Lecture Notes
in Physics

Edited by H. Araki, Kyoto, J. Ehlers, München, K. Hepp, Zürich
R. Kippenhahn, München, H. A. Weidenmüller, Heidelberg
J. Wess, Karlsruhe and J. Zittartz, Köln
Managing Editor: W. Beiglböck

297

A. Lawrence (Ed.)

Comets to Cosmology

Proceedings of the Third IRAS Conference
Held at Queen Mary College, University of London
July 6–10, 1987

Springer-Verlag

Berlin Heidelberg New York London Paris Tokyo

Editor

Andrew Lawrence
School of Mathematical Sciences, Queen Mary College
Mile End Road, London E1 4NS, England

ISBN 3-540-19052-X Springer-Verlag Berlin Heidelberg New York
ISBN 0-387-19052-X Springer-Verlag New York Berlin Heidelberg

© Springer-Verlag Berlin Heidelberg 1988
Printed in Germany

Printing: Druckhaus Beltz, Hemsbach/Bergstr.
Binding: J. Schäffer GmbH & Co. KG., Grünstadt
2153/3140-543210

DEDICATION

This volume is dedicated to the memory of Marc Aaronson
1950 - 1987

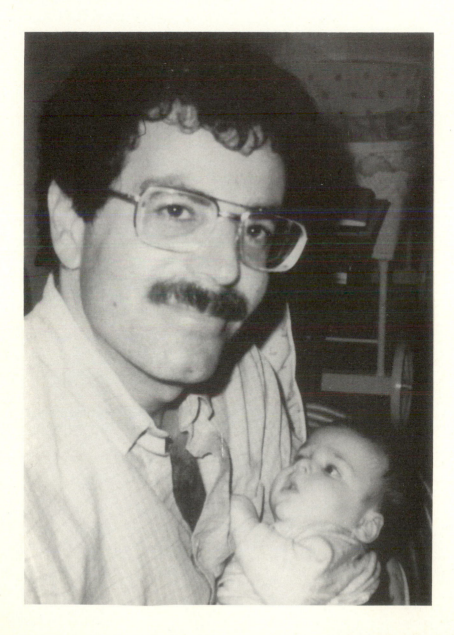

Editors Preface

The Third International IRAS conference was held at Queen Mary College (QMC), University of London, in the week of July 6–10, 1987. There was a feeling of maturity about this meeting in several ways. The Infra-Red Astronomical Satellite was developed, funded, built, and operated by three national scientific communities (the USA, the Netherlands, and the UK). With this latest conference, each of the three communities have had their opportunity to host a major meeting devoted to scientific results from IRAS data. (The first two were held at Noordwijk and Caltech, respectively). But this meeting was a very widely-based one, both geographically and scientifically. As well as the three original IRAS nations, we had contributions from twelve other countries, including some from every continent. (The IRAS survey products are available worldwide, and scientists everywhere are taking advantage of this rich mine of information). A large fraction of the papers presented were not simply discussion of discoveries from the IRAS data itself (lively though such work still is). Instead they represented projects which have either sprung directly from IRAS discoveries and data, but now involve a wide range of observational and theoretical work, or which are addressing the same questions upon which IRAS discoveries have shed light. The most striking thing of all, however, was the huge range of physical scales, and corresponding astronomical topics, addressed. The research presented concerned asteroids, comets, dust in the solar neighbourhood, young stars, old stars, the interstellar medium, Galactic Structure, normal galaxies, active galaxies, large scale structure, the distance scale, and the cosmic background at various wavelengths. It was of course for this reason that we named the conference, with blinding simplicity and style, "**Comets to Cosmology**."Carl Murray, inspired by this, designed a natty logo for our conference bags. There is always a danger that a meeting of such breadth will lose focus and fragment, but this didn't really seem to happen. Despite the diverse professional interests represented, almost everybody stayed the whole course, and the atmosphere was fruitful and friendly throughout.

There were two hundred and eighteen attendees at the meeting. Fifty three talks were given, and eighty nine poster papers. The poster papers remained on display for the whole week, to generate informal discussion during coffee and lunch breaks. There was a specially featured display on "Infrared facilities of the future" (COBE, ISO, SIRTF, SOFIA, NICMOS, and the Space Telescope). There was also an afternoon workshop session for IRAS data analysis pundits. A meeting of such size poses problems when it comes to producing a proceedings volume, especially as we were keen to keep the price of the volume as low as possible. In order to give the speakers enough room to write something sensibly approximating to their talks while keeping the volume a reasonable size, we had to make the sad decision not to include the text of poster papers. The interested reader will however find a list of all the titles and authors of poster papers, so that s/he can write and request preprints from the authors, where available.

The conference was masterminded by Michael Rowan-Robinson, with enormous quantities of nervous energy and dedication. Sally Wright, our Conference Secretary, was both administrative nerve centre and principal shock worker. Nobody ever doubted it would be a success with Michael and Sally in charge. Filled thus with confidence, the organising committee went about their tasks with vigour. In the closing stages however, the eventual success would have been impossible without the hard work of a large team of local staff members and students, many of whom had no IRAS research connection. We were worried by the problem of accomodation being so far from the College, but in the event, most participants seemed to be delighted to find themselves in the midst of a great city. Full advantage was taken of London entertainments, including Bill Fishman's famous tour of Jack the Ripper territory !

Finally, the meeting also marked a sad moment in recent astronomical history. Marc Aaronson was to have attended, with his family, and to have given a review talk on "The impact of IR astronomy on the distance scale". Shortly before the meeting, the news came that Marc had been killed in a tragic accident at Kitt Peak. We decided to set aside some time on the last day to remember Marc. First, L. Gougenheim gave a talk on the same subject as Marc's intended review. Then, Michael Rowan-Robinson and Roger Thompson spoke briefly, recalling both Marc's scientific achievements and his stature as a human being. We were delighted that Marc's wife Marianne, his daughter, and his father were able to come to London and attend our brief tribute, as well as taking part in some of the Conference social events. Marianne replied movingly to the tributes paid to Marc. Those of us who knew Marc only through his work were made aware that the community had lost more than just a good scientist.

Andy Lawrence
Queen Mary College
November 1987

TABLE OF CONTENTS

THE SOLAR SYSTEM

THE GALAXY

INFRARED GALAXIES

SCIENTIFIC ORGANISING COMMITTEE

M. Rowan-Robinson	QMC (Chairman)
P. Clegg	QMC
H.Habing	Leiden
M. Harwit	Cornell
R. Jennings	UC, London
M. Longair	ROE, Edinburgh
P. Marsden	Leeds
G. Neugebauer	Caltech
M.V. Penston	RGO, Herstmonceux
J.-L. Puget	Paris

LOCAL ORGANISING COMMITTEE

Chairman	M. Rowan-Robinson
Conference Secretary	Sally Wright
Conference Week	I.P. Williams
Word processing & printing	C. Murray , G. White
Entertainment	B.Carr , S. Harris
Rooms	R.Tavakol
Travel	A. Lawrence
Catering	Carey Adams
Accomodation	Sally Wright , K.Vedi
Advisory	I. Roxburgh , P. Clegg
	B.Stewart (RAL)

SPONSORS

The Third International IRAS conference was sponsored by several UK bodies : the Science and Engineering Research Council, The British National Space Centre, the Rutherford Appleton Laboratory, and the School of Mathematical Sciences and the Physics Department, Queen Mary College.

The Solar System

THE IMPACT OF IRAS ON ASTEROIDAL SCIENCE

S.F.Dermott, P.D.Nicholson, Y.Kim and B.Wolven

Center for Radiophysics and Space Research,
Cornell University,
Ithaca, NY 14853, U.S.A.

E.F. Tedesco

Jet Propulsion Laboratory,
California Institute of Technology,
Pasadena, CA 91109 U.S.A.

ABSTRACT

Asteroids are studied because, together with comets, they are one of
the few sources of information on conditions in the early solar
system. Asteroids have been classified into a small number of distinct
types and the distribution of these types with respect to their
distances from the sun gives information on the chemical structure of
the primitive solar nebula. Asteroids are also believed to be the
primary source of meteorites and thus the most abundant, diverse and
accessible source of extraterrestrial material. There have been
several major advances in asteroidal science in recent years. We now
have a clear understanding of the origin of some of the Kirkwood gaps
in the distribution of asteroidal semimajor axes and of the role that
these gaps have in the delivery of meteorites to the Earth. Secondly,
the number of meteorites available for study has increased since the
discovery of the meteorite fields on the Antarctic ice cap. As
expected, IRAS data on asteroidal albedos (the diameters and albedos
of 1,811 asteroids have now been measured) will vastly extend the
classification work. We anticipate that new insights into the
chemical structure of the asteroid belt will be revealed by plotting
the variations of the colors and albedos of asteroids in a given class
against their distance from the sun. Unexpected products of the IRAS
mission include the discovery of a small body in the Geminid meteor
stream and the discovery of structure in the zodiacal cloud. The IRAS
solar system dust bands provide a further connection with asteroids
since they may be collision products of asteroids, possibly members of
the major Hirayama families, implying that asteroids may be a
significant source of the particles in the zodiacal cloud.

(a)

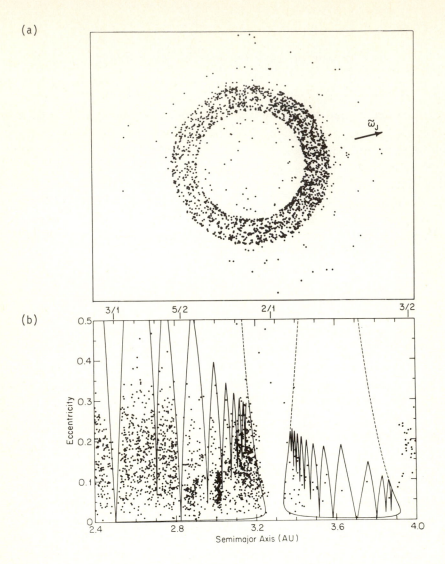

Figure 1. (a) Some of the dynamical structure of the asteroid belt is revealed in this plot of the semimajor axes (radial coordinate) and the longitudes of pericenter (angular coordinate). All the numbered asteroids listed in the 1979 TRIAD file (Bender, 1979) are represented. The arrow indicates Jupiter's longitude of pericenter (Dermott and Murray, 1984). (b) More of the dynamical structure is revealed by the distribution of asteroids in semimajor axis and eccentricity space. The solid lines represent the libration zones associated with the strongest jovian resonances. The locations of some of these resonances are marked by the ratio of the asteroidal and jovian orbital periods. The boundaries of the libration zones are largely clear of asteroids and neatly define the Kirkwood gaps. (Dermott and Murray, 1983).

1. INTRODUCTION

It is now widely believed that the various bodies in the solar system formed in an extensive disk of gas and dust concomitant with the formation of the sun. In one version of this model of the origin of the solar system, temperature and pressure gradients within the solar nebula determine the condensation sequence of the elements and their compounds and this sequence largely accounts for the major division in the solar system, that between the inner, rocky, terrestrial planets and the outer, gaseous planets. However, much of the detailed evidence of the early condensation and accretion processes is now either buried in planetary interiors or has been lost for ever because of post-accretional physical and chemical changes. The major sources of information on conditions in the early solar system are the small bodies that failed to accrete to planetary size, that is, asteroids, comets and the debris that they give rise to, namely, interplanetary dust particles and meteorites.

Asteroids are of particular interest since they are believed to be the primary source of meteorites, our most abundant, diverse and accessible source of extraterrestrial material, and because of their location in the transition zone between the terrestrial and jovian planets. A number of recent advances have placed asteroidal science in the forefront of planetary science. We now have a clear understanding of the origin of some of the Kirkwood gaps in the distribution of the asteroidal semimajor axes and the role that these gaps, particularly the 3:1 gap at 2.5 AU, play in the delivery of meteorites to the Earth. The nature of the Kirkwood gaps is partly revealed by their structure in semimajor axis and eccentricity space (Fig.1). The gaps correspond to those regions where one would expect to find asteroids librating about resonant configurations involving Jupiter (Dermott and Murray, 1983). Numerical experiments have shown that the motions of asteroids in those regions can be chaotic and that at sporadic intervals of ~ 10^6 years the eccentricities of the asteroids increase to values large enough for their orbits to cross that of Mars and, less frequently, that of the Earth (Wisdom, 1983, 1985).

Interest in meteorites has increased largely because of the development of techniques and laboratories to analyse lunar rocks but also because of the discovery by Japanese scientists in 1969 of rich meteorite fields on the Antarctic ice cap (Dodd, 1981). We can, perhaps, expect a similar surge of interest in the study of extraterrestrial dust particles due both to the recent flyby of Halley's comet and to the recent discovery of cosmic dust on the Greenland ice cap (Maurette et al., 1986, 1987). The Greenland deposits contain large abundances of the larger dust particles (> 100 micron) that are easier to analyse, they also seem to be less affected by terrestrial

(a)

(b)

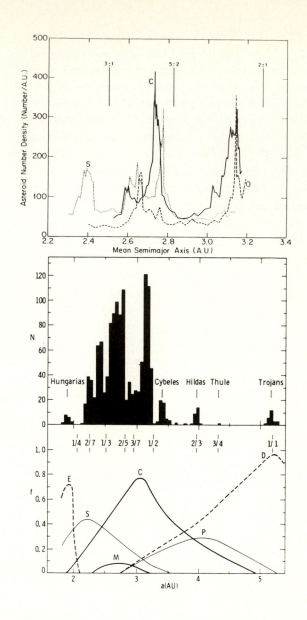

Figure 2. (a) The variation of the number density of asteroids with distance from the sun for the 236 bright asteroids in the bias-free set defined by Dermott and Murray (1983). The asteroids were classified as either S-type (dotted line) or C-type (solid line) according to the TRIAD classification scheme as modified by Tholen (1984). The remaining asteroids not in the S- or C-classes were classified as O (dashed line) or "other". (b) The distribution of asteroids in the S, C, M, E, P, and D classes in a sample of 1373 asteroids after correction for observational bias. f is the estimated fraction of asteroids at a given distance in each class (Gradie and Tedesco, 1982).

weathering than other deposits and, most importantly, a large fraction of the particles have not been melted by passage through the Earth's atmosphere and thus preserve their original mineralogy and petrology.

The IRAS data have a major role to play in these studies both through the extension of the asteroidal data base and because of the unexpected discovery, by IRAS, of the solar system dust bands.

2.   THE IRAS ASTEROID SURVEY

Averaged albedos and diameters derived from IRAS observations are available for 1,811 different asteroids (IRAS Asteroid and Comet Survey, 1986). However, it is important to realise that those data derived from low flux sightings are less reliable than those based on the means of several high flux detections (cf., Veeder, 1986; Tedesco et al., 1987; Tedesco et al., this book). The reasons for this include: (1) confusion of faint asteroids with background sources, (2) flux overestimation for the smaller asteroids, (3) real variations in flux from asteroids (due, for example, to variations in their cross-sectional areas), and (4) uncertainties in the absolute visual magnitudes. These problems are all more severe for faint asteroids.

The flux overestimation at 25 microns for faint (< 2 Jy) sources in the IRAS Point Source Catalog is as much as a factor of 3 (IRAS Circular, Nov. 1986) and increases with decreasing flux. The values and uncertainties given for the IRAS albedos and diameters do not take this overestimation into account. Yet as many as 698 (or 40%) of the asteroids with IRAS albedos have no sightings with flux densities greater than 1 Jy. As an example of one of the consequences of flux overestimation consider that the IRAS survey lists 1685 Toro and 1980 Tezcatlipoca as having low IRAS albedos. High quality ground-based observations contradict these and find instead that the albedos of both of these objects are moderate, in agreement with their taxonomic classifications which, in turn, are based entirely on visual data (Veeder et al., 1987).

IRAS source observations were accepted twice for 1685 Toro and once for 1980 Tezcatlipoca. The former was "sighted" 10 times and the latter 4 times. Of these 14 sightings all but 3 were rejected by the processing software. The two sightings of Toro which were not automatically rejected had flux densities at 25 microns of 0.56 Jy while that for Tezcatlipoca was 0.32 Jy. In view of the known problems with sightings having 25 micron flux densities less than 1 Jy it is not suprising that these fluxes yielded low albedos; the fluxes have been overestimated, the asteroids were probably observed near a peak in the noise or while confused with a background source. The key to recognizing such dubious results lies in looking at the 25 micron flux density together with the number of realized versus expected

(a)

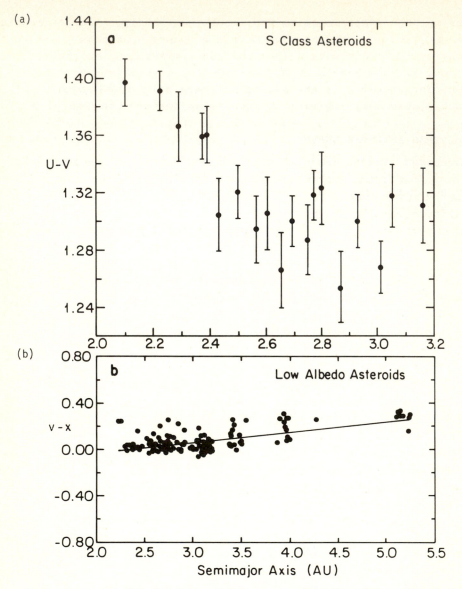

(b)

Figure 3. (a) Variation of the mean U – V color of 191 S-class asteroids with mean semimajor axis. The asteroids were sorted in order of increasing semimajor axis and then separated into independent samples of 10 asteroids (Dermott, Gradie and Murray, 1985). (b) Variation of the v – x color of low albedo (< 0.076) asteroids (those in the C, P and D classes) with semimajor axis. All asteroids with U – V < 0.90 were eliminated from the sample (this restriction removes the F-class asteroids). Tedesco (1987) concludes from this figure that the P-class merges smoothly into the C-class and is thus a subgroup of the C-class. The D-class (largely those asteroids with semimajor axis ~ 5 AU) do separate from the P-class. However, Tedesco notes that this separation could merely be due to a tendency for v – x to increase with semimajor axis.

sightings. If the flux density is below 1 Jy, and/or the number of expected sightings greatly exceeds the number realized, then the accuracies of the quoted albedo and diameter are probably much lower than their formal errors.

The IRAS albedos of the brighter asteroids will be used to extend the asteroidal classification schemes. Preliminary results have been announced by Barucci et al. (1987) and by Chapman (1987). Albedos are the backbone of all classification schemes and the availability of the IRAS albedos would appear to allow the separation of the asteroids into an even larger number of distinct classes than was previously the case. Using a statistical clustering technique, Barucci et al. classified 438 asteroids for which there are both eight color photometric data (Zellner et al., 1985) and IRAS albedos available. They were able to recognize a total of nine principal classes. Chapman has confirmed that the heliocentric distribution of asteroidal types is zoned (see Fig.2) and by comparing distributions of types within families with the general mix of asteroids at the same heliocentric distance, he has been able to evaluate the reality of some of the numerous "Hirayama" families proposed by various workers. Chapman has also confirmed an earlier result of Zellner (1979) that the different types have different size-frequency distributions, in particular, that there is a peak in the size-frequency distribution of C-types at 100 km.

Two comments are worth making on the distribution of asteroidal types. Figs.1 and 2 emphasize in several different ways that the structure of the asteroid belt has been determined by dynamical processes and that there has been a large amount of post-accretional mixing of the asteroids. The asteroids must have accreted in near-circular, near-coplanar orbits and it is hard to envisage any process that could have increased their eccentricities and inclinations without also changing their semimajor axes. The inner edge of the belt has been defined by the sweeping action of Mars while resonant interactions with Jupiter have both depleted the belt beyond 3.2 AU and sculpted the Kirkwood gaps. In Fig.2(a) we see that the distribution of asteroidal types is certainly zoned: S-types, for example, tend to dominate the inner edge of the main belt. However, in the same figure we also see that peaks in the number densities of the various types can occur at the same heliocentric distance. The origin of these peaks is obviously some, as yet obscure, dynamical process (associated with the major Kirkwood gaps), probably unrelated to the formation distances of the various asteroidal types. The observed zoned structure of the belt is best regarded as a weak reflection of an earlier primordial structure that, fortunately for us, has managed to survive a number of major dynamical processes that have shuffled the orbits. It would be wrong to regard the peaks in the distributions of, for example, the S- and C-types in Fig.2(b) as the formation locations of these asteroids. The implication is that S-types were formed closer to the sun than C-types, but, at present, that is all that can be stated with confidence. The groups of asteroids in the stable resonant locations, that is, the Hildas and Trojans, are special cases and the distributions of types within these groups deserve special consideration.

Secondly, the variation of the colors and albedos of asteroids within a given class with heliocentric distance should prove to be of interest. This has already been shown to be the case for S-class asteroids (Dermott et al., 1985) - see Fig.3(a). Tedesco (1987) points out that there appears to be a trend in the heliocentric distribution of the v - x colors of low albedo asteroids in the C, P and D classes - see Fig.3(b). This type of analysis may lead to the recognition of subclasses of asteroids, or, and this is somewhat disturbing, to the recognition that the colors of the asteroids have been modified by space weathering.

3. NEAR-EARTH ASTEROIDS

The major triumph of IRAS in this area was the discovery of a small object (1983TB, now 3200 Phaethon) in the middle of a (or the!) major meteor stream, the Geminids (Davies et al., 1984). While of great interest, this and other recent discoveries seem only to heighten the confusion that exists regarding the relationships between asteroids, comets, meteorites, meteors and near-earth asteroids (Davies, 1986).

Halliday (1987) has recently reported camera observations of a second large meteorite fall from the orbit of the Innisfree (brecciated LL) chondrite, although no object has yet been recovered. The detection of two related meteoritic events only 3 years apart and within 500 km of each other on the Earth's surface argues for the existence of a stream of meteorites that could contain as many as 5×10^8 objects with masses $\sim$ 10 kg. The Pioneer Venus orbiter magnetometer appears to have detected "comet wakes" in the orbit of the "asteroid" 2201 Oljato (Russell et al., 1984). There is no evidence from either IRAS or ground-based observations that Oljato possesses a coma, although the asteroid is very peculiar in many other ways (see Kerr, 1985).

3200 Phaethon has an albedo of 0.11   0.02, a diameter of 0.7 0.5 km and its thermal properties are consistent with those of solid rock (Green et al., 1985). There is no evidence of a coma or even a faint dust trail that may have enhanced its detectability (Davies, 1986; Davies et al., this meeting). The blue JHK colors and moderate albedo, when combined with the optical data, show Phaethon to be a unique object among the near-earth asteroids and dissimilar to the usually accepted model for an inert cometary nucleus (Green et al., 1985). These results offer some support to the hypothesis that the Geminid meteor stream is the result of collisions between asteroids.

4. THE IRAS SOLAR SYSTEM DUST BANDS

Support for the wider hypothesis that collisions between asteroids are a major source of all interplanetary dust particles may be provided by analysis of the structure of the IRAS solar system dust bands. During its all-sky survey, IRAS discovered three narrow bands of warm (165-200K) emission circling the sky at geocentric ecliptic latitudes -10, 0, and +10 degrees (Low et al., 1984, Neugebauer et al., 1984). Low et al. (1984) were the first to suggest that the bands may be associated with dust in the asteroid belt. We predicted (Dermott et al., 1984) that because of the secular perturbation of the dust grain orbits by the planets, the latitudes of the bands would vary with ecliptic longitude and we showed that measurement of the amplitude and phase of these variations might determine the orbital elements of the dust particles. We also suggested (Dermott et al., 1984) that the dust bands may be associated with the three most prominent Hirayama asteroid families and, by implication, that dust produced by the gradual comminution of asteroids in the main belt could be a significant contributor to the broader zodiacal dust cloud. As a consequence of the latter suggestion, we further predicted (Dermott et al., 1984) that the central dust band discovered by IRAS would be split and that the separation of the two components would be consistent with an origin associated with either the Koronis or Themis asteroid families (these two families have similar inclinations that may not be resolved in the binned data). The longitudinal variation of the latitudes of the dust bands and the appropriate splitting of the central dust band have both been confirmed by our subsequent analysis of the binned data in the IRAS Zodiacal History File (Dermott et al., 1986).

Examination of the unbinned high resolution data (IRAS Sky Flux Map Plate 81, HCON 3) by Sykes (1986) has revealed the existence of pairs of bands that he associates with the Themis, Koronis and Nysa asteroid families. However, Sykes and Greenberg (1986) are not convinced that the bands are a steady-state feature, favoring the alternative hypothesis that the observed bands are the results of comparatively recent collisions between random asteroids. Since 10% of all asteroids belong to the three most prominent asteroid families (Eos, Themis and Koronis), this may be a somewhat fine distinction (we could strengthen this argument by only considering asteroids with the

(a)

(b)

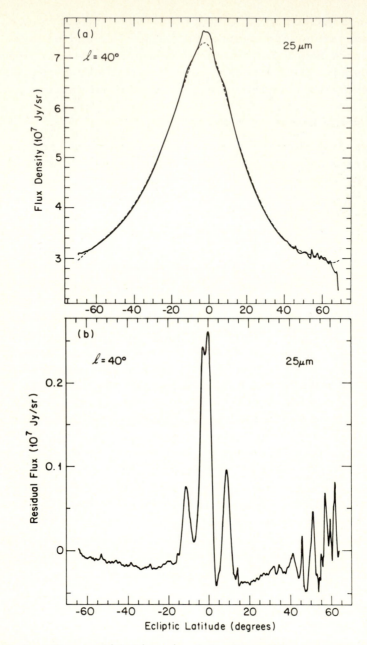

Figure 4. The solid line in (a) shows the average of four separate scans in the 25 micron waveband taken during SOP (Satellite Observing Plan) 394 on August 10 1983. Fourier methods were used to separate the smooth, large-scale background, shown as dashed curve in (a), from the narrower dust bands shown in (b). Note that the central dust band is split. The short-scale structure evident at high latitudes in (b) is due to the galaxy (Dermott, Nicholson and Wolven, 1986).

(a)

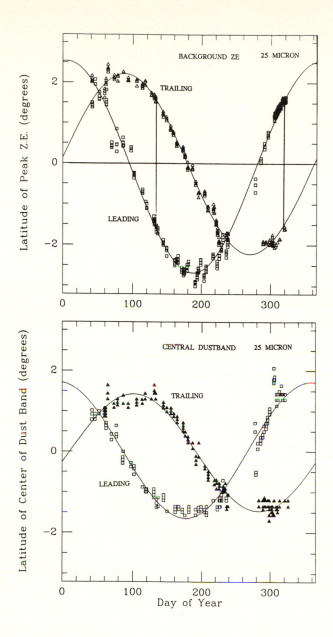

(b)

Figure 5. (a) Variation of the ecliptic latitude of the peak of the background zodiacal emission with the day of the of the year for observations in the 25 micron waveband. Data obtained when the satellite was in the leading or ascending phase of its motion round the Earth are represented by squares while data from the trailing or descending phase are represented by triangles. (b) A similar plot for the latitude of the center (at the half-power point) of the central dust band (the high peak close to the ecliptic plane shown in Figure 7(b)).

(a)

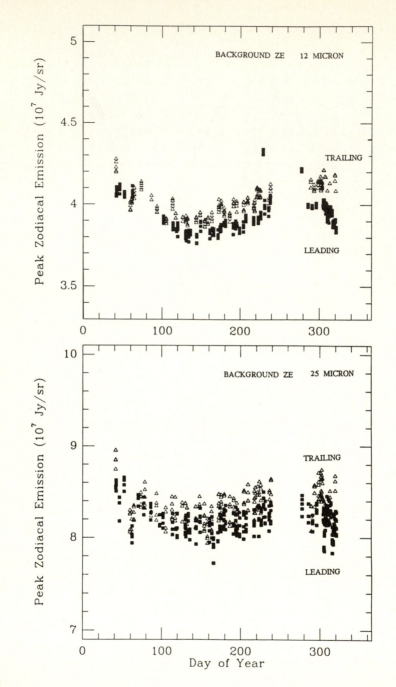

(b)

Figure 6. The variation of the peak zodiacal emission with the day of the year, after correction for the variation with elongation angle, for observations in (a) the 12 micron wavenband and (b) the 25 micron waveband. At all times of the year, observations in the trailing phase appear to be brighter than those in the leading phase.

appropriate inclinations: most asteroids with inclinations less than 2 degrees belong to either the Themis or Koronis families). It may also be irrelevent to the broader problem of the origin of interplanetary dust particles since they also argue that collisions between asteroids are sufficient to supply the steady-state zodiacal cloud.

We have now completed our analysis of most of the useful data in the IRAS Zodiacal History File at wavelengths of 12 and 25 microns. The raw data in a typical set of scans are shown in Fig.4(a). The high-frequency structure evident in these scans was separated from the large-scale, background structure using Fourier techniques. This involved using a Fast Fourier Transform to find the spatial frequency distribution of the signal strength and then dividing the signal into high- and low-frequency domains using a Parzen window. The features of interest in Fig.4 include the latitude of the peak of the background zodiacal emission (Fig.4(a)) and the latitudes of the peaks of the narrow dust bands (Fig.4(b)).

The very existence of the narrow dust bands implies that the dust particles in a given band have a common inclination (Low et al., 1984). If we assume that they also have other orbital elements in common, in particular, that they have a common semimajor axis, then we can apply Lagrange's secular perturbation theory to the band as a whole and from the band's observed latitude variation deduce its distance (Dermott et al., 1984, 1985, 1986). Similar arguments can be applied to the particles in the background zodiacal cloud, except that in this case, since the particles have a wide range of semimajor axes, the structure of the cloud will be determined by integrated quantities.

We have analysed nearly 1000 profiles similar to that shown in Fig.4. Some of our results are shown in Figs.5 and 6. Fig.5(a) shows the variation in latitude of the peak of the zodiacal background with the day of year, while Fig.5(b) is a similar plot for the center of the central (near-ecliptic) dust band. In all cases, the variations are well-matched by sinusoids with periods of one year. The data appear to be of high quality in that the sinusoidal variations are strong and allow accurate determinations of the orientations of the various planes of symmetry of the dust particle distributions. For elongation angles close to 90 degrees, the geocentric ecliptic latitudes of the peaks seen in the leading and trailing phases of the motion of the IRAS satellite around the Earth must be equal and opposite when the Earth crosses the symmetry plane of the dust. At these times, the peak latitudes are also equal in magnitude to the inclination of the plane of symmetry to the ecliptic. Thus, the determination of these locations (marked by vertical lines in Fig.5(a)) leads to a direct determination of both the inclination of

Table 1. Geometry of the zodiacal cloud with respect to the ecliptic

(a) 25 micron data   Inclination   Longitude
                            (degrees)   (degrees)
        Ascending node   1.59   51.9
        Descending node   1.53   231.2
                                           -179.3 = Difference

(b) 12 micron data   Inclination   Longitude
                            (degrees)   (degrees)
        Ascending node   1.36   47.8
        Descending node   1.62   229.4
                                           -181.6 = Difference

Table 2. Geometry of the central dust band with respect to the ecliptic

(a) 25 micron data   Inclination   Longitude
                            (degrees)   (degrees)
        Ascending node   1.21   52.4

(b) 12 micron data   Inclination   Longitude
                            (degrees)   (degrees)
        Ascending node   1.23   43.3

the symmetry plane and the longitudes of its nodes. Some of our results
are shown in Tables 1 and 2. Note that this geometrical method is
insensitive to problems associated with the flux calibration of the IRAS
data.

     Both the inclination and the nodes of the background zodiacal cloud
are what one would expect for dust particles near or inside the orbit of
Mars (the inclination and ascending node of Mars are 1.85 degrees and 49
degrees, respectively), and do not discriminate between models invoking
asteroidal or cometary origins for the dust. The inclination of the
central dust band is clearly less than that of the zodical cloud and is
exactly what one would expect for the Themis and Koronis families.
However, the node is the same as that of the background cloud and about
40 degrees less than the expected value for particles at a distance of
~ 3 AU. We are presently unsure of the seriousness of this discrepancy.

     We have observed that there are marked differences in the
amplitudes of the leading and trailing curves of the background ZE at
both 12 and 25 microns (only the 25 micron data are shown in Fig.5(a)).
Note that this is an asymmetry in the geometry of the cloud and cannot
be caused by calibration problems. We believe that this asymmetry may
reveal an important feature of the geometry of the zodiacal cloud which
has not previously been recognized in the literature, namely that the
sun is not at the center of symmetry of the zodiacal cloud (Dermott et
al., 1986).

We have found other asymmetries in the data which do lead us to question the IRAS flux calibration scheme which was based on a particular, and possibly inadequate, model of the zodiacal cloud. The peak brightness of the zodiacal background varies with ecliptic longitude because of the variable contribution of the dust bands (the plane of symmetry of the central dust band is clearly inclined to that of the background zodiacal emission). If the dust bands are filtered out, then we can investigate the longitudinal variation in the peak brightness of the large-scale, background emission. However, to do this, we first have to determine and remove the variations in brightness associated with elongation angle (Dermott et al., 1986). In Fig.6 we show the variation of the normalized peak brightness with the day of the year, separating those observations in the leading leg from those in the trailing leg. The distributions at both wavelengths are regular, remarkably similar and show a marked difference between the leading and trailing legs at all times of the year.

ACKNOWLEDGMENTS

A portion of the work described in this paper was carried out at the Jet Propulsion Laboratory, California Institute of Technology under contract to the National Aeronautics and Space Administration. At Cornell, this research was supported (in part) under the IRAS extended mission program by JPL contract 957290 and by NASA grant NAGW 392.

REFERENCES

Barucci,M.A., Capria,M.T.,Coradini,A., & Fulchignoni,M., 1987. Preprint.
Bender,D.F., 1979. In: Asteroids, ed. Gehrels,T., University of Arizona Press, Tucson.
Chapman,C.R., 1987. Preprint.
Davies,J.K., 1986. Mon. Not. R. astr. Soc. **221,** 19p.
Davies,J.K., Green,S.F., Stewart, B.C.,Meadows,A.J.,Aumann,H.H., 1984. Nature **309**, 315.
Dermott,S.F., & Murray,C.D., 1983. Nature **301**, 201.
Dermott,S.F., & Murray,C.D., 1984. In: IAU Colloquium No. 75 Planetary Rings, ed. Brahic,A., CEPAD, Paris.
Dermott,S.F., Gradie,J., & Murray,C.D, 1985. Icarus **62**, 289.
Dermott,S.F., Nicholson,P.D., Burns,J.A., & Houck,J.R., 1984. Nature **312**, 505.
Dermott,S.F., Nicholson,P.D., Burns,J.A., & Houck,J.R., 1985. In: IAU Colloquium No. 85 Properties and Interaction of Interplanetary Dust, eds. Giese,R.H., & Lamy,P., Reidel, Dordrecht, Holland.
Dermott,S.F., Nicholson,P.D., & Wolven,B., 1986. In: Asteroids, Comets, Meteors, II, eds. Lagerkvist,C-I., & Rickman,H., Uppsala.

Dodd,R.T., 1981. Meteorites, Cambridge Univ. Press, Cambridge.

Gradie,J., & Tedesco,E., 1982. Science **216**, 1405.

Green,S.F., Meadows,A.J., & Davies,J.K., 1985. Mon. Not. R. astr. Soc. **214**, 29p.

Halliday,I., 1987. Icarus **69**, 550.

Infrared Astronomical Satellite Asteroid and Comet Survey: Preprint Version No. 1, 1986. Ed. Matson,D.L., JPL Document No. D-3698.

Kerr,R.A. 1985. Science **227**, 930.

Low, F.J.,Beintema,D.A., Gautier,T.N., Gillet,F.D., Beichmann,C.A., Neugebauer,G., Young,E., Aumann, H.H., Boggess,N., Emerson,J.P., Habing,H.J.,Hauser,M.G., Houck, J.R., Rowan-Robinson,M., Soifer, B.T., Walker,R.G., & Wesseliu, P.R. 1984. Astrophys. J. **278**, L19.

Maurette,M., Hammer,C., Brownlee,D.E., Reah,N., & Thomsen,H.H., 1986. Science **233**, 869.

Maurette,M., Jehanno,C., Robin,E., & Hammer,C., 1987. Nature **328**, 699.

Neugebauer,G., Beichmann,C.A., Soifer,B.T., Aumann,H.H., Chester,T.J., Gautier,T.N.,Gillet,F.C., Hauser,M.G., Houck,J.R., Lonsdale,C.J., Low,F.J., & Young,E.T., 1984. Science **224**, 14.

Russell,C.T., Aroian,R., Arghavani,M.,& Nock,K., 1984. Science **226**, 43.

Sykes,M.V., 1986. Ph.D. thesis, University of Arizona.

Sykes,M.V., & Greenberg,R. 1986. Icarus **65**, 51.

Tedesco,E.F., Tholen,D.J., & Zellner,B., 1982. Astron. J. **87**, 1585.

Tedesco,E.F., 1987. Priv. Comm.

Tedesco,E.F., Matson,D.L., Veeder,G.J., & Lebofsky,L.A., 1987. Bull. Amer. Astron. Soc. **19**, in press.

Tholen,D., 1984. Ph.D. Thesis, University of Arizona.

Veeder,G.J., 1986. In Infrared Astronomical Satellite Asteroid and Comet Survey: Preprint Version No.1, ed. Matson,D.L., JPL Document No. D-3698.

Veeder,G.J., Matson,D.L. Tedesco,E.F., Lebofsky,L.A., & Gradie,J., 1987. Bull. Amer. Astron. Soc. **19**, in press.

Wisdom,J., 1983. Icarus **56**, 51.

Wisdom,J., 1985. Nature **315**, 731.

Zellner,B., 1979. In: Asteroids, ed. Gehrels,T., University of Arizona Press, Tucson.

Zellner,B., Tholen,D.J., & Tedesco,E.F., 1985. Icarus **61**, 355.

IRAS OBSERVATIONS OF ASTEROIDS

E.F. Tedesco, D.L. Matson, and G.J. Veeder

Jet Propulsion Laboratory
4800 Oak Grove Drive
Pasadena, CA 91109, U.S.A.
and
L.A. Lebofsky

Lunar and Planetary Laboratory
University of Arizona
Tucson, AZ 85721 U.S.A.

ABSTRACT

The IRAS Asteroid and Comet Survey was the largest, most uniform and least-biased survey ever conducted for asteroids and comets. The size and approach of this survey gave it marked advantages over earlier surveys. The large number of sightings over most of the sky provided excellent sampling of spatial distributions. The instrument and survey parameters were relatively constant throughout, yielding a uniform set of data. This was the first small bodies survey to observe thermal emission, rather than reflected light, thereby avoiding the severe albedo bias present in all previous surveys. Requirements of high reliability and completeness were also maintained. For the known asteroids a total of 11,449 sightings were obtained. Ultimately, 6,510 of these sightings were accepted into the catalog. These data yielded albedos and diameters for 1,811 different asteroids. Numerous previously unknown asteroids were also observed. There is an apparent difference in the albedo distribution of the asteroids as a function of size, however, this effect is probably an artifact due to a combination of factors such as faint source flux overestimation and confusion with faint background sources.

1. INTRODUCTION

The IRAS Asteroid and Comet Survey was conducted during the sky survey portion of the IRAS mission. With the exception of two five degree wide bands centered at ecliptic longitudes of 163° and 343° which went unscanned, 97% of the ecliptic plane was scanned four times; 72% was scanned six times. Because the same instrumentation and data reduction techniques were applied to each scan the resultant data is the most uniform obtained to date. Since each scan extended nearly from pole to pole the bias against sampling regions at high ecliptic latitudes was avoided. In addition, because this was the first survey to be conducted at wavelengths where thermal emission was the dominant source of photons, the severe bias against observing small, dark asteroids was avoided. For example, consider two asteroids of the same size but one with a geometric albedo of 0.03 and the other 0.30 both of which are observed under identical geometry. At visual wavelengths the lower albedo asteroid would have a flux one-tenth that of the brighter asteroid whereas at the IRAS wavelengths this ratio would be only 1.15!

In this paper we will limit ourselves to a discussion of what IRAS data have revealed about asteroids (IRAS Asteroid and Comet Survey, 1986). Because the IRAS asteroid data became available to the scientific community in December 1986, about six months before this conference, no science analysis of these data has yet been published.

2. DISTRIBUTION

What we present here is of a preliminary nature and is restricted to a discussion of the distributions of the 6,510 accepted asteroid sightings over heliocentric distance, albedo, and size. We use the following nomenclature throughout the remainder of this paper. By "albedo" we mean the visual (0.55 μm) geometric albedo. "Known asteroid" refers to one of the 3,318 asteroids which had been assigned a permanent number as of September 1985 or to one of the 135 unnumbered asteroids for which reliable orbital elements were known as of that date.

A "bright" asteroid is one with an albedo greater than 0.1 and a "dark" asteroid one with an albedo less than this.

2.1 Spatial Distribution

The main belt was well sampled as was the outer belt, including the Trojan groups. The limit on how many outer-belt asteroids were detected was not set by the sensitivity of the IRAS detectors but rather by the lack of completeness of the numbered asteroids, a limit set by the apparent visual brightness (e.g., from photographic surveys).

2.2 Size-Distance Distribution

The upper portion of Fig. 1 gives the diameter-distance distribution for known asteroids with IRAS-derived albedos less than 0.1. Note the bias against observing small, dark, asteroids at large heliocentric distances. This is due, in part, to the incompleteness in the known asteroids at these diameters. For example, there are numerous known asteroids with diameters less than 20 km at heliocentric distances between 2 and 3 AU but virtually none in this size range at heliocentric distances between 3 and 4 AU.

The lower portion of Fig. 1 is equivalent to the upper portion but for asteroids with IRAS-derived albedos greater than 0.1. A similar bias against observing small (bright) asteroids at large heliocentric distances is present here as well. Again, this is due, in part, to the incompleteness in the known asteroids at these diameters. Note the complete absence of bright known asteroids at distances beyond 4 AU. Because the albedo of a "bright" asteroid is about three times that of a "dark" asteroid, bright asteroids are more easily observed visually at any given distance. As can be seen from the upper portion of the figure, dark asteroids with diameters greater than 70 km are quite common at distances between 4 and 5 AU. If equal numbers of bright asteroids in this diameter range were present at these distances we would expect visual surveys to have discovered them down to diameters of about 30 km. The fact that IRAS observed none larger than this therefore cannot be attributed to the incompleteness of the known asteroids.

Conclusion: There are virtually no bright asteroids with diameters greater than about 30 km at heliocentric distances beyond 4 AU.

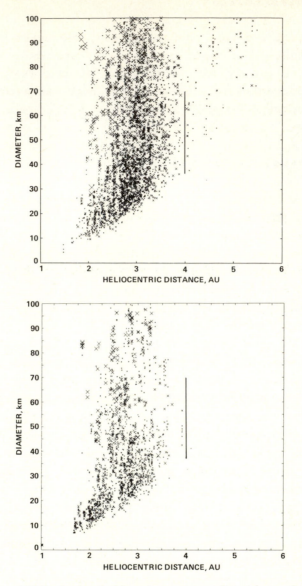

Fig. 1. Upper: IRAS-derived diameters as a function of heliocentric
        distance for accepted sightings with albedos < 0.1.
        Lower: IRAS-derived diameters as a function of heliocentric
        distance for accepted sightings with albedos > 0.1.

2.3 Albedo Distribution

Prior to IRAS it was established that the observed asteroid albedo histogram was bi-modal (cf., Morrison and Lebofsky, 1979). This conclusion was based on data for about 180 asteroids, much of it of low quality. There were more bright than dark asteroids in this sample, a consequence of the bias favoring observing asteroids which are bright at visual wavelengths.

The IRAS-derived albedo histogram for the larger asteroids (diameters greater than 40 km) presented in Fig. 2 is based on observations of 1,811 different asteroids, most of high quality. In this sample there are more dark than bright asteroids, a demonstration of the fact that IRAS lacked the severe albedo bias present in all earlier surveys.

On the other hand, as shown in Fig 3, the albedo histogram of known asteroids with IRAS-derived diameters less than 40 km is decidedly not bi-modal but rather resembles a Maxwellian curve.

The reason for the difference in these two distributions is currently being sought. Possible explanations include: 1) Confusion of faint asteroids with background sources, 2) Flux overestimation for the smaller asteroids, 3) Real variations in flux from asteroids (due, for example, to variations in their cross sectional areas), and 4) Uncertainties in the absolute visual magnitudes. These problems are all more severe for faint asteroids.

At this time we do not know whether this difference is real, i.e., this difference may not be physically significant.

3. UNKNOWN IRAS ASTEROIDS

Many of the 1,811 different asteroids with accepted sightings were observed multiple times. The average was 3.6 accepted sightings per asteroid. When the time between these multiple sightings is on the order of hours to days the sightings appear as "tracks". This property will be exploited to identify previously unknown asteroids in the IRAS asteroid data. By requiring that the individual sightings comprising each track have similar fluxes and, for the brighter sightings, similar flux ratios between bands, we will be able to assign a probability as to the reality of each track.

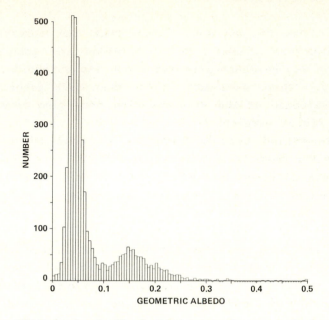

Fig. 2. IRAS-derived albedo histogram for accepted asteroid sightings with IRAS-derived diameters > 40 km.

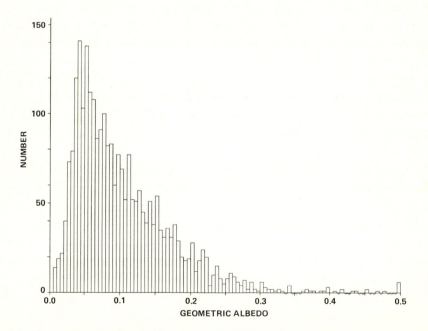

Fig. 3. IRAS-derived albedo histogram for accepted asteroid sightings with IRAS-derived diameter < 40 km.

To date we have identified on the order of 100 different tracks in the IRAS asteroid data.   Brian Marsden (Minor Planet Center) has identified at least one of these tracks with an as yet unnumbered asteroid discovered at visual wavelengths within a few months of the IRAS observations.   This gives us some confidence that the tracks we are finding are actually composed of sightings of real asteroids.

Before the sightings in a track can be used to estimate an albedo and diameter the distance and visual magnitude of the asteroid must be known.   In the case of those few percent of tracks which are linked with asteroids for which such visual observations are available those parameters can be obtained from them.   In the vast majority of cases, however, these estimates must come from the IRAS observations themselves.

Precisely how this can be done is beyond the scope of this paper. For the brighter candidates, i.e., those observed at 12, 25, and 60 um, we can obtain a crude estimate of the heliocentric distance from the position in a two-color diagram as shown in Fig. 4 as well as from their apparent rate of motion.

4. CONCLUSION

IRAS has increased the data base of asteroids with reliable albedos and diameters by an order of magnitude.   Although more work needs to be done on assessing the reliability of the results for the smaller asteroids that work is well under way.   The job of exploiting the IRAS asteroid data base for previously unknown asteroids will be a challenging one.   That many such asteroids are present is firmly established.   The task facing us now is how to reliably extract as many such sightings as possible.

ACKNOWLEDGMENTS

This work was carried out at the Jet Propulsion Laboratory, California Institute of Technology, under contract with the National Aeronautics and Space Administration and the Air Force Office of Scientific Research.

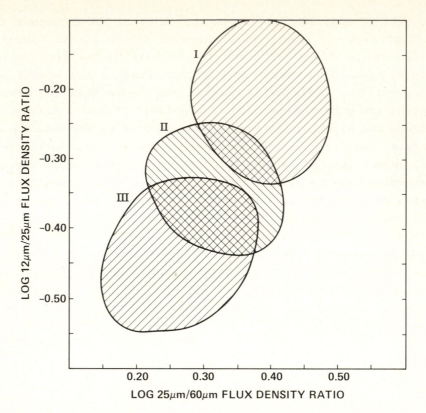

Fig. 4. Color-Color regions for known asteroids at three heliocentric distance ranges. I - less than 2.5 AU, II - between 2.5 and 3.0 AU, and III - greater than 3.0 AU.

REFERENCES

Infrared Astronomical Satellite Asteroid and Comet Survey: Preprint Version No. 1 (D.L. Matson, ed.) JPL Document No. D-3698.

MODELS FOR INFRARED EMISSION FROM ZODIACAL DUST

Michael G. Hauser
Laboratory for Astronomy and Solar Physics
Goddard Space Flight Center
Greenbelt, Maryland 20771, U. S. A.

ABSTRACT

The large-scale brightness of most of the sky at wavelengths from 10 μm to 30 μm is dominated by thermal emission from interplanetary dust. Measurement of this 'zodiacal emission' has long been of interest for many reasons. The IRAS survey has provided comprehensive brightness measurements in the infrared, covering the celestial sphere with solar elongation angles ranging from 60° to 120°. These data have stimulated numerous approaches to segregation and modeling of the zodiacal emission. Current models provide a reasonable basis for study of the interplanetary dust, but are severely stressed when one wishes to know the 'residual' sky brightness. The nature, status, and implications of various models are reviewed.

1. INTRODUCTION

The IRAS survey data show clearly that emission from interplanetary dust dominates the infrared sky brightness over most of the sky at 12 and 25 μm, and is a substantial contributor at 60 and 100 μm as well (Hauser et al., 1984). Study of this emission has already revealed several new features of the interplanetary dust (IPD), such as the apparent bands of emission near the ecliptic plane (Low et al., 1984) and dust trails in cometary orbits (Sykes et al., 1986). Because of the high sensitivity, multi-wavelength mapping of the celestial sphere for most of a year, with solar elongation angles ranging from 60° to 120°, the IRAS data provide a major new tool for study of the structure and origins of the dust cloud as well as of the character of the dust (see reviews by Hauser and Houck, 1986 and Beichman, 1987).

The IRAS data have triggered many new attempts to model infrared emission from the IPD. These models have had numerous motivations in addition to study of the IPD itself, including: to facilitate studies of extended galactic sources by removal of the bright zodiacal emission foreground; to search for a possible residual

extragalactic infrared background; to permit comparison of the solar
system dust cloud with those found in other stellar systems; to aid
in establishing a consistent zero point calibration of the IRAS data
over the year; and to provide a basis for determining natural
sensitivity limits for future space infrared astronomy measurements.
The fidelity of the model required, and consequently the approach
employed to obtain it, generally depends upon the particular
motivation. Because of the dominant contribution of the zodiacal
emission to the sky brightness at the shorter IRAS wavelengths,
studies of the residual galactic or extragalactic brightness at these
wavelengths typically demand the greatest precision, a few percent or
better, in determining zodiacal emission. On the other hand, at 100
μm where the Galaxy is dominant, accurate determination of the
zodiacal component itself is more difficult.

Significant challenges arise in seeking models of such fidelity.
Fundamental among these is the difficulty of distinguishing emission
originating beyond the solar system from the IPD contribution. Many
models are based upon assumptions of simple properties for the IPD
cloud, such as axial symmetry, symmetry about a plane, lack of small-
scale features, or spatial homogeneity of dust properties. Since, as
we shall discuss, none of these assumptions are totally satisfied,
such models are inherently limited. Remaining uncertainties in the
calibration of the IRAS data, particularly in the absolute zero point
and its consistency over the year, also impose limitations. Finally,
the large volume of IRAS data presents a significant computational
burden to serious modeling efforts.

My aim here is to review the nature and attributes of current
models of zodiacal emission, primarily those based upon IRAS data,
and to summarize some of the implications for the IPD cloud. I will
focus on properties of the main cloud; the zodiacal emission bands
and dust trails are addressed by others at this Conference.

2.   REVIEW OF MODEL APPROACHES

A number of techniques have been used to isolate and model the
zodiacal emission, including: (1) analysis of time variation in
observed properties of selected subsets of data; (2) angular
frequency decomposition; (3) averaging observations of regions where
galactic emission is not prominent; (4) fitting parameterized
physical models to the data; and (5) inversion techniques. Though
some models incorporate aspects of several of these techniques, I
shall try to illustrate each in what follows.

2.1 Time variation

Analyses of temporal variations in the observed diffuse emission have the advantages that the variable component is certainly zodiacal emission and that the results often are relatively insensitive to photometric uncertainties. Methods applied so far have the disadvantage that they determine limited specific properties of the IPD cloud rather than a full photometric model. Two examples used to study the geometry of the IPD cloud have been determination of the annual variation of the difference between the North and South ecliptic polar brightnesses, and determination of the annual variation of the apparent symmetry axis of the emission (the ecliptic latitude about which the brightness distribution is symmetric in scans at fixed elongation). Both analyses assume that the IPD is symmetric about a plane. The polar difference analysis, carried out numerous times on the various revisions of the IRAS Zodiacal History File (Hauser et al., 1984; Rickard et al., 1985; Rickard and Hauser 1987), yields Ω, the ecliptic longitude of the ascending node of this plane, without further assumptions about cloud properties. Determination of the inclination angle of the plane relative to the ecliptic plane, ι, requires some knowledge of the density distribution normal to the ecliptic plane. Results are summarized in Table 1 and discussed in section 4.

Table 1. Symmetry surface of the interplanetary dust cloud

Inclination (deg)	Ascending Node deg	Reference
1.5	55	Hauser and Gautier (1984)
1.47 ± 0.10	50 ± 4	Dermott et al. (1986)
1.58 ± 0.1	43 ± 5	Vrtilek and Hauser (1987)
1.8 ± 0.1	75.7 ± 1.0	Rickard and Hauser (1987)
1.40 ± .05	67 ± 2	Good (1987)
1.6 − 3	77 − 110	Murdock and Price (1985)
3.0 ± 0.3	87 ± 4	Leinert et al. (1980)

The seasonal variation of the location of the apparent symmetry axis has been studied with several approaches. Hauser and Gautier (1984) identified the symmetry axis for each IRAS scan with the mid-latitude between points of equal intensity north and south of the ecliptic plane, and estimated ι and Ω by comparing with results calculated for simple models of the radial variation of dust density. Less model-dependent results have recently been obtained by Dermott, Nicholson, and Wolven (1986; and private communication) and Vrtilek and Hauser (1987). The former group found the symmetry latitude by first separating the high angular frequency zodiacal band contribution from the low angular frequency main zodiacal cloud contribution using Fourier analysis of the scan data. The symmetry latitude was then found from the main cloud profile. The latter group fit a smooth parameterized empirical function to the lower envelope of the data for scans near 90° elongation, obtaining the symmetry latitude as one of the parameters. Dermott et al. noted that the times at which scans looking ahead of (leading) and behind (trailing) the direction of the Earth's orbital motion yield symmetry latitudes of equal magnitude but opposite sign are the times at which the Earth is in the ascending or descending line of nodes. The symmetry latitude at these times is the inclination of the symmetry plane. Results from the two methods applied to the IRAS 25 μm data are very consistent (see Table 1). Both groups also found that the amplitudes of the annual variation of the apparent symmetry latitude in the leading and trailing directions are not the same, suggesting that the dust distribution is not strictly heliocentric. A more quantitative interpretation of this effect requires analysis of detailed cloud models.

2.2 Angular frequency decomposition

Many investigators have assumed that the zodiacal emission varies only slowly over the sky, with the exception of the relatively low contrast zodiacal band features and comet trails, and have identified the zodiacal emission with the low frequency part of the IRAS scan data. This is clearly subject to significant systematic error, since, for example, any isotropic extragalactic background or large-scale galactic gradient would erroneously be called zodiacal emission. Nevertheless, at those wavelengths where zodiacal emission dominates anyway, the errors are presumably small.

The Fourier decomposition and lower envelope empirical function fitting described above are examples of this approach.   A comprehensive model of zodiacal emission in the IRAS data based on this assumption was reported by Good, Gautier, and Hauser (1984) (see also discussion by Hauser and Houck 1986).   This model, in which ellipses were fit to the lower envelope of each scan when represented as polar coordinate plots with intensity I and inclination angle i (angle between the ecliptic plane and the plane defined by the line of sight and the observer-Sun line) as radial and angular coordinates respectively, yielded typical residuals of several percent and systematic shape errors at low ecliptic latitude.   It also did not include the zodiacal bands, and often was not successful in fitting the 100 μm data.   Characteristics of this model are summarized in Table 2.

2.3 Averaging techniques

An alternate approach to separating galactic and IPD emission is to average data over regions of the sky where galactic emission is weakest.   Jongeneelen, Deul, and Burton (1985) took this approach in order to remove zodiacal emission at 60 and 100 μm from the IRAS quick-look data product Spline I.   They incorporated a number of simplifying assumptions about symmetries of the zodiacal emission in their analysis, but their residual maps were suitable for study of global interstellar emission, particularly at low galactic latitude (Table 2).

A more complete empirical model for the zodiacal emission as seen by IRAS based upon few a priori assumptions and yielding very small residuals has recently been developed by Boulanger (private communication; see Boulanger and Perault 1987 for a brief description).   The observed zodiacal emission can be represented as a function of time t (due to the Earth's motion within the cloud), elongation angle ε (the angle between the line of sight and the Sun), and ecliptic latitude β.   It is explicitly assumed that the zodiacal emission varies smoothly with ε and time.   The model consists of a set of average scan profiles (ASP), that is, intensity (at each IRAS wavelength) as a function of β at finite intervals of t and ε, such that the intensity at arbitrary (t, ε, β) can be found accurately by tri-linear interpolation.   Note that the angular frequencies

Table 2. Models of zodiacal emission based on IRAS data.

Reference	Method	Bands Covered*	Time Covered*	Elongation Covered (deg)	Zodi Bands incl.?	Isotropic Bkgd. incl.?	Residuals	Systematic Errors	Data Source
Good, Gautier, and Hauser (1984)	lower envelope ellipse in I-i space (4 parameters per scan)	12	H1-H3	60-120	No	Yes	~several %	Systematic shape errors; secular trends in parameters	ZOHF
		25	"	"	"	"	"		"
		60	"	"	"	"	"	No fit for many scans	"
		100	"	"	"	"	"		"
Jongeneelen, Deul, and Burton (1985)	average profile, elongation dep.	60	H1-H3	60-120	Yes	Yes	1.6 MJy/sr	Band remnants. Large scale artifacts	Spline I
	scaled to 60 μm	100	H1-H3	60-120	Yes	Yes	0.8 MJy/sr	Large scale artifacts	Spline I
Good (1987)	parameterized physical model (12 parameters)	12	H1-H3	60-120	No	No	~few % zodi	Ad hoc constant 2.6 MJy/sr	ZOHF
		25	"	"	"	"	"	9.1 MJy/sr	"
		60	"	"	"	"	"	1 MJy/sr	"
		100	"	"	"	"	--	--	"
Boulanger and Perault (1987)	average scan profiles	12	H1-H2	81-99	Yes	Yes	0.1 MJy/sr (|b|>50°) 1 MJy/sr (b ~ 0°)	Includes extended gal. emission	ZOHF
	"	25	H1-H2	81-99	Yes	Yes	0.2 MJy/sr (|b|>50°) 2 MJy/sr (b ~ 0°)	Includes extended gal. emission	ZOHF
	"	60	H1-H2	81-99	Yes	Yes	0.4 MJy/sr (|b|>50°) ~0.5 MJy/sr (b ~ 0°)	Less accurate near gal. plane	ZOHF
	"	100	H1-H2	81-99	Yes	Yes	0.5 MJy/sr	Less accurate near gal. plane	ZOHF

*H1, H2, and H3 denote IRAS Hours-confirmed sky coverages 1, 2, and 3 respectively.

represented in the model are determined by the sampling rate in β, which can readily be chosen to include the zodiacal emission bands in the model.

To separate galactic and zodiacal emission contributions, Boulanger determined the ASP's from the IRAS Zodiacal Observation History File (ZOHF) in an iterative fashion. The sky regions over which data were averaged were first restricted to exclude galactic latitudes less than 25° and the Magellanic Clouds. After completing determination of ASP's and finding maps of residual sky emission, additional regions of substantial (presumably galactic) emission were identified and excluded in a subsequent repeat of the averaging process. Large-scale galactic emission at 60 μm was largely eliminated from the ASP's by subtracting a component proportional to velocity-integrated HI emission at each point. The 100 μm ASP's were obtained by scaling those at 60 μm using an average ratio of 60 to 100 μm zodiacal intensity (found after subtraction of the HI-correlated galactic emission from each). In practice, it was also necessary to find separate ASP's for IRAS scans taken in the Earth-leading and trailing directions, since significant differences exist between these profiles.

The residual sky brightness maps obtained by subtracting the Boulanger model from the IRAS scans appear to be the least-biased and highest fidelity view we now have of galactic infrared emission on a large scale (Table 2). In order to minimize the effects of residual small calibration errors in individual scans at 12 and 25 μm, where galactic emission is small compared to zodiacal emission, Boulanger used a lower envelope fit of the ASP to each scan to obtain a gain and offset correction for that scan before subtracting. It is evident that any isotropic component in the sky brightness remains in the zodiacal emission model determined in this fashion. It is also true that spatially slowly varying 12 or 25 μm galactic emission at high galactic latitude remains in the zodiacal model.

2.4 Parameterized physical models

A very different approach to determining zodiacal emission is to construct a physical model of the IPD cloud, and then to evaluate unknown model parameters by fitting calculated brightnesses to the observational data. This has the virtues of yielding a model which can be represented with a relatively small number of parameters and which can directly be used to obtain meaningful physical properties of the cloud. It has the disadvantages of explicitly incorporating assumptions about the nature and distribution of the dust and of being computationally intensive, both to determine the parameters and to evaluate emission at an arbitrary direction and time from the model.

Good (1987) has recently generated a model of this type using the IRAS data (Good, Hauser, and Gautier (1986) reported an early version of this model; see also papers by Rowan-Robinson and Reach and Heiles at this Conference). He assumed that IPD particles have spatially homogeneous properties, and adopted a modified fan model for the dust density distribution (Giese, Kneissel, and Rittich (1986) have reviewed such models):

$$n(r,z) = n_o (R_o/r)^\alpha \ \exp \ [-\beta \ (|z|/r)^\gamma] \qquad (R_o = 1 \ \text{AU})$$

where r and z are cylindrical coordinates relative to the dust symmetry plane, and n_o is the dust number density at 1 AU. The symmetry plane was characterized by its inclination ι and longitude of the ascending node Ω. He adopted a power law radial variation of dust temperature T,

$$T(r) = T_o (R_o/r)^\delta$$

where T_o is the temperature at 1 AU, and a dust emissivity law of the form

$$e(\lambda) = \begin{cases} 1 & \text{for } \lambda \leq \lambda_o \\ \lambda_o/\lambda & \text{for } \lambda > \lambda_o. \end{cases}$$

The overall brightness scale determines the infrared volumetric absorption coefficient at 1 AU, ρ_o, which is the product of n_o, the mean particle cross-sectional area, and the radiative absorption efficiency at wavelength λ_o. To obtain reasonable fits, Good found it necessary to express the zodiacal emission as the integrated line-of-sight radiance of this model plus an additive constant C_λ for each spectral band. The model was fitted to the lower envelope of the data for 200 IRAS scans selected to span the time and elongation coverage of the mission. Preliminary values for the parameters are summarized in Table 3.

Table 3.   Parameters of a physical IPD cloud model
(Good 1987)

Dust Density	Dust Temperature
$\rho_O = 1.4 \times 10^{-20}$ cm^{-1}	$T_O = 263$ K
$\alpha = 1.7$	$\delta = 0.27$
$\beta = 4.05$	Dust emissivity
$\gamma = 1.16$	$\lambda_O = 36$ μm
Symmetry plane orientation	Brightness offsets
$\iota = 1.4° \pm .05°$	$C_{12} = 2.6$ MJy sr^{-1}
$\Omega = 67° \pm 2°$	$C_{25} = 9.1$ MJy sr^{-1}
	$C_{60} = 1.0$ MJy sr^{-1}

2.5 Inversion methods

Classical inversion methods used to invert the brightness integral using optical zodiacal light data can also be applied to infrared data to obtain the volume emissivity function. This has been discussed by Hong and Um (1987), who applied their analysis in the ecliptic plane to the Zodiacal Infrared Photometer rocket data of Murdock and Price (1985) at 11 and 21 μm. They find that the radial dust density distribution can not be represented by a single power law applicable to both wavelengths and all elongations, suggesting that the IPD particle properties are not spatially homogeneous.

3. SUMMARY OF MODEL ATTRIBUTES

The characteristics of the global zodiacal emission models discussed here are summarized in Table 2. The relative merits of the models depend upon the intended application. For the general purpose of describing the infrared sky brightness, e.g., for planning future experiments, all are quite sufficient. All share the disadvantage that they are not fully described in the published literature, and are not yet widely available. However, they do shed light on the merit of various approaches, and will be discussed from that point of view.

The most complete representation of the time, elongation angle, and spectral range encompassed in the IRAS data is provided by the models of Good, Gautier, and Hauser (1984) and Good (1987) (note that the first of these was evaluated for a now-obsolete version of the ZOHF). The least model-dependent approach, and hence the least biased for studies of the zodiacal emission itself, is that described by Boulanger and Perault (1987). However, this model does not include the HCON 3 data, which limits its time and elongation angle coverage.

The Boulanger and Perault model is the only one of these models which covers all four spectral bands and which represents the zodiacal band emission, making it also the most useful for study of the residual galactic emission. The values in the 'Residuals' column of Table 2 represent estimates of the rms noise in the residual maps due to lack of fidelity in the zodiacal model. These should be compared with typical zodiacal emission at the ecliptic plane near 90° elongation of 40, 85, 28, and 10 MJy/sr at 12, 25, 60, and 100 μm respectively, and galactic emission of .05, .08, .2, and 1 MJy/sr in the same bands at the galactic poles (Boulanger and Perault 1987). Whereas the residual noise in the maps of Boulanger and Perault is impressively only ~0.3% of the peak zodiacal emission at high galactic latitude at 12 and 25 μm, it is still somewhat larger than the mean galactic brightness at the galactic poles. This clearly demonstrates the challenge of determining the extra-solar system sky brightness in the infrared. However, these residuals are certainly small relative to galactic emission near the galactic plane.

The model by Good (1987) is the only one of these four which could in principle discriminate an isotropic background from zodiacal emission. Though appreciably non-zero additive constants were found at 12, 25, and 60 μm, it is not clear whether these constants indicate real isotropic emission or calibration problems. This model does more readily provide insight into physical conditions in the IPD cloud than the others, as discussed below.

4. SUMMARY OF INTERPLANETARY DUST CLOUD PROPERTIES

While it is not my intent to undertake a thorough review of IPD cloud properties as we now understand them, some comments on the implications of the analyses discussed above seem in order.

I have already noted the evidence for lack of axisymmetry in the IPD cloud from the asymmetries in annual variation of the latitude of peak emission in the Earth-leading and trailing directions. There is also evidence supporting the suggestion (Misconi 1980) that there is not a plane of symmetry for the IPD, but rather a warped surface with different orientation in different parts of the solar system. This may be seen in the differences of the symmetry plane parameters shown in Table 1. For example, the analyses of Hauser and Gautier (1984), Dermott et al. (1986), and Vrtilek and Hauser (1987), which depend upon cloud properties near the ecliptic plane somewhat exterior to 1 AU, yield ι and Ω values close to those of the Mars orbital plane ($\iota = 1.8°$, $\Omega = 49°$). On the other hand, the Helios data of Leinert et al. (1980), which sampled the cloud interior to 1 AU, yield values more reminiscent of the Venus orbital plane ($\iota = 3.4°$, $\Omega = 76°$). The ecliptic polar analysis based upon IRAS data (Rickard and Hauser 1987) and the IRAS global model of Good (1987), which sample intermediate regions, yield intermediate values of ascending node.

The steep radial decrease of dust density and shallow radial gradient of dust temperature found by Good (1987) (Table 3) suggest a decrease in particle albedo with heliocentric distance. This is in accord with the conclusions of Dumont and Levasseur-Regourd (1986), based upon analysis of IRAS data, and of Lumme and Bowell (1985), based upon optical intensity and polarization data. The volumetric absorption coefficient at 1 AU found by Good, $\rho_0 = (1.4\pm0.05) \times 10^{-20}$ cm^{-1} for wavelengths less than 30 μm compares favorably with that of $(0.5 - 0.7) \times 10^{-20}$ cm^{-1} deduced by Hong and Um (1987) from the Murdock and Price (1985) data when one allows for the 40% calibration discrepancy between the two experiments.

Finally, the Good (1987) model can be integrated to yield an estimate of the total IPD cloud mass of 3×10^{18} g, assuming a mean particle radius of 15 μm, particle density of 3 g cm^{-3}, and maximum cloud radius of 3.3 AU. This is somewhat lower than the estimate of 15×10^{18} g derived under the same assumptions from optical data by Lumme and Bowell (1985). Both estimates are very much smaller than the estimated range of ~$6 \times 10^{25} - 2 \times 10^{30}$ g for the dust shell mass of the Vega system (Aumann et al. 1984), underscoring the difference in nature of the two dust systems.

5.   CONCLUSIONS AND FUTURE PROSPECTS

We have seen that considerable effort has already been expended on extracting, modeling, and interpreting the zodiacal emission. Though none of these efforts has yet produced a definitive model best for all purposes, considerable progress has been made and much insight gained. Simple assumptions often made regarding the IPD cloud, such

as axisymmetry, symmetry with respect to a plane, spatial homogeneity
of particle properties, and lack of small scale structure, have all
been shown to have their limitations. The problem of unequivocal
separation of zodiacal emission from other large-scale components of
infrared emission remains a challenge, particularly in determining
large-scale galactic emission accurately at 12 and 25 μm, zodiacal
emission accurately at 60 and 100 μm, and extragalactic emission at
any infrared wavelength. Progress with both modeling approaches and
calibration consistency suggest that IRAS data will ultimately yield
zodiacal emission models superior to present versions.

In the future, we can look forward to major new zodiacal
emission data from the Diffuse Infrared Background Experiment (DIRBE)
on NASA's Cosmic Background Explorer mission (Mather 1982), now
scheduled for launch in early 1989. With absolute photometry from 1
to 300 μm, including linear polarization measurements at 1.2, 2.2,
and 3.2 μm, and hundreds of observations of each sky position over
the course of a year, the DIRBE data promise to enhance substantially
our ability to extract and model the various large-scale emission
components in the infrared sky.

Acknowledgments

I have been greatly assisted in the preparation of this review by the
insights and sharing of pre-publication results of many colleagues,
especially F. Boulanger, S. Dermott, J. Good, L. J. Rickard, and J.
Vrtilek. The cheerful and efficient assistance of the IPAC
librarian, R. Hernandez, was an essential aid to compilation of the
literature. This work has been supported in part by NASA's IRAS
Extended Mission Program.

References

Aumann, H. H., et al., 1984. Astrophys. J. (Letters), 278, L23.
Beichman, C. A., 1987. Ann. Rev. Astron. Astrophys., 25, 521.
Boulanger, F., and Perault, M., 1987. Preprint subm. to
    Astrophys. J..
Dermott, S. F., Nicholson, P. D., and Wolven, B., 1986.
    Asteroids, Comets, Meteors II, ed. C-I. Lagerkvist et al.,
    (Uppsala) p. 583.
Dumont, R. and Levasseur-Regourd, A. C., 1986. Light on Dark
    Matter, ed. F. P. Israel (D. Reidel Publ. Co., Dordrecht),
    p. 45.

Giese, R. H., Kneissel, B., and Rittich, U., 1986. _Icarus_, <u>68</u>, 395.

Good, J. C., 1987. Preprint.

Good, J. C., Gautier, T. N., and Hauser, M. G., 1984. _Bull. Am. Astr. Soc._, <u>16</u>, 921.

Good, J. C., Hauser, M. G., and Gautier, T. N., 1986. _Adv. Space Res._, <u>6</u>, 83.

Hauser, M. G., _et al._, 1984. _Astrophys. J. (Letters)_, 278, L15.

Hauser, M. G., and Gautier, T. N., 1984. _Bull. Am. Astr. Soc._, <u>16</u>, 495.

Hauser, M. G., and Houck, J. R., 1986. _Light on Dark Matter_, ed. F. P. Israel (D. Reidel Publ. Co., Dordrecht), p. 39.

Hong, S. S., and Um, I. K., 1987. _Astrophys. J._, <u>320</u>, 928.

Jongeneelen, A. A. W., Deul, E. R., and Burton, W. B., 1985. Preprint.

Leinert, C., Hanner, M., Richter, I., and Pitz, E., 1980. _Astron. Astrophys._, <u>82</u>, 328.

Low, F. J., _et al._, 1984. _Astrophys. J. (Letters)_, <u>278</u>, L19.

Lumme, K., and Bowell, E., 1985. _Icarus_, <u>62</u>, 54.

Mather, J. C., 1982. _Opt. Eng._, <u>21</u>, 769.

Misconi, N. Y., 1980. _Solid Particles in the Solar System_, eds. I. Halliday and B. A. McIntosh (D. Reidel Publ. Co., Dordrecht), p.49.

Murdock, T. L., and Price, S. D., 1985. _Astron. J._, <u>90</u>, 375.

Rickard, L. J., Dwek, E., White, R. A., and Hauser, M. G., 1985. _Bull. Am. Astr. Soc._, <u>17</u>, 591.

Rickard, L. J., and Hauser, M. G., 1987. In preparation.

Sykes, M., _et al._, 1986. _Science_, <u>232</u>, 1115.

Vrtilek, J., and Hauser, M. G., 1987. In preparation.

SEPARATING THE SOLAR SYSTEM AND GALACTIC CONTRIBUTIONS TO THE DIFFUSE INFRARED BACKGROUND

William Reach and Carl Heiles

Astronomy Department
University of California
Berkeley, CA 94720, USA

ABSTRACT

A model for the zodiacal emission detected by IRAS is presented. The model allows for power-law variations with heliocentric distance of the grain temperature and density, and includes an outer edge with the same ellipsoidal shape as the density contours. Results of a preliminary least-squares fits suggest that the outer edge is located near the edge of the asteroid belt, that the grains absorb and radiate like blackbodies, and that the density is consistent with Poynting-Robertson drag on grains originating in the asteroid belt.

1. INTRODUCTION

The diffuse IR emission is due to at least two distinct components: Galactic and zodiacal (Solar System) emission. The zodiacal emission (ZE) dominates in the 12 and 25μm bands, and at 60μm the zodiacal and Galactic plane brightnesses are comparable. At 100μm the Galactic plane clearly dominates, but intermediate and high latitude Galactic features are comparable in brightness to the ZE. Reliable color temperatures from the IRAS diffuse emission maps are difficult to determine, particularly over large angular scales. Thus an accurate model for the infrared background is needed for practical reasons, over and above its physical significance.

We believe that the ZE and Galactic emission can be accurately separated by a global least-squares fit to a model of the zodiacal dust distribution together with the galactic HI. An important feature of this approach is that all four IRAS bands, and both the Galactic and ZE models, will be used simultaneously to calculate the fit. The residuals can then be examined for correlations with ecliptic coordinates or with regions of strong Galactic emission, to test for inadequacies of the fit. Previously undetected astronomical backgrounds which may appear in the residuals include Galactic emission associated with the ionized component of the ISM, Solar System emission from comets in the hypothetical inner Oort cloud, and a cosmological background radiation. In this paper we present preliminary results on the ZE model.

2. Characterization of the Zodiacal Emission

To understand the angular distribution of the infrared background, it is advantageous to consider particular wavelengths and coordinate systems. Figure 1 presents two examples of 25μm maps in different coordinate systems. It is clear that the appropriate coordinates for displaying the ZE are not the standard ecliptic longitude and latitude, but rather the solar elongation and ecliptic latitude. Thus it is not feasible to attempt to model the infrared background without knowledge of the location of the sun. Further, two geometric effects, caused by the fact that IRAS observations occurred as the Earth traversed nearly an entire orbit, are also evident in the data, as discussed below. These effects can all be modelled if, in addition to the direction of the line of sight, the time of observation is specified. We have used the time-ordered Zodiacal Observation History File, with $1/2°$ resolution, as our data base.

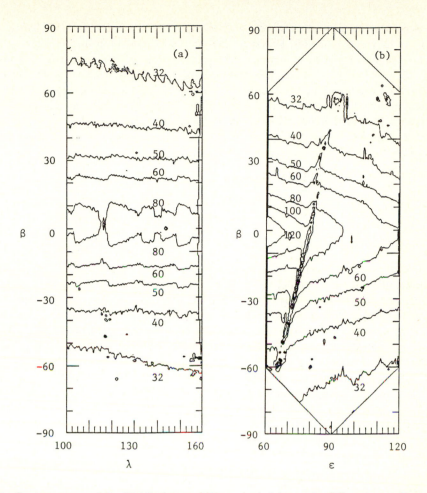

Figure 1. Contour maps of the IRAS-observed 25μm sky brightness in different coordinate systems. *(a)* Ecliptic longitude, λ, and ecliptic latitude β. *(b)* Solar elongation angle (between the solar direction and the line of sight), ϵ, and ecliptic latitude. Contours in both figures are labelled in units of MJy/Str. The variation of brightness with the coordinates in *b* is real, while some of the variation in *a* is not. Note in particular the sawtooth pattern in the highest contour level in *a*; the jogs are due to abrupt changes of the elongation angle, which is held constant for each scan, from one orbit to another. Also note the slope of the lowest contour in *a*; this is due to the changing height of the satellite with respect to the dust symmetry plane. Contours in *b* are smooth and vary in a natural way, except near the galactic plane, which runs from $(\epsilon, \beta) = (60°, -60°)$ to $(100°, 70°)$, passing through the galactic center near $(75°, -15°)$. Map *a* covers the galactic anticenter, where the galactic 25μm emission is very weak.

A table of the optically-observed zodiacal light (ZL), in coordinates similar to those shown in Figure 1b, was presented by Levasseur-Regourd and Dumont (1980). Since the spectrum of the ZL is the same as the solar spectrum, the ZL is interpreted as sunlight scattered by interplanetary dust. The similarity of the angular distributions of the ZL and ZE suggests a common origin. Indeed, the existence of the dust required to produce the ZL implies that there must be an infrared background due to emission of the fraction of the incident sunlight which is absorbed. Thus we interpret the ZE as thermal emission from the same dust grains which produce the ZL. This fact allows us to use the results of optical modelling as an important stepping stone from which to begin to model the infrared background.

3. A Model for the Zodiacal Emission

The observed intensity is equal to the integral over the line of sight of the infrared emissivity,

$$I_\nu = \int_0^{d_{max}} dx \, n \, \sigma_\nu \, B_\nu(T) \tag{1}$$

where n is the dust density, σ_ν is its absorption cross section, $B_\nu(T)$ is the Planck function evaluated at the temperature of the dust, and the upper limit to the integral, d_{max}, represents the outermost edge of the dust distribution for the specified line of sight. The temperature was approximated by a power law in heliocentric distance, r. The expected dependence for large blackbody grains is $r^{-1/2}$, with a local value of 280 K at 1 AU. The dust density distributions which have been successful in modelling the ZL are usually separated into one function of heliocentric distance multiplied by another function of heliocentric latitude. We have found that an acceptable fit is obtained by using a power law in heliocentric distance and an ellipsoid-shaped function of the heliocentric latitude, i.e.

$$n = n_o r^{-\nu} [1 + (\gamma \sin \beta_{\odot z})^2]^{-\nu/2} \tag{2}$$

The latitude, $\beta_{\odot z}$, in this expression is zero in the dust symmetry plane, the location of which is discussed below. From space-borne observations of the ZL in the inner solar system, the radial power law index is known to be about $\nu = 1.3$ (Lienert, 1981). Also, the ratio of the major to minor axes of the dust distribution is specified by γ, which is near $\gamma = 6.5$ to reproduce optical observations (Giese, 1986).

Integrating this density distribution out to a heliocentric distance r, the total number of particles in the cloud is found to be proportional to $r^{1.7}$. Clearly the density cannot continue a $r^{-1.3}$ decrease for all r, since the total number of particles implied diverges so rapidly. In order to take this into account, we have introduced an outer edge to the density distribution, which terminates the intensity integral at an upper limit, d_{max}. As a first attempt, the upper limit was taken to be the same for all lines of sight. We found that with this assumption, a negative constant had to be added to the model in order to fit the IRAS data. As an alternative, d_{max} for each line of sight was calculated by assuming that the outer edge of the cloud has the ellipsoidal shape of the density contours. Using this assumption, the choice of major axis for the cloud which minimized the error in a simultaneous least-squares fit to the $12\mu m$ and $25\mu m$ data was 3.6 AU. Optimal values for the other parameters were found by minimizing the total error calculated when offsets from the estimated values were applied. The density radial power-law index was found to be $\nu = 1.3 \pm .15$; the local temperature was found to be 274 ± 2 K; and the power-law index of the temperature variation was found to be 0.50 ± 0.05. No evidence was found for a wavelength dependence of the absorption cross-section.

4. Geometry of the Zodiacal Dust Distribution

The fact that the ZE and ZL isophotes are not parallel to the ecliptic (see *e.g.* Figure 1*a*) is due to the fact that the surface of maximum dust density is not parallel to the ecliptic. The simplest assumption is that the surface is a plane, called the "symmetry plane", which intersects the ecliptic in a line that passes through the sun. The effect of the inclination of the symmetry plane is most clearly seen at high ecliptic latitudes, where the brightness is observed to have a clear annual variation. Sinusoidal curves fit to the variations were used to provide relative calibration of the IRAS extended emission data. It is evident that the north and south ecliptic polar brightness variations are well fit by a sinusoid, but the amplitude of the north polar brightness variation is consistently 27% brighter than that at the south pole.

We have found that this effect is predicted if the eccentricity of the Earth's orbit is taken into account when the brightness integral is evaluated. All IRAS-observed ZE brightnesses are modulated annually by this effect, with a phase set by the perihelion of the terrestrial orbit. Using the *difference* between the north and south polar brightness nearly eliminates the effect. We have fit the observed annual variation of this difference to sinusoids to find the ascending longitude of the intersection between the symmetry plane and the ecliptic; we obtain $\Omega = 79°$. Using model calculations of the amplitude of the annual variation, we find that the inclination of the symmetry plane with respect to the ecliptic is $i = 1.8°$.

5. Discussion

The ZE model presented here differs from previous models by allowing an outer edge of the dust distribution. The value of 3.6 AU obtained for the outer edge in the ecliptic has a straightforward interpretation, because it corresponds to the outer edge of the main asteroid belt. Some marginal direct evidence for a steepening of the density distribution in the asteroid belt was obtained by the optical cameras aboard Pioneer 10, but the ZL brightness was near the minimum detection level for $r > 2.5$ AU. Indeed, no evidence exists for significant amounts of dust in the outer Solar System.

The value obtained for the density radial power-law index, $\nu = 1.1$, is marginally consistent with the Helios value (1.3) and the theoretically-predicted value (1.0) for a steady-state equilibrium with Poynting-Robertson drag. Together with the outer edge, our results support an asteroidal origin for the dust which produces the ZE. The possibility of a radial gradient in the absorption cross-section, which would modify ν in equation (2), seems to be ruled out by the fact that the temperature gradient is so close to the theoretical prediction for blackbody grains. The temperature results suggest that the ZE is due to larger particles than those responsible for the ZL. Thus it may not be reasonable to calculate the albedo of the dust using the ratio of the observed ZE and ZL emissivities. We obtain an albedo of 0.09 from such an estimate, which is probably darker than the actual grains responsible for the ZE.

REFERENCES

Giese, R.H., Kneissel, B., and Rittich, U. 1986, *Icarus*, **68**,395.
Levasseur-Regourd, A.C. and Dumont, R. 1980, *Astron. Astrophys.*, **84**, 277.
Lienert, C., Richter, I., Pitz, E. and Planck, B. 1981, *Astron. Astrophys.*, **103**, 177.

ZODIACAL DUST PROPERTIES AS DEDUCED BY INVERSION OF IRAS OBSERVATIONS

A.-Ch. LEVASSEUR-REGOURD

Université Paris VI - Service d'Aéronomie

BP 3, 91371 Verrières le Buisson, France

R. DUMONT

Observatoire de Bordeaux, 33270 - France

ABSTRACT

The method of the nodes of lesser uncertainty, which allowed in zodiacal photometry to retrieve local information, is extended to the thermal case. Results about temperature T, global volume intensity $\mathcal{R}$, albedo of the dust $\mathcal{A}$, and their gradients with heliocentric distance r (in AU) are derived in the ecliptic from the available IRAS observations. Typically T is found to be of the order of 250 $r^{-0.32}$ K and $\mathcal{A}$ of the order of 0.09 $r^{-0.65}$.

1. INTRODUCTION

Many progresses have been made during the last 20 years in the study of the complex of interplanetary dust grains in the solar system, i.e. the zodiacal dust (Fechtig et al., 1981, Weinberg, 1985). It has been possible to collect grains in the Earth environment, to study their impacts on board space probes or to perform accurate measurements of the solar light they scatter and of their thermal emission. In a first approximation, it can be said that the grains are mainly fluffy particles, their size is in a 1-100μm range, there is a maximum of concentration near the ecliptic plane, and the density increases with decreasing distance to the Sun.

The grains are spiralling around the Sun under Poynting-Robertson effect, unless they are too small and are blown out by the radiation pressure. They are believed to originate from comets and/or asteroids. The question of the homogeneity, smoothness, and stability of the zodiacal dust has been disputed for a long time (Levasseur and Blamont, 1975, 1976, Dumont and Levasseur-Regourd, 1978). IRAS results do confirm that the zodiacal cloud is neither completely smooth (Low et al., 1984), nor really homogeneous (Levasseur-Regourd and Dumont, 1986). Instead of fitting a general model with IRAS observations, we derive our results from an inversion method adapted for the thermal emission from a method developped for the optical scattering case (see Dumont, 1983 or Dumont and Levasseur-Regourd, 1985).

2. INTERPLANETARY DUST ; THERMAL EMISSION

The quantity observed along the line of sight is the infrared brightnesses $I(\nu)$ (basic unit MJy sr^{-1} ; dimension MT^{-2})

$$I(\nu) = \int_{-\infty}^{x_0} \mathcal{E}(\nu) \, dx = F \int_{-\infty}^{x_0} \frac{\mathcal{R}(\nu)}{r^2} \, dx \qquad (1)$$

where $\mathcal{E}(\nu)$ is the monochromatic intensity of infrared emission, dx the elemental section of the line of sight, F the solar constant at 1 AU, r the heliocentric distance, and $\mathcal{R}(\nu)$ the "remote" (=relative monochromatic thermal emission) volume intensity.

We know that $\mathcal{R}(\nu) \rightarrow 0$ when $r \rightarrow \infty$ (fall of density + temperature) and that it exists a rotational symmetry (absence of seasonal dependance, after correction for oscillations of the Earth on either side of the symmetry plane ; see Dumont and Levasseur-Regourd, 1978, or Dermott et al., 1986). With one degree of freedom, the simplest mathematical model for $\mathcal{R}(\nu)$, in agreement with density laws and temperatures is found to be:

$$F\mathcal{R}(\nu) = a/r + b/r^2 \qquad (2)$$

Calling ϵ the elongation, r_0 the Sun-Earth distance, m the Sun-line of sight distance (=$r_0 \sin \epsilon$), equations (1) and (2) lead to $I = aA + bB$, where :
$A(m,\epsilon) = (1 + \cos \epsilon)/m^2$ $B(m,\epsilon) = (\pi - \epsilon + \sin \epsilon \cos \epsilon)/2m^3$

For each elongation, there exists one peculiar point where, once $I(\nu)$ is measured, $\mathcal{R}(\nu)$ is retrieved with less uncertainty than elsewhere. All the curves $F\mathcal{R}(\nu)$ satisfying the observed integral $I(\nu)$ focus, or at least constrict in one point: for abscissa $r_1 = A(m,\epsilon)/B(m,\epsilon)$
and ordinate $F\mathcal{R}(\nu) = I(\nu) \ B(m,\epsilon)/A^2(m,\epsilon)$ at r_1.

Note that numerous gradients of the temperature (from $r^{-0.3}$ to $r^{-0.7}$) and of the space density (from $r^{-0.6}$ to $r^{-1.4}$) have been combined to derive $F\mathcal{R}(\nu)$ for typical elongations ; the r.m.s. deviation at the node remained weak and the method was found to be moderately model dependant.

3. TEMPERATURE OF THE DUST AT 1 AU ; GRADIENT IN THE ECLIPTIC

IRAS allows to derive $I(\nu)$ for 12 μm and 25 μm at various elongations (see Hauser et al., 1984, Hauser and Houck, 1986 or Zodiacal History File). From the nodes of lesser uncertainty method, and for given heliocentric distances $r_1(m,\epsilon)$, $\mathcal{R}(12\mu m)$ and $\mathcal{R}(25\mu m)$ are therefore obtained. Assuming the absorption cross section to be the same at all wavelengths (greybody case), Planck law gives, from the ratio $\mathcal{R}(12\mu m)/\mathcal{R}(25\mu m)$ at r_1, the local temperature $T(r_1)$.

We typically obtain T = 250 K at 1 AU (see also Dumont and Levasseur-Regourd, 1986). In the ecliptic, at least in a 1 AU - 1.3 AU range,
$$T(r) = 250 \, r^{-0.32\pm0.08} \text{ K}$$

Note that ZIP data (see Murdock and Price, 1985), would lead to a larger value for T ($\simeq$ 40 K discrepancy), but to the same gradient r=0.32.

Such a gradient ($\neq$ 0.5) conflicts with the greybody assumption. However the results remain valid if the volumetric absorption cross section is assumed to have two distinct values C_{abs}^{vis} in the visible and C_{abs}^{IR} in the infrared (see Röser and Staude, 1978). The deviation of the temperature gradient from -0.5 comes from the distinct heliocentric changes of the two C_{abs}.

4. RELATIVE GLOBAL VOLUME INTENSITY; ITS GRADIENT IN THE ECLIPTIC

Given the local temperature T, $\mathcal{E} = \int_{o}^{\infty} \mathcal{E}(\nu) \, d\nu$ and $\mathcal{R} = \int_{o}^{\infty} \mathcal{R}(\nu) \, d\nu$,

so called the relative global volume intensity, can be derived :

$$\mathcal{E} = \frac{\pi^4}{15}\left(\frac{h}{k}\right)^4 \nu^{-3} \left(\exp \frac{h\nu}{kT} - 1\right) T^4 \, \mathcal{E}(\nu) = \beta(\nu,T)\,\mathcal{E}(\nu)$$

$$\mathcal{R} = \beta(\nu, T)\,\mathcal{E}(\nu) \; r^2/F$$

IRAS data allow to obtain $\mathcal{R}$ and its gradient, again in good agreement for the gradient with ZIP

$$\mathcal{R} \simeq 13 \quad 10^{-9} \; r^{-0.7} \; AU^{-1}$$

In the non gray case, this gradient $\rho = -0.7$ is demonstrated to be also the gradient of C_{abs}^{vis}, while the heliocentric gradient of C_{abs}^{IR} is found to be twice steeper and equal to 1.4.

5. ALBEDO AND ITS GRADIENT IN THE ECLIPTIC

The local albedo $\mathcal{A}$ of the zodiacal dust is accessible if the thermal energy reemitted in the IR ($\mathcal{R}$) and the energy scattered in the visual domain ($\mathcal{D}$) are known

$$\mathcal{A} = \int_{4\pi} \mathcal{D} \, d\omega \; [4\pi\mathcal{R} + \int_{4\pi} \mathcal{D} \, d\omega]$$

The directional scattering coefficient $\mathcal{D}$ depends on location and scattering angle θ. From all observational sources of the visual brightness, we have obtained (Dumont and Levasseur-Regourd,1985) $\mathcal{D}$ at r = 1 AU, θ = 90° and its heliocentric gradient δ

$$\mathcal{D}_{90} = (1.3 \overset{+}{-} 0.2) \; 10^{-9} \; r^{(-1.4 \pm 0.3)} \; AU^{-1}$$

The albedo at 90° scattering angle, and its heliocentric gradient α can therefore be obtained

$$\mathcal{A}_{90} = \mathcal{D}_{90} / \mathcal{R} + \mathcal{D}_{90} \qquad\qquad \alpha = (1 - \mathcal{A}_{90})(\delta - \rho)$$

The albedo and its heliocentric gradient are obtained from a combination of IRAS and zodiacal light observations.Typically,

Weinberg (*ground based*) 64 gives :

Roach and Gordon (*ground based*)73

Frey et al (*balloon*) 74

Levasseur-Regourd and Dumont (*s/c + ground 80*)

$\tilde{A}_{90} = 0.107 \ r^{-0.28}$

$\tilde{A}_{90} = 0.091 \ r^{-0.87}$

$\tilde{A}_{90} = 0.079 \ r^{-0.79}$

$\tilde{A}_{90} = 0.082 \ r^{-0.65}$

The albedo at $r = 1AU$, $\theta = 90°$ is of the order of 9% ; its heliocentric gradient, definitely negative, could be in a -0.5 to -0.8 range. Such a small value of the albedo at 1 AU, and an even smaller one farther from the Sun ($\simeq 4\%$ at 4 AU), rule out the possibility for grains to be icy. It is rather in favour of dark fluffy particles breaking off and evaporating while spiraling towards the Sun.

6. CONCLUSION

Despite the great difficulty of retrieving local information from brightnesses integrated along the line of sight, the method of the nodes of lesser uncertainty allows to retrieve various parameters of the zodiacal cloud in the ecliptic. By such an approach, further analyses of IRAS data should allow to derive unambiguously a precise description of the physical properties and of the evolution of interplanetary grains.

References

Dermott, S.F., Nicholson, P.D., & Wolven, B., 1986 in *Asteroids, comets, meteors II*, Uppsala Universiteit, 583

Dumont, R., 1983, *Planet. Space Sci.*, **31**, 1381

Dumont, R & Levasseur-Regourd A-C., 1978, *Astron. Astrophys.* 64,9

Dumont, R & Levasseur-Regourd, A-C., 1985, *Planet. Space Sci.*,**33**, 1

Fechtig, H., Leinert, C. & Grün, E., 1981 *Landolt-Bornstein VI*, **2a**, 228

Frey, A., Hofmann, W., Lemke, D. & Thum, C. 1974, *Astron. Astrophys.*36,447

Levasseur-Regourd, A-C. & Blamont, J.E., 1975, *Space Research*, **XV**, 573

Levasseur-Regourd, A-C. & Blamont, J.E.,1976,*Lectures notes in Physics*, 48,58

Levasseur-Regourd, A-C & Dumont, R., 1980, *Astron. Astrophys.* **84**, 277

Levasseur-Regourd, A-C.,& Dumont, R., 1986, *Advances Space Res.*, **6**, 7, 87

Low F.J., Beintema, D.A., Gautier, T.N., Gillett, F.C., Beichman, C.A., Neugebauer, G., Young, E., Aumann, H.H., Boggess, N., Emerson, J.P., Habing, H.J., Hauser, M.G., Houck, J.R., Rowan-Robinson, M., Soifer, B.T., Walker, R.G., Wesselius, P.R., 1984, *Ap. J.*, **278**, L19

Murdock, T.L. & Price, S.D., 1985, *Astron. J*1,**90**, 375

Roach, F.E. & Gordon, J.L., 1973, *The Light of the Night Sky*, Reidel ed.

Röser, S. & Staude, H.J., 1978, *Astron. Astrophys.* **67**, 381

Weinberg, J.L., 1964, *Ann. Astrophys.* **27**, 718

Weinberg, J.L., 1985, in *Properties and interactions of interplanetary dust*, Reidel,1

THE INFRARED SPECTRUM OF COMET P/HALLEY

Thérèse ENCRENAZ
Observatoire de Paris, Section de Meudon
92195 Meudon-Cedex, France

ABSTRACT

Infrared observations of Comet P/Halley have provided the first unambiguous detection of parent molecules in a comet. The most abundant parent molecule is H_2O ; CO and CO_2 are also present in minor abundances. The IR spectrum of P/Halley has also revealed for the first time a 3 μm signature showing some analogy with the unidentified IR interstellar features. The cometary signature is interpreted by the presence of hydrocarbons, both in the saturated and the unsaturated form, with a total number of carbon atoms equal to about 30 % of H_2O. With this result it is possible to define a cometary composition which is in global agreement with the composition found for interstellar dust.

1 - INTRODUCTION

Infrared observations of comets have been performed for many years. However, before Comet Halley's apparition in 1985-86, they were mostly restricted to broad-band photometry (Ney, 1982). All infrared cometary spectra show the 2 characteristic components (Fig. 1) due respectively to reflected and scattered sunlight (below 3 μm) and to thermal emission of cometary dust (above 3 μm). Depending upon the heliocentric distance R of the comet, which affects the grain temperature through a $R^{-1/2}$ law, the maximum of the observed thermal emission ranges between about 6 μm (R $\cong$ 0.4 AU) and 14 μm (R $\cong$ 2 AU).

Signatures of ices and grains have been searched for at low or medium spectral resolution (R $\cong$ 50). A broad emission due to silicates has been observed in most of the comets, at 10 μm and 18 μm. Many attempts have been made to detect ice at 3 μm but, in spite of a tentative identification on Comet Cernis (Hanner, 1984) this research has not been fully conclusive at this time.

Before Comet Halley's last apparition, only one comet, the non-periodic comet West (1976 VI) was bright enough for high resolution spectroscopy. A Fourier-Transform spectrum was recorded between 0.9 μm and 2.5 μm by Johnson et al (1983) with a resolving power of about 3000, showing in particular the red system of CN.

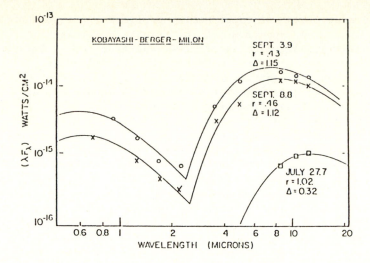

Fig.1 The infrared spectrum of Comet Kobayaschi-Berger-Milon at several heliocentric distances (from E.P. Ney, 1982).

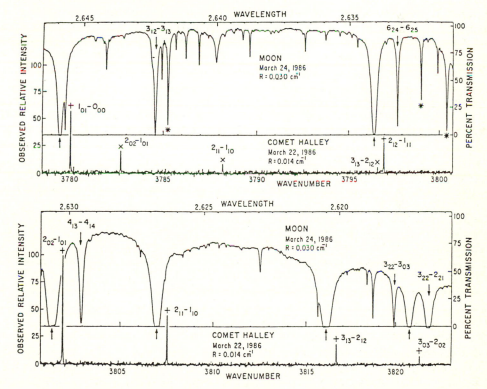

Fig. 2 High-resolution spectrum of Comet Halley at 2.7 μm, showing individual components of the ν_3 H_2O band (from Mumma et al, 1986).

The 1986 apparition of Comet Halley provided a unique opportunity to explore the infrared spectrum of a bright comet; at high spectral resolution, from the ground, from an aircraft and from space. Observations from above the Earth's atmosphere were especially needed to search for cometary parent molecules, directly outgassed from the nucleus. These molecules, which provide a key information about the nature of the cometary material itself, have strong transitions in the infrared and millimeter ranges, while the UV and visible ranges are, in contrast, best suited for the study of cometary radicals and ions.

Several types of infrared experiments have been performed on Comet Halley : (1) photometry and spectrophotometry, from ground-based telescopes ; (2) high resolution spectroscopy, from the Kuiper Airborne Observatory ; (3) medium resolution spectroscopy from the Vega probes. In addition high resolution spectroscopy has been performed from Mauna Kea Observatory in the near infrared, and spectrophotometry has been performed in the far infrared from the KAO and from the Lear Jet. Thanks to all these means, the spectrum of Comet Halley has been measured, from 1 μm to about 100 μm with many repeated photometric measurements over about 2 years (1985-86). Information has been obtained upon the nature of parent molecules, their production rate, their distribution as a function of the nuclear distance, the nature and properties of cometary grains, and their spatial distribution. Apart from the detection of new parent molecules in Comet Halley, the IR spectrum has also revealed a 3 μm emission feature showing some analogy with the "unidentified IR interstellar features". This unexpected result has been interpreted by the presence of carbonaceous material in the vicinity of the comet nucleus, but the exact identification of the 3 μm feature in Comet Halley is still an open question.

This paper analyses the major results obtained on the nature of parent molecules in P/Halley from IR spectroscopy (Section 2). A study of the 3 μm emission feature and its possible interpretation is given in Section 3. On the basis of these results, a comparison is shown between the abundances in Comet Halley, as derived from molecular observations, and the abundances of interstellar dust, as derived from IR observations of dense clouds (Section 4).

2 - DETECTION OF PARENT MOLECULES IN COMET HALLEY

In the near infrared range, cometary molecules are expected to be excited by resonent fluorescence from the solar infrared radiation field. Calculations have been performed before Halley's apparition to estimate which parent molecules could be detectable (Yamamoto, 1982 ; Crovisier and Encrenaz, 1983 ; Crovisier and Le Bourlot, 1983 ; Crovisier, 1984 ; Weaver and Mumma, 1984). Although not directly observed, H_2O was believed to be the most abundant constituant, on the basis of the presence of H_2O^+, and the H and OH abundances. This assumption has been confirmed with the first unambiguous detection of the H_2O ν_3 band (Fig. 2) at 2.7 μm from the Kuiper Airborne Observatory (Mumma et al, 1986). The same band was also observed with the IKS infrared experiment aboard Vega 1 (Fig. 3) which explored the

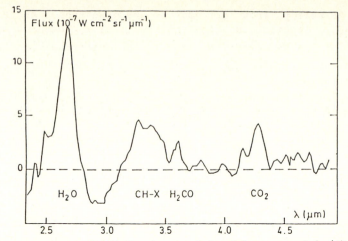

Fig. 3 The spectrum of P/Halley between 2.5 and 5 μm, recorded with the IKS-Vega spectrometer (from Moroz et al, 1987).

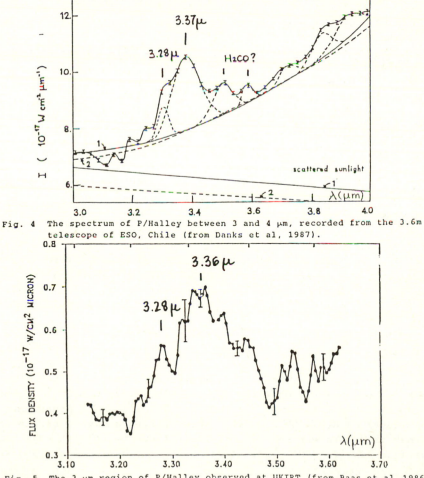

Fig. 4 The spectrum of P/Halley between 3 and 4 μm, recorded from the 3.6m telescope of ESO, Chile (from Danks et al, 1987).

Fig. 5 The 3 μm region of P/Halley observed at UKIRT (from Baas et al, 1986).

spectrum of P/Halley from 2.5 to 5 µm, and from 6 to 12 µm. The IKS instrument also detected CO_2 for the first time, with a mixing ratio $CO_2/H_2O = 2\ 10^{-2}$ (Moroz et al, 1987). The distribution of H_2O and densities as a function of nuclear distance r follows the r^{-2} law expected for parent molecules.

Other parent molecules were also tentatively identified : (1) H_2CO, at 3.5 µm and 3.6 µm from IKS and from ground-based observations (Moroz et al, 1987 ; Knacke et al, 1986 ; Danks et al, 1987) ; (2) CO, marginally present at 4.7 µm in the IKS data, and also observed in the UV range (Woods et al, 1986). If the IKS identifications are real, they correspond to the following mixing ratios : $H_2CO/H_2O \sim 10^{-2}$, $CO/H_2O \sim 0.2$.

Two species are absent from the list of parent molecules detected in P/Halley : CH_4 and NH_3. Upper limits of 0.04 and 0.10 have been derived respectively (Drapatz et al, 1986 ; Krankowsky et al, 1986). HCN has been detected from ground-based millimeter observations (Despois et al, 1986) with a mixing ratio $HCN/H_2O \sim 10^{-3}$. This amount is apparently not sufficient to account for the CN abundance derived from visible observations (Krasnopolsky et al, 1986).

3. THE 3 µm EMISSION FEATURE

The 3 µm feature, first revealed by the IKS experiment aboard the Vega 1 probe, is observed for the first time in a comet (Combes et al, 1986 ; Moroz et al, 1987). The density distribution of the emitter follows the distribution of a parent molecule. This detection has been confirmed by several ground-based observations (Knacke et al, 1986 ; Wikramasinghe and Allen, 1986 ; Baas et al, 1986 ; Danks et al, 1987). Examples are shown in Fig. 3, 4 and 5. It can be seen that, as the spectral resolution increases, the 2 components at 3.28 µm and 3.36 µm become well separated. Both components appear in emission.

These features appear at the same wavelengths as the interstellar features, but the latter appear in different conditions in the interstellar medium. Indeed the 2 features are not present in the same kinds of objects. The 3.28 µm is narrow and appears in emission near UV sources (Willner et al, 1977), associated with other strong emissions at 6.2, 7.7, 8.6 and 11.3 µm (Fig. 6). In contrast, the 3.4 µm feature is broader (Fig. 7), appears in absorption in dense clouds (Butchart et al, 1986) and is sometimes associated to weaker features, in particular at 6.8 µm. In the case of Comet Halley we see for the first time the 3.28 µm and the 3.4µm features both in emission.

Another striking point is the absence of associated features beyond 6 µm. Indeed, the spectrum of Comet Halley between 6 and 12 µm only shows the strong silicate emission band, observed from the ground (Tokunaga et al, 1986 ; Bouchet et al, 1987 ; Hanner et al, 1987), the KAO (Campins et al, 1986 ; Fig. 8) and with the IKS Vega 1 experiment. A preliminary reduction of the IKS data (Combes et al, 1986) had revealed a strong emission centered at 7.5 µm but a more careful analysis of the data has shown that this feature was due to an instrumental effect. The corrected IKS spectrum is in full agreement with the ground-based and KAO data

(Combes et al, 1987).

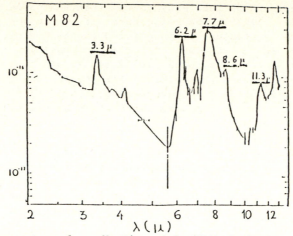

Fig. 6 The spectrum of a reflection nebula (M82) showing the "unidentified IR features" at 3.3 μm, 6.2 μm, 7.7 μm, 8.6 μm and 11.3 μm (from Willner et al, 1977).

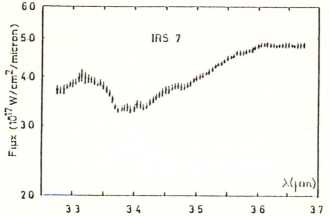

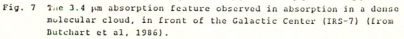

Fig. 7 The 3.4 μm absorption feature observed in absorption in a dense molecular cloud, in front of the Galactic Center (IRS-7) (from Butchart et al, 1986).

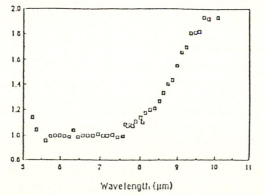

Fig. 8 The spectrum of P/Halley between 6 and 10 μm as observed from the KAO (Campins et al, 1986).

It has been known for many years that the 3.28 μm and 3.37 μm features were most likely the signature of C-H bonds in carbonaceous material. The 3.28 μm feature is associated to the C-H stretching mode of unsaturated hydrocarbons (-CH=CH-, CH=CH2) or aromatics (= C-H). In the case of interstellar features, Leger and Puget (1984) have pointed out the remarkable agreement between the interstellar spectra and the IR laboratory spectra of polycyclic aromatic hydrocarbons (PAH). As first suggested by Sellgren (1984), the excitation mechanism would be transient heating by a single UV photon.

The 3.37 μm signature is associated to the C-H stretching mode of saturated hydrocarbons. In dense clouds, these chains probably form grains larger than PAH, which cannot be heated by a single UV photon, and are observed in absorption in front of an IR source. The 3.37 μm interstellar feature shows some analogy with the organic material obtained in the laboratory from the irradiation of ices by UV radiation or high energy particles(Fig.9).

In the case of Comet Halley, the 3.28 μm and 3.36 μm emission features are likely to be due to the same material than in the interstellar medium, i e both unsaturated and saturated hydrocarbons, but the emission mechanism has to be different. The fact that both cometary features are in emission and the absence of associated features beyond 6 μm can be simply explained if we assume that the hydrocarbon molecules are excited by solar resonance scattering (or resonent fluorescence), as in the case of the other cometary parent molecules. As the pumping rate is strongly dependent upon the solar flux (Crovisier and Encrenaz, 1983), this mechanism decreases very rapidly towards longer wavelengths and cannot be efficient beyond 6 μm.

With this interpretation, it is possible to derive the abundance of carbon in the material responsible for the 3 μm feature. The intensities of the two 3 μm feature are actually functions of the number of C-H stretches in the material. By taking a mean value of the intensities measured in the various cometary spectra (Figs. 3, 4, 5), a value of 10 % can be derived for the ratio $[C-H]/[H_2O]$, for each class of hydrocarbons. In order to derive the carbon abundance, an assumption has to be made for $[C]/[C-H]$ and $[H]/[C-H]$ in hydrocarbons. For saturated hydrocarbons, assumed to be composed of CH_2 -chains, it is reasonable to take $[C]/[C-H] = 1$ and $[H]/[C-H] = 2$. For unsaturated hydrocarbons we tentatively assume $[C]/[C-H] = 2$, $[H]/[C-H] = 1$. Then the total number of carbon atoms in Comet Halley (derived from the 3 μm signature) is about 30 % of the number of H_2O molecules. This number is surprisingly high, compared to the amount of other carbon molecules detected in comet Halley ($CO_2/H_2O = 2 \ 10^{-2}$; $H_2CO/H_2O \sim 10^{-2}$; $CO/H_2O \cong 0.15$; $HCN/H_2O \sim 10^{-3}$). It means that, for molecular abundances, the most important contribution to the total carbon content comes from hydrocarbons (Encrenaz et al, 1987).

4 - ABUNDANCES IN COMET HALLEY AND IN THE INTERSTELLAR DUST

Table 1 summarizes the molecular abundances derived in Comet Halley, relatively to H_2O. Upper limits are given for CH_4 and NH_3. These

Table 1

Relative abundances of molecular species in Comet Halley, normalized to H_2O

Molecule	Mixing Ratio (per volume)
H_2O	1
CO	0.15
CO_2	0.02
Number of carbon atoms (saturated)	0.10
Number of carbon atoms (aromatics hydrocarbons)	0.20
CH_4	< 0.04
NH_3	< 0.10
H_2CO	< 0.01
HCN	0.001

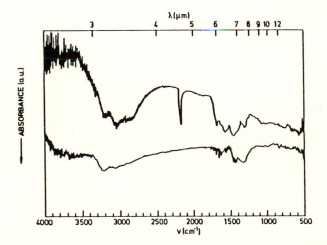

Fig. 9   An example of organic residue obtained in the laboratory from the irradiation of ices, showing the 3.4 μm absorption feature. The ice is a mixture of H_2O, CO, CH_4 and NH_3 (from d'Hendecourt, 1984).

Table 2

Nature	Fraction in mass to total mass of gas (measured)	Fraction of carbon (1)	Fraction of oxygen (2)
Silicates SiO_2 + Mg,Fe..	4. 10^{-3}		18 %
Ices H_2O	3.3 10^{-3}		34 %
NH_3	1.4 10^{-3}		
CO (gas + ice)	1.1 10^{-3}	15 %	8 %
PAH	4.7 10^{-4}	15 %	
Very small grains	5.2 10^{-4}	17 % (if dominated by carbon)	
Organic refractory	1.45 10^{-3}	24 %	6%
Amorphous carbon/ graphite	1.3 10^{-3}	29 %	
TOTAL		100 %	66 %

(1) relative to total carbon, assuming cosmic abundances (Ref. 35)
(2) relative to total oxygen, assuming cosmic abundances (Ref. 35)

Table 3

Relative Abundances in Interstellar Dust
and in Comet Halley (derived from the gaseous phase)
normalized to oxygen

	Cosmic Abundance	Interstellar Dust (1)	(2)	Comet Halley
H	1.51 10^3	2.13	1.45	1.93
O	1.00	1.00	1.00	1.00
C	0.50	0.50	0.50	0.39
N	0.14	0.14	0.14	< 0.08
Si	0.05	0.05	0.05	—

numbers can be converted in elements abundances ; the result is H:O:C:N = 1.93 : 1.00 : 0.39 : < 0.08. These numbers refer to the gaseous phase (except for hydrocarbons which might be partly in the form of small grains) so that the material which is trapped in cometary grains in not considered in this table.

A similar compilation can be made for the composition of dense molecular clouds, as observed from IR interstellar spectra, mostly from the solid phase (ices and refractory grains). This work has been made by Puget (1987) and the results are shown in Table 2. These results show that the observed carbon and nitrogen abundances fit the cosmic values reasonably well, while, in contrast, the oxygen observations can account for 66 % of the cosmic value. It is thus reasonable to assume that about 34 % of the total oxygen is in the gaseous form, or in some undetected ice. Two extreme assumptions can be made: (1) the missing oxygen is in form of gaseous H_2O; (2) the mixing oxygen is in form of O_2 (ice or gas). These 2 assumptions lead to 2 different compositions of the interstellar dust : in the first case, H:O:C:N= 2.13 : 1.00 : 0.50 : 0.14 ; in the second case, H:O:C:N = 1.45 : 1.00 : 0.50 : 0.14 (Encrenaz et al, 1987).

We can now compare the compositions which have been derived for Comet Halley and for the interstellar dust (under the 2 assumptions mentioned above). This comparison is shown in Table 3.

A first comment is that the H/O ratio in Comet Halley ranges between the 2 values derived for interstellar dust under the 2 extreme cases. In all cases the H/O ratio is in the order of 1.5 to 2, which means that H_2O is one of the major sources of H and O. Thus, the amount of condensable hydrogen seems to be comparable in interstellar and cometary material.

A second remark concerns the abundances of C and N in Comet Halley. Table 3 shows that there is a depletion of C and N relatively to the interstellar (and cosmic) values. A reasonable explanation is that C and N are partly trapped in grains in Comet Halley. This result is in qualitative agreement with the conclusions of the Vega and Giotto mass spectrometer experiments (Kissel et al, 1986) which indicate a large fraction of light elements ("CHON") particles. It has been also suggested that CO, CN and C_2 in Comet Halley could partly originate from grains (Eberhardt et al, 1986 ; A'Hearn et al, 1986 a,b). Assuming that about 20 % of cometary oxygen is trapped in grains, ($SiO2$, MgO, FeO...), we derive that about 40 % of carbon and at least 50 % of nitrogen are trapped in grains in Comet Halley. These numbers do not seem unreasonable, from the studies of the cometary dust composition; unfortunately it seems difficult to derive accurate element abundances from the study of cometary grains.

5 - <u>CONCLUSIONS</u>

Infrared spectroscopy of Comet Halley has provided a major step in our knowledge of the cometary chemical composition. For the first time, parent molecules have been unambiguously detected. Furthermore, besides the expected H_2O and CO_2 signatures, hydrocarbons have been detected in large amounts. From what wa know now from both the gaseous and the solid phase, it seems possible to define a cometary composition which could be similar to the composition of interstellar dust, as observed in dense molecular clouds. Moreover, the cometary icy material seems to have been irradiated by UV radiation or high energy particles, in the same way as the interstellar dust. All these results suggest an analogy in the global nature of cometary and interstellar material at the time of the comet's formation, and reinforce the already suspected "primitive" nature of comets.

References

- A'Hearn, M.F., Birch, P.V. and Klingesmith, D.A., 1986. ESA SP 250, Vol I, 483.
- A'Hearn, M.F., Hoban, S., Birch, P.V., Bowers, C., Martin, R. and Klingesmith, D.A., 1986. Nature, 324, 649.
 Baas, F., Geballe, T.R. and Walther, D.M., 1986. Astrophys. J. 311, L97.
- Bouchet, P., Chalabaeev, A., Danks, A., Encrenaz, T., Epchtein, N. and Le Bertre, T., 1987. Astron. Astrophys. 174, 288.
- Butchart, I., Mc Fadzean, A.D., Whittet, D.C., Geballe, T.R. and Greenberg, J.M., 1986, Astron. Astrophys. 154, L5.
- Campins, H., Bregman, J.D., Witteborn, F.C., Wooden, D.H., Rank, D.M., Allamandola, L.J., Cohen, M. and Tielens, A.G., 1986. ESA SP-250, Vol II, 121.
- Combes, M. et al, 1986. Nature, 321, 266.
- Combes, M. et al, 1987. In preparation.
- Crovisier, J., 1984. Astron. Astrophys., 130, 361.
- Crovisier, J. and Le Bourlot, J., 1983. Astron. Astrophys., 130, 61.
- Crovisier, J. and Encrenaz, T., 1983, Astron. Astrophys., 126, 170.
- Danks A.C., Encrenaz, T., Bouchet, P., Le Bertre, T., and Chalabeaev, 1987. Astron. Astrophys. in press.
- Drapatz, S., Larson, H.P. and Davis, D.S., 1986. ESA SP-250, 347.
- Despois, D., Crovisier, J., Bockelée-Morvan, D., Schraml, J., Forveille, T and Gérard, E., 1986. Astron. Astrophys., 160, L11.
- Eberhardt, P. et al, 1986. ESA SP-250, Vol. I, 383.
- Encrenaz, T., Puget, J.L., Bibring J.P., Combes, M., Crovisier, J., Emerich, C., d'Hendecourt, L. and Rocard, F., 1987. Proceedings of the Symposium on the Diversity and Similarity of Comets, Brussels, April 1987, in press.
- Hanner, M.S., 1984. Astrophys. J., 277, L75.
- Hanner, M.S., Tokunaga, A.T., Golish, W.F., Griep, D.M. and Kaminski, C.D., 1987. Astron. Astrophys, in press.
- d'Hendecourt, L.B., 1984. Ph.D Thesis, Leiden University.
- Johnson, J.R., Fink, U. and Larson, H.P., 1983. Astrophs. J., 270, 769.
- Kissel, J. et al, 1986. Nature, 321, 280.
- Knacke, R.F., Brooke, T.Y. and Joyce, R.R., 1986. Astrophys. J., 310, L49.
- Krankowsky, D. et al. 1986. Nature, 321, 326.
- Krasnopolsky, V. et al, 1986. Nature, 321, 269.

- Léger, A. and Puget, J.L., 1984. Astron. Astrophys., 137, L5.
- Mumma, M.J., Weaver, H.A., Larson, H.P., Davis, H.S., and Williams, M., 1986. Science, 232, 1523.
- Moroz, V.I. et al, 1987. Astron. Astrophys., in press.
- Ney, E.P., 1982. In "Comets", L. Wilkening ed., U. of Arizona Press, p. 323.
- Puget, J.L., 1987. Private Communication.
- Sellgren, K., 1984. Astrophys. J., 277, 623.
- Tokunaga, A.T., Golish, W.F., Griep, D.M., Kaminski, C.D. and Hanner, M.S., 1986. Astron. J., 92, 1183.
- Wickramasinghe, D.T. and Allen, D.A., 1986. Nature, 323, 44.
- Weaver, H.A. and Mumma, M.J., 1984. Astrophys. J., 276, 782.
- Willner, S.P., Soifer, B.T., Russell, R.W., Joyce, R.R and Gillett, F.C., 1977. Astrophys. J., 217, L121.
- Yamamoto, T., 1982. Astron. Astrophys., 109, 326.

HALLEY'S COMET

I.P. Williams
Astronomy Unit, School of Mathematical Sciences,
Queen Mary College,
Mile End Road,
London E1 4NS

ABSTRACT

At its 1986 apparition, Halley's Comet became by far the most studied comet in history with five space probes encountering it and a multitude of ground based telescopes observations carried out by both amature and professional astronomers. In this overview we will discuss only the part of the data set that is of most interest ot infrared astromers, namely the nucleus and the dust. The molecular emission has been dealt with in another chapter.

1. INTRODUCTION

In ancient times comets were objects of great interest primarily because of the belief that they either fortold or were the cause of great disasters. This was very fortunate for our understanding of comets, for the dates of the appearance of many comets have been carefully noted, and this has allowed much valuable information regarding the orbital evolution of Halley's comet to be gathered [Yeomans & Kiang 1981]. The 1910 apparition was significant in that it produced the first photograph of the comet, but the recent apparition also caused great interest as there was a hope of observing the nucleus for the first time and so confirming its existence. The main reason for the study of comets is to gain an insight into the structure and evolution of the nucleus. It is generally believed that the nucleus of comets may be composed of primordial material, essentially unaltered since the epoch of formation of the planets and so an understanding of the nucleus will throw light on the problem of the formation of the planets and ultimately on the process of star formation. Clearly, observing the nucleus is therefore of prime importance.

The results from the recent apparation obtained both from space probes and ground based telescopes can be roughly be divided into four categories, plasma, gas, dust and nucleus. The plasma experiments produced many results regarding shock fronts, ionization, disconnection events and magnetic fields but in general these are of little interest to infrared astronomers. In general, the plasma results also confirmed the pre-encounter theories so that no startling new discoveries were made. The plasma physcisists had got their calculations correct. We will not therefore discuss this aspect further. The results of the observations of gas has been reviewed in this book by Encrenaz (1988) and so will not be discussed further here. There were many results obtained from the study of the nucleus and the dust and space will permit only a brief account of some of the more important discoveries.

2. THE NUCLEUS

Comet Halley was recovered on 16th October 1982 by Jewitt and Danielson (1982) at a visual magnitude of 24.5 while still at a distance of 11 A.U. from the Sun using the 5.1 m telescope at Mount Palomar. At this time all the radiation would have been solar radiation, scattered or re-emitted from the nucleus but because of extreme faintness no information could be obtained beyond the fact that Halley was still in existence and close to the position predicted by orbital calculations. By September 1984, while at a distance in excess of 6 A.U. a weak coma was observed (Spinrad et al, 1984) indicating that some outgassing had commenced. On its emergence from behind the Sun in February 1985, CN emission bands were identified by Wyckoff et al (1985) with the comet still at a heliocentric distance of nearly 5.A.U. In between these two dates, the first detection in the infrared was made by Birkett et al. (1986) using the U.K. Infrared telescope in Hawaii. At the time, because of the previous indications of some outgassing, it was difficult to ascertain whether the infrared emission was from the nucleus, or the ejected dust or any combination. As the comet approached the Sun, this situation worsened and so it was not until the spacecraft encounters occured that any data on the nucleus which did not depend on the theoretical modelling of dust ejection was obtained.

The first space probe encounter was by Planet A (Suisei) the first of the pair of Japanese spacecraft (the second being Sakigake). This indicated a number of discrete sources of emission on the nucleus. The results also suggested that the nucleus was covered with some hard substance, with the jets of predominantly H_2O being emitted from vents or fractures (Kaneda et al. 1986). Vega 1 and 2 improved on this picture by showing that the nucleus was an elongated irregular body of rough dimensions 14×7.5 km [Sagdeev et al. 1986]. The GIOTTO encounter, passing within 605 km of the nucleus produced images of the nucleus by means of the multicolour camera. Photographs of the nucleus may be seen for example in Keller et al. (1986) and Reitsema et al. (1986). The projected size is 14.9 x 8.2 km but the actual size is somewhat larger and Wilhelm et al. (1986) calculated, using a tri-axial ellipsoid fit, that the real dimensions may be 16 x 9 x 10 km.

The geometric albedio is very low of the order of .044 (Whipple 1986) so that the nucleus is very dark, considerably darker than had hitherto been suspected. In the past, all estimates of cometary radii have been obtained from the absolute magnitude by assuming an average albedo for a 'dirty ice' body. If an albedo similar to that of Halley were to be universal for comets, then the previous estimates for cometary radii have to be increased by a factor of between 2 and 3. The emission is predominantly from jets or vents which occupy no more than about 15 percent of the total surface area. One hemisphere apears to be much more active than the other. Activity is also only present from the Sunward side which indicates that the nucleus both has a low thermal capacity and low thermal conducitivity.

We thus have a picture of the nucleus as a dark irregular elongated object emitting gas (mostly H_2O) containing small grains, the emission occuring in well defined jets. It is pertinent to ask whether this tells us all we need to know about the structure of comets. This clearly is not the case for a steam locomotive would also fit the above description and it is meaningful to ask, now that space encounters are past, whether we can tell the difference.

The one physical property not determined by the fly-by missions was the mass of the nucleus. [The deceleration measured for GIOTTO was dominated by collisions with ejected material rather than the gravitational field due to the nucleus]. It is possible to obtain a reliable estimate of the mass using variations of the following basic principle. Astrometric observations of the comet show that a delay ΔT of 4.1 days in the time of perihelion passage occurs due to non-gravitational effects that is after account has been taken of all the planetary gravtational perturbations, [Yeomans, 1977]. These non-gravitated effects are of course due to the asymmetric emission of dust and gas mentioned above. Estimating the rate of loss of momentum due to out gassing gives the force on the nucleus, while the 4.1 day delay is related to the deceleration. By division, the mass can thus be inferred. More details of the method can for example be found in Rickmann (1986). A mass in the range 5 to $13 \times 10^{16}g$ is estimated for the Halley nucleus which implies a density in the range 0.08 to 0.24 g cm^{-3} ; the nucleus is not thus a steam locomotive! The low density implies a very loosely packed material much like terrestrial snow. This type of structure also has the advantage of explaining the low conductivity and low thermal capacity required to explain why emission only occurs on the sunward side.

3. THE EJECTED DUST

One of the most noteworthy results from the study of Halley dust has been the discovery of variability in its composition with considerably more material based on combinations of Carbon, Hydrogen, Oxygen and Nitrogen (the so called CHON particles) than had hitherto been suspected in addition to the more usual grain compositions. This has opened up the possibility that some gas molecules may originate from these grains rather than all being emitted from the nucleus. Another aspect of great interest was the physical side of the problem namely the size and size distribution of grains found near Halley. It is important to distinguish between the insitu measurements for the mass spectrum [i.e. what was actually measured by the spacecraft] and what this may imply for the mass spectrum at the nucleus. The two will differ as both the initial velocity achieved by the grains and the retardation forces due to solar radiation are dependent upon grain sizes. Consequently the very large grains do not move far from the nucleus as they have low ejection speeds while very small grains with both a high initial velocity and small deceleration as the efficiency factor, Q, for radiation is small, can travel very far from the nucleus. An addition of about 0.2 to the index is called for in the case of larger grains but the exact correction is of course model dependent and in what follows only the insitu measurements are given. It should also be noted that the cumulative mass index is used throughout, that is the number of grains with mass greater than m is assumed to be proportional to m$^{-\alpha}$ [note that this is not what is always used by authors cited, but conversion is standard and straight forward. Most of the data comes from spacecraft encounters but some grain masses can also be obtained from a study of the two meteor streams associated with Halley, namely the Orionids and the η Aquarids. It is convenient to discuss the results as five separate mass ranges, though the exact location of the boundaries between the mass ranges is somewhat arbitrary.

(i) Very Small Grains, m $< 10^{-13}$g

The discovery of these very small grains was one of the more exciting aspects of the Halley encounter these being essentially undetectable from Earth, [Vaisberg et al. 1986a, Mazets et al. 1986a, McDonnell et al. 1986a]. The index of the mass spectrum for this size range was found to be very variable with distance from the comet, with an average value for α of 0.2 from Vaisberg et al. (1986b), in agreement with values given by Mazets et al. (1986a).

(ii) Small Grains, $10^{-13}\text{g} < \text{m} < 10^{-10}\text{g}$.

Some variation, both with nuclear distance and size within the range was found by Simpson et al. (1986) though somewhat surprisingly, the change of spectrum slope with distance was less for the smaller grains. They found values of the index in the range 1.0 to 1.9. All these values appear to be in conflict with values of α around 0.5 - 0.7 given by Vaisberg et al. (1986a.b) Mazets et al. (1986a.b) McDonnell et al. (1986b).

(iii) Medium Grains, $10^{-10}\text{g} < \text{m} < 10^{-5}\text{g}$.

McDonnell et al. (1987b) give a value for the index α of 0.85 which is very much in line with a value of 0.8 given by Mazetz et al. (1986a). This is probably the best determined of all the indices.

(iv) Large Grains, $10^{-5}\text{g} < \text{m} < 10^{-1}\text{g}$.

The mass spectrum now comes from the DIDSY experiment alone. The largest mass detected was .035 g though and impact with a grain as large as 1 g can be inferred [McDonnell private communication]. The index for this range has a value of 0.54 (McDonnell, 1987). A large asymmetry in number was found between pre and post encounter data. Radio meteors also fall within this mass The index for meteors in the Ovionid and Aquarid streams for a similar mass range in 0.7 (Hughes, 1986).

(v) Very Large Grains, $\text{m} > 10^{-1}\text{g}$.

The only source of information here is the visual meteors found in the two meteor streams and Hughes (1986) found an index of 1.27 for these.

In general there appears to be a general increse in the index α , with increasing grain size, starting at around 0.2 for very small grains, increasing to 1.3 for the very large. The only exception to this rule appears to be the large grains where an index of around 0.7 was found, rather than values just above one for a continuous trend. It maybe that in this range the spacecraft data is biased by statistics of small number while the difficulty of both mass and number estimation for the small meteors is also considerable. It also has to be remembered that the difference between the in situe and nuclear index may be largest here.

4. THE SUNWARD SPIKE

One other feature of the Halley dust environment that deserves comment is the Sunward Spike which was observed between April 28th and June 7th. [See Sekanina et al. 1986 for a list of observations]. As already mentioned, the larger grains are ejected with a small velocity relative to the nucleus and consequently never move far from the orbit of the comet. The anti-tail is seen when the comet is close to the ecliptic and the dust, in orbit close to the comet, is seen due to projection effects as an anti-tail. Such an anti-tail was indeed expected in February 1986 and was observed (Sekanina and Larson, 1986). By March 1986 it was fading rapidly (West et al. 1986). The Sunward Spike is not this anti-tail but is a genuine Spike pointing towards the sun with a length of 7 x 10^5 km. In order to achieve such a separation from the nucleus in the Sunward direction it is necessary to have both a high initial velocity and a low deceleration by solar radiation. Both conditions are only satisfied by grains less than 0.1 μm in size the acceleration being small because of their reduced efficiency to absorb solar radiation. The appearance of the spike thus also indicates the presence of the very small grains detected by the space probes.

The appearance of the Spike is suggestive of a thin dust sheet observed almost edge on. Since it is composed of small grains with no forces acting on them, their location in a plane must signify their ejection in a plane. The Halley nucleus is precessing about the angular momentum vector (Sekanina, 1987) and a plane normal to this vector is the only plane in which one might expect to find ejected grains. The direction normal to the plane containing the small grains turns out to be at Right Ascension 17° , declination −63° , very close to the position previously determined for the rotation pole of the nucleus. The angle between the Earth-Comet line and the deduced angular momentum vector was changing rapidly in April but slowly in June which explains the sudden appearance but slow fading of the Spike.

5. CONCLUSIONS

The recent exploration of Halley has produced many exciting results, many of interest to infrared astronomers. It was found that a single nucleus exists, which is an irregular elongated body of very low albedo and low bulk density. Most of the dust and gas is ejected from the Sunward side of the nucleus from well defined vents indicating also a low conductivity for the nucleus. One hemisphere appears to be more active than the other. Two unexpected results from the study of the ejected dust were the composition (many CHON particles) and the extreme small size of some of the grains. Because of their very small size the solar radiation pressure does not excert a significant force on these grains and so they can be found at much larger distances from the nucleus than had hitherto been suspected and in particular, when seen at the correct angle can generate a Sunward Spike.

ACKNOWLEDGEMENTS

I would like to thank McDonnell and his DIDSY team and Seknina for a number of preprints and helpful communications and the Science and Engineering Research Council for granting me facilities which allowed me to observe Halley.

REFERENCES

Birkett, C.M. *et al* , 1985. *I.A.U. Circular 4025*.

Encrenaz, Th., 1988. In *"Comets to Cosmology"*, proceedings of 3rd IRAS conference, ed. A.Lawrence. (Springer-Verlag, Berlin).

Hughes, D.W., 1986. *E.S.A. Workshop on Comet Sample Return*, SP-249.

Jewitt, D.C. and Danielson, G.E., 1982. *I.H.W. Newsletter*, No. 2.

Kaneda, E., *et al* , 1986. *Nature*, **321**, 297.

Keller, H.U. *et al* , 1986, in Exploration of Halley's Comet, *E.S.A.* SP-250, 347.

McDonnell, J.A.M. *et al* , 1986a, nature, 321, 338.

McDonnell, J.A.M. *et al* , 1986b, in Exploration of Halley's Comet, *E.S.A.* SP-250, 25.

McDonnell, J.A.M. *et al* , 1987, Astron. Astrophys. (in Press).

Mazets, E.p. *et al* , 1986a, Nature 321, 276.

Mazets, E.P. *et al* , 1986b, in Exploration of Halley's Comet, *E.S.A.* SP-250, 3.

Reitsema, H.J. *et al* , 1986, in Exploration of Halley's Comet, *E.S.A.* SP-250, 351.

Rickmann, H., 1987, E.S.A. Workshop on Comet Nucleus Sample Return (in press).

Sagdeev, R.Z. *et al* , 1986, Nature, 321, 262.

Sekanina, Z., 1987, Nature, 325, 326.

Sekanina, Z. and Larson, S.M., 1986, Astron. Jl. 92, 462.

Sekanina, Z. *et al* , 1986, in Exploration of Halley's Comet, *E.S.A.* SP-250, 177.

Simpson, J.A. *et al* , 1986, in Exploration of Halley's Comet, *E.S.A.* SP-250, 11.

Spinrad, H., Pjorgovski, S. and Belton, M.J.S., 1984, I.A.U. Circular, 3996.

Vaisberg, O. *et al* , 1986a, Nature, 321, 274.

Vaisberg, O., Smirnov, V. and Omelchenko, A., 1986b, in Exploration of Halley's Comet, *E.S.A.* SP-250, 17.

West, R.M. *et al* , 1986, Nature, 321, 363.

Whipple, F.L., 1986, in Exploration of Halley's Comet, *E.S.A.* SP-250, 281.

Wilhelm, K. *et al* , 1986, in Explortion of Halley's Comet, *E.S.A.* SP-250, 367.

Wyckoff, S. *et al* , 1985, Nature, 316, 241.

Yeomans, D.K., 1977, Astron. Jl. 82, 435.

Yeomans, D.K. and Kiang, T., 1981, Mon.Not.R.astr.Soc., 197, 633.

THE ALBEDO OF LARGE REFRACTORY PARTICLES FROM P/TEMPEL 2

Mark V. Sykes

Steward Observatory
University of Arizona
Tucson, AZ 85721

ABSTRACT

Thermal emission from a portion of the Tempel 2 dust trail indicates that the large refractory particles ejected from the nucleus have a bolometric Bond albedo of ~0.05. It is inferred that the nucleus itself is dark. Upper limits on the number density of trail particles are calulated. The mean free path between trail particles suggests that the Comet Rendevous/Asteroid Flyby mission is unlikely to be affected if it flies within the trail ahead of the comet in its orbit, but may collide with trail particles (having diameters of a few hundred microns) behind the comet.

1. INTRODUCTION

Cometary dust trails were first observed by the Infrared Astronomical Satellite (IRAS) and consist of debris continuously tracing out a portion of a comet's orbit both behind and often ahead of the comet's orbital position (Sykes *et al.*, 1986a). Seen from the earth, these trails extend from a few degrees to many tens of degrees of sky at thermal wavelengths. Their narrowness and proximity to the projected orbit of of their parent bodies argues for particle sizes in the submillimeter range and larger with low (meters/sec) relative velocities with respect to the comet nuclei from which they derive. Simple dynamical analysis of several trails show that significant surface area resides in particles several millimeters and larger in diameter (Sykes *et al.*, 1986b).

With the exception of P/Schwassmann-Wachmann 1, the comets so far identified as having associated trails all have perihelia within 3 astronomical units (AU) of the sun. When ejected from the comet nucleus dust trail particles are likely to have significant ice components. This ice, however, is unstable so close to the sun, and will sublime on short timescales. Trail particles thus provide a unique opportunity to study the refractory component of a comet nucleus independent of the gas and micron-sized dust which dominate comae and tails.

At the time of the IRAS mission, the brightest and largest (in apparent length) dust trail was associated with the short-period comet Tempel 2. It was first observed at 25 μm as a string of point sources which were found to be continuous (Davies *et al.*, 1984). Examination of the IRAS sky flux maps showed it to be a narrow contrail-like feature clearly and continuously seen at 12, 25, and 60 μm and extending across 7 adjacent 16.5 × 16.5 degree plates (HCON 1). It is seen very faintly at 100 μm.

Because each successive scan which is incorporated into a plate is generally eastward of the previous scan, the satellite effectively tracks the trail motion. This results in a far greater apparent

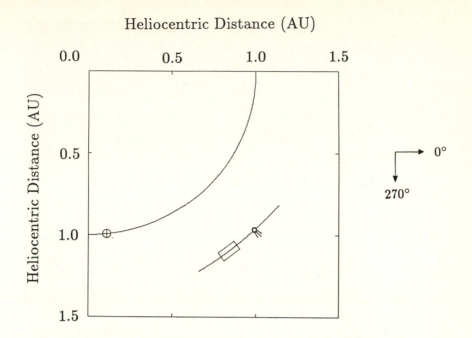

Figure 1. Geometry of P/Tempel 2 and its observed dust trail and the earth on 29 June, 1983. Ecliptic longitude is shown on the right. The box encloses the portion of the trail seen in Plate 95, HCON 1. The solar elongation of the observation is ~100°.

length in the sky flux maps than would be the case if each map was a "snapshot" of the sky. The trail appears to fade away ~4° in mean anomaly forward of the comet's orbital position and gets lost in galactic emission when it is observed ~8° in mean anomaly behind the comet. Some plates sample the P/Tempel 2 orbit more than 20° in mean anomaly behind the comet's orbital position, but show no evidence of trail emission. The location of the observed trail in late June, 1983, is shown in Fig. 1. It was almost 100,000,000 km in length and subtended more than 30° of sky as seen from earth.

2. ANALYSIS AND RESULTS

For purposes of this study, a single skyflux plate (Plate 95, HCON 1) was chosen for analysis because of its relatively low scan-to-scan noise ("striping") and the very small time interval (9 days) over which the scans comprising the plate were taken (Fig. 2). This latter minimizes parallactic effects which can result in a very "choppy" trail. The times of observation were 19 June to 28 June, 1983, when the trail segment in Plate 95 HCON 1 had a heliocentric distance of 1.38 AU and geocentric distance ranging between 0.82 and 0.94 AU. The trail is viewed at an angle between 5 and 8 degrees away from its orbital plane.

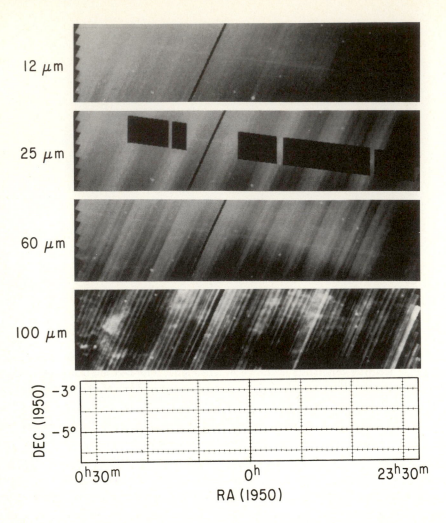

Figure 2. The Tempel 2 dust trail as seen in all four IRAS bandpasses in IRAS Skyflux Plate 95, HCON 1. The trail extends from the upper left to lower right of each image. The blacked out areas in the 25 micron field show the portions of the trail and background emissions which were coadded.

Verticle cuts were made through the trail which were 1.37° (41 pixels) in length, centered on the trail. Adjacent columns were sampled over the length of the trail on the plate, but were rejected if the column contained a bright source in any bandpass (primarily 12, 25, and 60 μm). An example of the sampled region is shown in Fig. 2.

The sampled columns were then registered with respect to the trail and averaged. The resultant coadded profiles are shown in Fig. 2. Trail emission is clearly seen in all four IRAS passbands, but the trail profile is largely due to the blurring of the true profile (having a width of ~3 pixels) in the process of coaddition. At each wavelength, the trail profile was manually removed and the modulation in the background was estimated by fitting a parabola to the remaining points. The background was then subtracted from the trail profile and the result was integrated to get the total trail flux density within a cut through the trail. The uncertainty of the background fit was taken to be the uncertainty of trail flux density at each point in its profile. The total uncertainty of the trail flux is then that value multiplied by the square-root of the number of profile points summed. The results are shown in Table 1.

Table 1.

Wavelength (microns)	Flux density* (Jy/sr)	σ (Jy/sr)
12	7.47×10^5	4.0×10^4
25	1.34×10^6	9.5×10^4
60	5.37×10^5	3.0×10^4
100	2.83×10^5	3.0×10^4

* This is the mean dust trail surface brightness density compressed into one 2×2 arcminute pixel along a North-South line. The correction factor to get the uncompressed value is 0.62.

Since dust trail particles are submillimeter in size and larger, their thermal emissivities are assumed to be spectrally flat over the IRAS passbands. Assuming the particles to be rapidly rotating gray spheres in radiative equilibrium with sunlight, the coadded observations are best fit by a graybody of temperature 233 K (Fig. 4). A rapidly rotating spherical blackbody at the heliocentric distance of the trail particles would have an equilibrium temperature of 237 K. The uncertainty of 11 K is derived by propogating the uncertainties of the fluxes through the blackbody radiation equation. The dominant term in this case is that of the uncertainty in the 100 μm flux. Neglecting the 100 μm term results in an increase of one degree in particle temperature and a reduction of the temperature uncertainty to ± 5 K. Temperatures exceeding the blackbody temperature for that heliocentric distance of 237 K are rejected as they would require the presence of particles very small by comparison with the wavelength observed. Dynamical considerations would require that they be tens of nanometers in size and break off from larger dust trail particles well away from the comet nucleus where coupling with the gas outflow would accelerate them to speeds of ~1 km/s relative to the nucleus (which is much larger than the relative velocities of trail particles). Of course, if future analysis of the remainder of the Tempel 2 dust trail were to result in a firm temperature in excess of 237 K, then such particles would have to be considered. The nominal bolometric Bond albedo of the dust particles is thus 0.05, which says that the particles are likely very dark. For a Lambertian scatterer, this would correspond to a geometric albedo of 0.03.

The uncertainty in temperature translates to a range of bolometric albedos from totally black to 0.23 (using all for wavelengths) or 0.13 (neglecting the 100 μm point). Careful future coaddition of more than an order of magnitude additional Tempel 2 dust trail observations should substantially reduce uncertainties in particle temperatures and albedos.

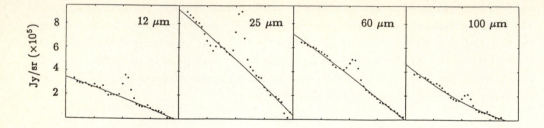

Figure 3. Coadded and averaged trail profile along a North-South cut through the trail. The solid line is a parabolic fit to the background emission. Each point corresponds to the flux density of a 2 arcminute pixel. The true trail width is only ~3 arcminutes. The wider profile results from blurring due to the method of coaddition.

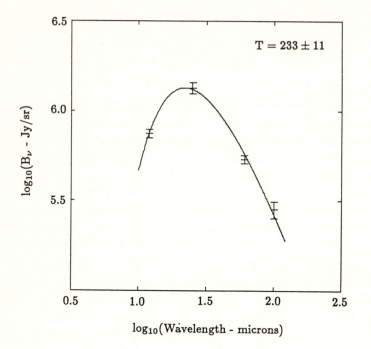

Figure 4. The total trail flux densities in all four IRAS passbands have been fit to a graybody curve using a method of least squares. Neglecting the 100 micron point results in an increase in particle temperature of one degree, while reducing its uncertainty by a factor of two.

3. THE TRAIL ENVIRONMENT

P/Tempel 2 is the proposed target for the Comet Rendevous/Asteroid Flyby mission (CRAF). Since the spacecraft is expected to remain in the vicinity of the comet over most of an orbit, an understanding of the nature of the environment is important, especially with respect to dust contamination. Given the large size of the trail particles, there is also the concern of possible damage if the spacecraft is struck at large relative velocities. Since the trail has been spatially resolved (Sykes *et al.*, 1986b), an upper limit to the volume density of trail particles can be estimated by assuming a minimum patrticle size as well as a bolometric albedo. The mean free path of the spacecraft travelling through or within the trail can then be calculated.

The Tempel 2 dust trail was observed to have a resolved width of ~125,000 km (*ibid.*). This is assumed to hold for all locations. At the trail, each skyflux map pixel (2 arcminutes on a side) has a width of ~78,000 km. Thus in order to get the true average surface brightness density of the trail, the values in Table 1 must be multiplied by 0.62 in each band. The number density, N, assuming all particles have a radius a is

$$N = F_\nu / (B_\nu \pi a^2 D) \tag{1}$$

where,

$$B_\nu = \frac{2hc}{\lambda^3 [\exp(h\nu/kT) - 1]} \tag{2}$$

and F_ν is the observed surface brightness density, and D is the depth of the trail along the line of sight (which is assumed to bve equal to its width – perpendicular to the line of sight). The temperature of the particles, T, is related to the blackbody equilibrium temperature at the heliocentric of the trail by the particle albedo, A_b:

$$T = 237(1 - A_b)^{1/4} \tag{3}$$

Number density profiles have been calculated assuming a range of particle radii and bolometric albedos, and are plotted in Fig. 5. The values of N were are averaged over the values obtained for observations in all four IRAS passbands.

CRAF is assumed to have a projected surface area, A_{CRAF}, of 10 m^2. The mean free path of the spacecraft is given by

$$\Delta = 1/(N A_{CRAF}) \tag{6}$$

which is also plotted in Fig. 5. It turns out that both N and Δ are only weakly dependent on the bolometric albedo over the range of uncertainty of the albedo previously determined.

Forward of the comet's orbital position the trail particles are several millimeters or larger in radius (Sykes *et al.*, 1986b). The spacecraft would have to travel more than 10^8 km within the trail to have a reasonable chance of being hit. Since the trail particles move with respect to the nucleus at only meters per second (Eaton *et al.*, 1984; Sykes *et al.*, 1986b), a static position relative to the nucleus would be "safe" indefinitely. Behind the comet's orbital position the situation changes

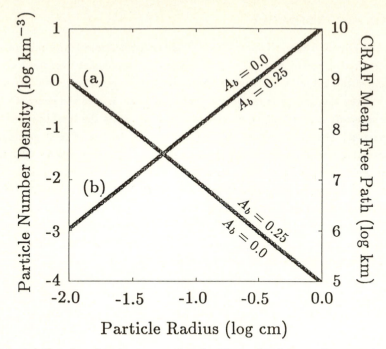

Figure 5. (a) Number density curves for the Tempel 2 dust trail, assuming all trail particles to be the same size. The bolometric albedo of the particles are allowed to vary from 0.0 to 0.25, with little effect. (b) Curves for the mean free path of CRAF within the trail with the same conditions as in (a).

as smaller, submillimeter size particles probably dominate. At worst, CRAF would have to move through only 10^6 km of trail to have a chance of collision. At a relative velocity of $\sim$1 km/s, this would correspond to a travel time of several months within the trail. These estimates assume that the spacecraft is well away from the dust coma surrounding the nucleus.

ACKNOWLEDGMENTS

This work has been supported in part by Contract No. F19628-87-K-0045 from the Air Force Geophysics Laboratory and Contract No. 958047 from the IRAS Guest Investigator Program.

REFERENCES

Davies, J., S. Green, B. Stewart, A. Meadows, and H. Aumann, 1984. *Nature* **309**, 315.
Eaton, N., J.K. Davies, and S.F. Green, 1984. *Mon. Not. R. Astr. Soc.* **211**, 15P.
Sykes, M.V., L.A. Lebofsky, D.M. Hunten, F.J. Low, 1986a. *Science* **232**, 1115.
Sykes, M.V., D.M. Hunten, F.J. Low, 1986b. *Adv. Space Res.* **6**, 67.

INFRARED STUDIES OF SOLAR SYSTEM BODIES

Dale P. Cruikshank

Institute for Astronomy, University of Hawaii
Honolulu, Hawaii 96822 USA

ABSTRACT

Recent progress in near-infrared reflectance spectroscopy of several solar system bodies is reviewed.

1. INTRODUCTION

Progress in understanding the planets and minor bodies of the solar system owes much to infrared techniques used at ground-based observatories, airborne platforms, and aboard Earth-orbital and deep space probes. Information on the surface mineralogical and the atmospheric gaseous compositions of the planets and their satellites is contained in the near-infrared spectral region, encompassing that wavelength range where reflected sunlight dominates the flux from the bodies, normally between 0.8 and 3 μm. Longer wavelengths, where intrinsic thermal emission from planets, satellites, comets, and asteroids are, contain additional compositional information, but also yield basic data on temperatures, thermal structures of the uppermost surface layers, and overall dimensions of the objects. Of special interest is the thermal signature of Jupiter's satellite Io, because of the strong component due to surface hot spots as a manifestation of active volcanism.

IRAS observations have given very important data on the asteroid population, the zodiacal material, comets, Pluto, and other solar system bodies; most of this is reviewed elsewhere in this volume. The IRAS data are complementary to those obtained from ground-based and airborne observatories, and together form an emerging picture of the compositions, dimensions, and thermophysical properties of the Earth's neighbors in the solar system.

This is a brief and incomplete review of some current work on planetary objects in the infrared, with emphasis on near-infrared reflectance observational studies.

2. SPECTROPHOTOMETRY OF PLANETS AND SATELLITES

The passively scattered solar radiation received from planets and their satellites contains absorptions at the visible and near-infrared wavelengths where electronic and molecular transitions occur. At wavelengths longward to about 2.5 μm, transitions of d-shell electrons in transition metal ions and electron exchange between ions are largely responsible for the absorptions. Molecular oscillations result in absorptions longward of about 1 μm, with water, carbon dioxide, methane, ammonia, and various other molecules being of importance in the outer solar system.

Reflectance spectroscopy of planets and satellites observed at relatively low spectral resolution (1–5%) has given a general picture of the distribution of volatile materials in the outer solar system, while in the inner solar system it has revealed the presence of several mineral types in the asteroids and on Mercury, the Moon, and Mars. In this brief review, only some of the most recent work published and some in progress can be considered.

2.1 Pluto and Charon

The presence of methane on Pluto was established from filter photometry in two near-infrared wavelengths chosen to distinguish the ices of methane, water, and ammonia (Cruikshank et al. 1976) and later confirmed in a spectrum obtained by Soifer et al. (1980). Subsequent work (e.g., Buie and Fink 1987) has shown the methane band system at 0.89 μm and has demonstrated that the band strength is variable with the planet's 6.4-day rotation. Cruikshank and R. H. Brown (unpublished) noted that the band strengths of methane in the near-infrared also vary with Pluto's rotation. Thus, the absorption bands must arise largely from solid methane on the planet's surface rather than from a gaseous atmosphere, though a tenuous atmosphere consistent with the vapor pressure equilibrium may exist.

For two years, 1987 and 1988, Pluto and its satellite Charon have been in an epoch of mutual transits and eclipses. Photometric and spectroscopic observations during these total events has permitted the determination of the dimensions of both the planet and satellite (e.g., Tholen et al. 1987), as well as the spectrophotometric signature of water ice on Charon (Buie et al. 1987). The radius of Pluto determined from eclipse and transit photometry is 1145 $\pm$ 46 km, and that of Charon is 642 $\pm$ 34 km. Measurements of the dimensions of Charon's orbit by a variety of techniques gives the mass of the planet-satellite system, which together with the sizes of the bodies, yields the system density of 1.84 $\pm$ 0.19 g cc^{-1} (Tholen et al. 1987).

2.2 Triton

Near-infrared spectrophotometry of Triton shows six methane bands in the region 0.8–2.5 μm, plus another weak band at 2.15 um that is attributed to molecular nitrogen (Cruikshank and Apt 1984; Cruikshank et al. 1984). The early data were obtained with a circular variable interference filter (CVF), but the development of the infrared array detector spectrometer at the NASA Infrared Telescope Facility (IRTF) by A. T. Tokunaga and colleagues makes it possible to observe the spectrum of Triton and other faint planetary objects with higher signal precision and greater resolution than previously was possible. New observations of Triton with the Cooled-Grating Array Spectrometer (CGAS) on the IRTF show structure in the strongest of the methane bands and give the band center and shape of the nitrogen feature with greater precision than possible with the CVF.

Triton is revealed as a volatile-rich satellite on which the extreme seasonal cycle probably results in the mass migration of methane and nitrogen gas and condensates. The opportunity for direct *in situ* observations from Voyager 2 in 1989 will offer an extraordinary opportunity to compare the hard-won Earth-based telescopic data with a little world revealed to the spacecraft.

2.3 Io

Reflectance spectroscopy of Io shows a band complex of sulfur dioxide in the vicinity of 4.1 μm. Recent work on this topic by Howell et al. (1987) gives details of the strengths of the combination and overtone bands of the normal isotopes of sulfur and oxygen, as well as indications of the abundances of the isotopes ^{33}S, ^{34}S, and ^{18}O. There are no indications of measurable isotopic anomalies among those studied so far. The sulfur dioxide on Io appears to be mostly in the form of condensed frost or snow in a mixture with the surface soil materials, whatever they might be. The source of the sulfur dioxide is the system of active volcanic vents.

2.4 Non-Water Volatiles on Planetary Satellites

In addition to the water-ice surfaces of the three Galilean satellites Europa, Ganymede, and Callisto, the surfaces of the large satellites of Saturn and Uranus are also dominated spectroscopically by the presence of water ice. The spectra are quite different from one another in overall reflectance level and in the shape of the near-infrared reflectances from 0.8 to 2.5 μm (see the review by Clark et al. 1986). In particular, there are hints of the presence of non-water ice components in the spectrum in the cases of Europa (Jupiter), Enceladus (Saturn), and Ariel (Uranus), particularly in the 2.1–2.4 μm region. In search of other volatile components of the ices, Brown et al. (1988) have begun to explore the spectra of these bodies with higher resolution than that afforded by the CVF spectrometers used for the original work. In preliminary work on Europa, Brown et al. report the possibility of ammonium hydroxide as a minor constituent of the water ice surface of this high-albedo and recently resurfaced satellite.

Ammonia is a particularly important potential constituent of those planetary satellites whose surfaces show evidence of relatively recent activity, because ammonia has the effect of lowering the melting temperature of the interior ices of these bodies, thus making recent eruptive activity more easily understood. Methane is another potential component of the icy surfaces of some planetary satellites because it may have been incorporated in the condensing water ice as a clathrate compound during the cooling of the solar nebula. There is, as yet, no firm evidence for the presence of methane clathrate on planets or their satellites.

3. ASTEROIDS AND COMETS

3.1 Asteroids

The reflectance spectra of asteroids from 0.8 to 2.5 μm show various minerals of importance for understanding the mineralogical evolution of the asteroids and their associations with the meteorites (e.g., Gaffey and McCord 1979; Larson and Veeder 1979; Cruikshank and Hartmann 1984). Spectrophotometric work has been extended to longer wavelengths, in particular the 2.8–3.6 μm region, in studies of the water of hydration in certain classes of asteroids and in search for the diagnostic C–H stretching mode band in hydrocarbons.

Lebofsky found bound water in the surface minerals of some asteroids (Lebofsky 1980; Lebofsky et al. 1981). In the C-type asteroids, the amount of bound water is highly variable (Feierberg et al. 1985), and while it is presumed to be present in asteroids of other types, particularly the D and P types in the outer regions of the asteroid system, the characteristic 3-μm absorption band has not yet been detected.

On the smooth wing of the bound water absorption band seen in certain primitive carbonaceous meteorites in the laboratory lies the 3.4-μm C–H stretching mode band arising from the hydrocarbons that are abundant in such meteorites as Murchison and Murray. The 3.4-μm band has been tentatively identified in the spectrum of the asteroid 130 Elektra by Cruikshank and Brown (1987) from IRTF spectra obtained with the CGAS in 1986. Asteroid 130 Elektra is a C-type object with bound water in the surface minerals. The 3.4-μm band attributed to C–H is only about 4% deep as seen in diffuse reflectance against the combined thermal radiation and reflected sunlight from the asteroid.

3.2 Comets

The reflectance spectra of comets tend to be featureless between 0.8 and 2.5 μm, more or less independent of the level of activity of the comet (e.g. Hartmann et al. 1987). Jewitt et al. (1982) found evidence for an absorption band at 2.2 μm in the spectrum of Comet Bowell (1980b), but they could not identify it. In the 3-μm region, a number of studies of Comet P/Halley revealed the 3.4-μm C–H band complex in emission (e.g., Wickramasinghe and Allen 1986). This band and other emission features in the spectrum of Comet P/Halley were seen by the spectrometers aboard the Vega spacecraft that flew by the comet's nucleus in 1986. Continued studies of comets from ground-based telescopes in the region of the 3.4-μm emission features will give crucially important statistical information on the compositions of the old comets that have made many passes through the inner solar system in comparison with "new" comets making their first approach to the Sun.

4. CONCLUSIONS

Near-infrared reflectance spectroscopy of solar system bodies, particularly those in the outer solar system, continue to give fundamental information on the surface compositions of these bodies because of the diagnostic spectral features in the relevant minerals and ices that occur in the accessible spectral region. These discoveries establish a backdrop against which the next generation of spectral studies from spacecraft such as Galileo, Mars Observer, Comet Rendezvous Asteroid Flyby, Cassini, and others can be planned and eventually realized.

5. ACKNOWLEDGMENTS

This work is supported in part by NASA Grant NGL 12-001-057 to the University of Hawaii. The author acknowledges with thanks partial support for his participation in the IRAS conference awarded by the Organizing Committee.

References

Brown, R. H., D. P. Cruikshank, A. T. Tokunaga, R. G. Smith, and R. N. Clark (1988). Search for volatiles on icy satellites, I: Europa. *Icarus* (in press).

Buie, M. W., D. P. Cruikshank, L. A. Lebofsky, and E. F. Tedesco (1987). Water frost on Charon. *Nature* (in press).

Buie, M. W., and U. Fink (1987). Methane absorption variations in the spectrum of Pluto. *Icarus* **70**, 483–498.

Clark, R. N., F. P. Fanale, and M. J. Gaffey (1986). Surface composition of natural satellites. In *Satellites* (J. A. Burns and M. S. Matthews, Eds.), pp. 437–491. Univ. of Arizona Press, Tucson.

Cruikshank, D. P., and J. Apt (1984). Methane on Triton: Physical state and distribution. *Icarus* **58**, 306–311.

Cruikshank, D. P., and R. H. Brown (1987). Organic matter on asteroid 130 Elektra. *Science* (in press).

Cruikshank, D. P., R. H. Brown, and R. N. Clark (1984). Nitrogen on Triton. *Icarus* **58**, 293–305.

Cruikshank, D. P., and W. K. Hartmann (1984). The meteorite-asteroid connection: Two olivine-rich asteroids. *Science* **223**, 281–283.

Cruikshank, D. P., C. B. Pilcher, and D. Morrison (1976). Pluto: Evidence for methane frost. *Science* **194**, 835–837.

Feierberg, M. A., L. A. Lebofsky, and D. J. Tholen (1985). The nature of C-class asteroids from 3-μm spectrophotometry. *Icarus* **63**, 183–191.

Gaffey, M. J., and T. B. McCord (1979). Mineralogical and petrological characterizations of asteroid surface materials. In *Asteroids* (T. Gehrels, Ed.), pp. 688–723. Univ. of Arizona Press, Tucson.

Hartmann, W. K., D. J. Tholen, and D. P. Cruikshank (1987). The relationship of active comets, "extinct" comets, and dark asteroids. *Icarus* **69**, 33–50.

Howell, R. R., D. B. Nash, T. R. Geballe, and D. P. Cruikshank (1987). High resolution infrared spectroscopy of Io and possible surface materials. *Icarus* (in press).

Jewitt, D. C., B. T. Soifer, G. Neugebauer, K. Matthews, and G. E. Danielson (1982). Visual and infrared observations of the distant comets P/Stefan-Oterma (1980g), Panther (1980u), and Bowell (1980b). *Astron. J.* **87**, 1854–1866.

Larson, H. P., and G. J. Veeder (1979). Infrared spectral reflectances of asteroid surfaces. In *Asteroids* (T. Gehrels, Ed.), pp. 724–744. Univ. of Arizona Press, Tucson.

Lebofsky, L. A. (1980). Infrared reflectance spectra of asteroids: A search for water of hydration. *Astron. J.* **85**, 573–585.

Lebofsky, L. A., M. A. Feierberg, A. T. Tokunaga, H. P. Larson, and J. R. Johnson (1981). The 1.7 to 4.2 μm spectrum of asteroid 1 Ceres: Evidence for structural water in clay minerals. *Icarus* **48**, 453–459.

Soifer, B. T., G. Neugebauer, and K. Matthews (1981). The 1.5–2.5 μm spectrum of Pluto. *Astron. J.* **85**, 166–167.

Tholen, D. J., M. W. Buie, R. P. Binzel, and M. L. Frueh (1987). Improved orbital and physical parameters for the Pluto-Charon system. *Science* **237**, 512–514.

Wickramasinghe, D. T., and D. A. Allen (1986). Discovery of organic grains in comet Halley. *Nature* **323**, 44–46.

The Galaxy

THE LARGE SCALE DISTRIBUTION OF INFRARED RADIATION IN OUR GALAXY

Harm. J. Habing

Observatory Leiden,

PO Box 9513, 2300 RA LEIDEN, the Netherlands

1. LARGE SCALE DISTRIBUTIONS: AN OLD PASTIME FOR ASTRONOMERS

Characterisation of the large scale distribution of astronomical objects has been astronomers entertainment for a long time: already Herschel tried to fathom the depth of the Universe by counting stars. To obtain more convincing results than Herschel Kapteyn used his considerable diplomatic, organisational and mathematical skills to derive a spatial distribution via his "Plan of Selected Areas". His efforts failed for a reason feared by Kapteyn but demonstrated only after his death: interstellar extinction. Successful and lasting interpretations of systematic large scale surveys have been made only at wavelengths where extinction is of minor or no significance: at radio wavelengths and at gamma wavelengths. These interpretations show clearly that the best and most direct information in such surveys is on the interstellar medium -only one of the components of our Galaxy. The surveys give some insight in the stellar component but only in an indirect way -see the interesting work on the distribution of ionizing stars in our Galaxy by Guesten and Mezger (1982).

Now we have the IRAS results in front of us: Complete and reliable data superior to that of all earlier surveys. I will not offend anybody when I say that we will soon forget all previous broad-band infrared surveys in the same wavelength interval as IRAS (18 to 150μm). Only surveys outside this interval remain significant; for example the Japanese balloon borne surveys at 2.4μm (Maihara et al., 1978; Hayakawa et al., 1978) or the Goddard balloon survey at 150, 250 and 350μm (Hauser et al., 1984). The former measure the stellar distribution in the Galaxy, and the latter interstellar matter. It is thus quite possible that the four IRAS surveys (one in each band) contain information on the distribution of the interstellar medium and of the stellar component in the Galaxy. In fact, I will argue that the 12μm and 25μm surveys contain very good information on the stellar component. And yet, I guess that even at 12μm the point sources contribute only a small fraction of the total emission - a qualitative inspection of 12μm sky flux maps leads me to this statement.

This review divides naturally into two parts: first, a discussion of the infrared emission by interstellar matter and, second, a discussion of the stellar component.

2. THE EXTENDED INFRARED EMISSION

Before IRAS was launched there was a paradigma accepted by most (but not all) scientists involved in its preparations: the only high-quality product to come out of the mission was going to be a catalogue of point sources: extended structures could not be measured, the detectors would be too unstable. As it turned out, the few enthousiasts who before launch were willing to bet on extended structures (Mike Hauser was the most pronounced team member), proved to be right: the stability of the detectors was excellent and very good maps of the extended emission have been constructed. And yet, the present "sky flux" maps are not the best possible; significant improvements are still being obtained, in the U.S. at IPAC, the IRAS data processing center at the California Institute of Technology, and in the Netherlands in a project called GEISHA at the Space Research Laboratory in Groningen.

Analyses of the IRAS "Sky flux" maps have up to now been preliminary, even though it is over 2½ year after the data were published. The major cause for delay has not been the quality of the maps, but the presence of significant zodiacal emission at all four IRAS wavelengths. Models of the zodiacal light have now become so well developed that the interested researchers begin to subtract the "zody" from the IRAS sky flux maps with some confidence. In the last few weeks before this conference I obtained preprints of a number of more or less "definitive" analyses (Sodritski et al., 1987; Boulanger and Pérault, 1987; Pérault et al., 1987; Burton and Deul, 1987). Time of preparation was too short to attempt a synopsis of these papers. Therefore I will isolate a few important points of discussion. I will use two references, bench marks, to "calibrate" the discussion: the first reference is to the important paper by Cox et al. (1986), that summarizes and discusses mainly pre-IRAS data on the Galaxy. The second reference is to the thesis on M31 by Walterbos (1986) who included IRAS data; this thesis allows very useful comparisons between M31 and our Galaxy. Table 1 is the first example of what I termed "calibration": it compares the infrared emission from the solar neighbourhood as derived by Boulanger and Pérault (1987) with that of the inner Galaxy derived by Cox et al. and of M31 (Walterbos).

Table 1

Normalized spectrum of the diffuse emission

λ	λF_λ		
	CKM[1]	BP[2]	W[3]
12μm	4	38	51
25μm	8	27	21
60μm	50	34	36
100μm	100	100	100

[1] Cox et al. (1986): pre-IRAS data; inner parts of our Galaxy
[2] Boulanger and Pérault (1987): IRAS data; local, solar neighbourhood
[3] Walterbos (1986): M31

Table 1 shows that the spectrum of our local neighbourhood compares quite well with that of M31 and, at least qualitatively, with the pre-IRAS views. The surprise is, as you all know, the secondary maximum at 12μm; that has already been discussed by Cox et al. (who give references to earlier work), but it is stronger than they thought.

Practically all infrared emission recorded by IRAS is thermal emission by dust grains. Symbolically this is represented by the following: $j = \sum n_i \, \varepsilon_i(T)$, where j is the emissivity per volume element, ε_i the emissivity of a grain of type i at temperature T and n_i the number density of particles of type i. Of crucial importance is the temperature T, which is determined by the absorption mainly of UV photons from the surrounding interstellar radiation field. The emissivity j is thus determined (1) by the particle density and (2) by the local density of the UV radiation field: IRAS measured a product of the two; to derive either one extra information is needed.

Consider now the particles or grains. As is well known, there are two different kinds of particles, each identified by an emission/absorption feature: at 9.7μm (silicate type particles) and at 11.3μm (carbon rich particles). The distribution of the particle sizes is an important datum, because many particles are smaller than the wavelengths at which they emit; for example, the exponential MRN-distribution (Mathis et al., 1977) extends from 10nm to 250nm. New information from the IRAS observations is the presence of large numbers of small particles with sizes down to 1, or even 0.3nm: these particles are required to explain the emission at 12 and 25μm, and perhaps, part of the 60μm emission. The existence of such small particles had been discussed already a long time ago but the first convincing evidence came shortly before the IRAS emission from the detection of 2μm continuum in reflection nebulae (Sellgren, 1984).

Small particles add the following properties infrared: (1) they are poorer emitters in the infrared, but remain good absorbers in the UV; thus they will be hotter than larger grains; (2) the smallest particles do not reach a time-constant temperature in the interstellar radiation field: the absorption of a single photon will elevate temporarily their temperature; nevertheless a sufficiently large number of such particles will show an equilibrium distribution of temperatures, with a tail of high temperatures (200-300k); (3) the smallest particles are probably better described as macro molecules (the suggestion by Léger and Puget (1984) is that they are PAH's or polycyclic aromatic hydrocarbons); the particles will show quantum mechanical effects, such as discrete band structures. The upturn of the spectrum at the short wavelength end of the spectrum (table 1) is now generally attributed to the small particles -clearly they occur not only in our Galaxy, but also in M31.

The analysis of extended emission usually progresses as follows: from the observed distribution of the emission one derives some local emissivity $j(\lambda)$ at wavelength λ and compares this with other local properties, for example n(H), the density of (atomic or molecular) hydrogen. Another frequently discussed parameter is $j(100μm)/j(60μm)$, which is used to derive the grain temperature T. I noticed that

all observers agree that j(100μm)/j(60μm) is very constant in our Galaxy and also
over the face of M31; however, the ratio j(100μm)/n(H) decreases by a factor 8-10
when one moves away from the galactic center. Taken at face value this would suggest
that the temperature of the dust grains remains constant, but that the number of dust
grains per hydrogen atom decreases with increasing R (R is the galactocentric
distance). This, however, is not necessarily true; when there are a large number of
small grains then, as Walterbos showed (see his table 4 at his page 165), a decrease
of just the interstellar radiation field will have the same effect as a decrease of
the relative particle density.

The general outcome of the analyses in our Galaxy and in M31 is that at 100μm
the emission is from the larger grains; at 60μm most of the radiation is still from
large grains, but there may already be a significant contribution by the very small
grains; at 25μm and at 12μm the small grains contribute essentially all the
radiation.

Table 2

Contribution by percentage of various components of the interstellar medium to the
infrared flux density

	CKM[1]	BP[1]	W[1]
very cold dust 14k (molecular clouds)	4%	10%	(7%)?
cold dust 15-25k (HI region)	33%	70%	54%
warm dust 30-40k (HII regions)	50%	20%	8%
hot dust 250-350k (very small grains and circumstellar dust)	13%	included in 70%)	31%

[1] CKM: Cox et al. (1986); BP: Boulanger and Pérault (1987);
W: Walterbos (1986)

Table 2 shows the break-down of the contributions by various components of the
interstellar medium to the extended infrared emission. It is evident that the results
concerning the local neighbourhood (BP) and concerning M31 (W) agree quite well: most
emission is from clouds of atomic hydrogen; HII regions give a small contribution in
M31 compared to that in our Galaxy; this is probably real and reflects the fact that
M31 performs rather poorly in the formation of stars. In the Cox et al. paper warm
dust in HII regions contributes 50% to the total infrared emission; in the local
neighbourhood Boulanger and Pérault estimate this fraction to the only 20%. Two
causes for this discrepancy are possible: in the inner Galaxy HII regions are more
important, or Cox et al. underestimated the contribution by the dust in HI regions.
It seems to me that the second explanation is the more probable (see also Pérault et
al.).

Finally I give a summary of the most important conclusions reached so far. Please remember that all analyses are new and my summary may be a little too early!

a. Most of the infrared emission from extended sources is from regions with atomic hydrogen (HI region). However, the heating of the grains is through photons that have escaped from HII regions!

b. PAH's and the smallest grains appear to occur especially (exclusively?) in HI regions.

c. Discrete HII regions dominate locally, but not over the whole Galaxy.

d. Molecular clouds are relatively unimportant contributors to the IR emission between 12 and 150μm.

e. The emissivity at 100μm per hydrogen atom (i.e. j(100μm/n(H)) decreases strongly with increasing galactocentric distance, but the ratio j(100μm)/j(60μm) stays remarkably constant. This may be the result of an decrease in the radiation field with R or in the dust abundance -or a decrease in both.

f. For M31 the same conclusions can be drawn, except that M31 has a lower star formation activity.

g. To equate the IR emission of a galaxy with star formation activity is dangerous, because it oversimplifies the relation between the presence of young stars and interstellar matter.

3. POINT SOURCES AND GALACTIC STRUCTURE

I have already given my opinion (more a guess than the conclusion of a serious investigation) that most of the radiation detected by IRAS is from extended sources, already at 12μm and much more so at the other three wavelengths. But such a conclusion does of course not imply that point sources are useless for the study of galactic structure. In fact, I think that the opposite is true: the distribution of the point sources is highly informative! To see this, one has to make the right selection of point sources; especially useful is the distribution of IRAS point sources with their 25μm flux density about equal to their 12μm flux (for example the range 0.8<f(25)/f(12)<3.8: See Habing et al., 1985 and Habing, 1988). After selection of point sources in this way a beautiful picture of our Galaxy emerges (Fig. 1). The explanation why this narrow selection "window" gives such a nice result is well established: point sources at 12μm are practically all stars, most with circumstellar shells of gas and dust. The ratio of the 25μm to 12μm flux density is an indication of the optical depth in the shell; if the ratio is close to 1 the shell is optically thick around 10μm, and all the stellar radiation is converted into infrared emission, usually with a peak at 12μm. In fact, when $f_\nu(25\mu m) \simeq f_\nu(12\mu m)$ then the IRAS band at 12μm contains as much as 30% of the total stellar flux. And since these stars are all giants and thus are very luminous, they become very strong sources at 12μm and at 25μm. Interstellar extinction plays only a minor role and the stars have been detected by IRAS over more than 10kpc distance. Thus the IRAS data permit us to

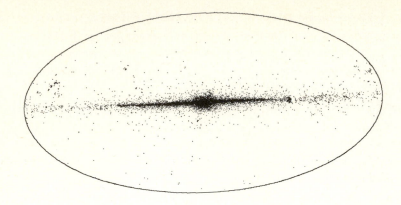

Figure 1:   Distribution on the sky, in galactic coordinates, of IRAS point sources
with a good detection at 12μm and at 25μm and with a flux density at 25μm
between 0.82 and 3.8 times the flux density at 12μm.

derive the space distribution and the luminosity distribution of our Galaxy -see
further down. How informative are such distributions?

The space distribution throughout the Galaxy and the luminosity distribution can
be interpreted in the best way once we know what these stars are. Where do they
belong in the Hertzsprung-Russell diagram? Much effort has gone into this question,
and generally accepted conclusions have been reached -see for an extensive discussion
Van der Veen and Habing (1987). All stars with strong circumstellar emission appear
to be Asymptotic Giant Branch stars, objects that have become red giants for the
second time, this time with a degenerate carbon-oxygen core and alternatingly burning
hydrogen into helium and helium into carbon in two adjacent shells. Such stars are
long period variables with periods between 250 and 2000 days; Mira variables are
identified with those of shorter periods. All stars loose mass at a significant rate,
but those with the longer periods (≥500 days) have such large mass loss rates that
the shell is optically thick even at 9.7μm. Estimates have been made of the main
sequence mass and these suggest that all stars with a mass between 1 and 5 M_Θ range
(this could also be between 0.9 and 8 M_Θ) ultimately pass through this particular
phase of evolution. Because the phase lasts briefly, there are always only a very few
stars in this phase -the duration is ,say, 10^4yr for stars with the largest mass loss
rate and the longest periods, and 10^5yr for the Mira's.

The apparent distribution of the AGB stars in the plane of the sky is a
convolution of their space distribution in the Galaxy and their luminosity
distribution. Specify a quantity $n(1,b,f_\nu)df_\nu$, where n is the number of stars per
square degree in direction (1,b) with a flux density (at 12μm)
between f_ν and $f_\nu + df_\nu$. It is easy to show that n equals the spatial
distribution, ρ, of the stars convolved with the luminosity distribution, ψ. By
determining n in a large number of directions one can then try to unravel ρ and ψ. I

have recently made such an attempt (Habing, 1988). Assuming cylindrical symmetry for the density distribution ρ one can indeed find distributions ρ and ψ that together give an acceptable description of n. For ρ I found an exponential distribution in R (galactocentric distance) with scaleheight h_R = 4.5±0.5kpc, and a similar distribution in z(h_z=0.3kpc). The luminosity distribution is broad, but shows a clear maximum at 6000 L_Θ, and virtually no stars at L<1500 L_Θ and at L>16,000 L_Θ. The analysis is hampered by confusion effects in the galactic plane, especially near the galactic centre: the count in such a direction is significantly smaller than the actual number of sources. A very convincing conclusion is that most of the stars are found in the galactic disk within 90° from the galactic center (i.e. 270°<1<90°); this implies that the Sun is at the edge of this disk of stars which, as I argued before, consists of objects shortly ago evolved from stars with main-sequence masses between 1 and 5 M_Θ -quite ordinary stars. A puzzle is that in my analysis the observations show a small excess of rather faint stars compared to model predictions. These stars seem to be real enough, but their galactic distribution is thicker in z and the scale length in R is larger; are they related to the stars in the so-called thick disk of the Galaxy (see Freeman, 1987)? I am convinced that my analysis can be much improved by treating in more detail confusion effects and by making counts to a deeper level than the IRAS point source catalog permits -which is possible via the cleaned-up and better organized IRAS data bases now available.

Shortly after the IRAS mission the presence of the Bulge of our Galaxy in figures like figure 1 created some interest. But after this first wave of enthousiasm very little has been published, although several groups are working on the data. The existence of stars with L between 4000 and 6000 L_Θ (by necessity AGB stars; stars on the first red giant branch have $L \lesssim 3000 L_\Theta$) remains a fact; the most direct interpretation is that their main sequence mass is over 1.0 M_Θ, and that they are younger than the globular clusters, although the bulge is hypothesized to be among the oldest structures in the Galaxy. Undoubtedly the last word has not been said on this subject.

ACKNOWLEDGEMENTS

The processing of IRAS at the University of Leiden has been made possible in part through a special grant, 78-218, from the Netherlands Foundation for the Advancement of Pure Research (ZWO).

REFERENCES

Boulanger, F.,Pérault, M. 1987, Astrophys. J. (submitted)

Burton, W.B., Deul, E.R. 1987, in the "Galaxy", eds. R. Carswell and G. Gilmore
     (Reidel, Dordrecht) p. 141

Cox, P., Kruegel, R., Mezger, P.G. 1986, Astron. Astrophys. 155, 380

Freeman, K., 1987, Ann. Rev. Astron. Astrophys. 25, 603

Guesten, R., Mezger, P.G. 1982, Vistas in Astronomy 26, 159

Habing, H.J. 1988, submitted to Astron. Astrophys.

Habing, H.J., Olnon, F.M., Chester, T., Gillett, F., Rowan-Robinson, M., Neugebauer,
     G. 1985, Astron. Astrophys. 152, L1

Hauser, M.G., Silverberg, R.F., Stier, M.T., Kelsall, T., Gezari, D.Y., Dwek, E.,
     Walser, D., Mather, J.C., Cheung, L.H. 1984, Astrophys. J. 285, 74

Hayakama, S., Ito, K., Matsumoto, T., Murakami, H., Uyama, K. 1978, Publ. Astron.
     Soc. Japan 30, 369

Léger, A., Puget, J.L. 1984, Astron. Astrophys. Lett. 137, L5

Mathis, J.S., Rumpl, W., Nordsieck, K.H. 1977, Ap.J. 217, 425

Maihara, T., Oda, N., Sugiyama, T., Okuda, H. 1978, Publ. Astron. Soc. Japan, 30, 1

Pérault, M., Boulanger, F., Puget, J.L., Falgarone E. 1987, preprint

Sellgren, K. 1984, Astrophys. J. 277, 623

Sodroski, T.J., Dwek, E., Hauser, M.G., Kerr, F.J. 1987, Astron. J. (in press)

Van der Veen, W.E.C.J., Habing, H.J. 1987, Astron. Astrophys. (in press)

Walterbos, R. 1986, "Stars, Gas and Dust in the Andromeda Galaxy", thesis, University
     of Leiden

LARGE SCALE STRUCTURE OF DUST AND GAS IN THE GALAXY

E.R. Deul
Sterrewacht Leiden
P.O. Box 9513, 2300 AA Leiden
The Netherlands

ABSTRACT. Both the large scale and intermediate scale structure of the dust and gas in the Galaxy are discussed. A set of clean–sky maps are obtained using a physical model that describes at the four IRAS wavelengths the contaminating contribution of the zodiacal emission. Direct comparison between the $100\,\mu m$ and HI column density maps shows a good correlation at high galactic latitudes. After radial unfolding the 12, 25, 60 and $100\,\mu m$ radial profiles are strikingly similar and show both the molecular ring and a steep increase toward the galactic center where the $12\,\mu m$ emission is relatively weak. The profiles clearly show a changing $100\,\mu m$/HI ratio which can be explained by a changing interstellar radiation field. The intermediate scale infrared structure can be separated in velocity space using high resolution HI data. There is evidence for a correlation between the HI material at anomalous velocities, with wide profiles and the corresponding dust, showing higher $60/100\,\mu m$ flux ratios than the material at normal galactic velocities.

1. INTRODUCTION

The distribution of dust and gas in our Galaxy can be studied by comparing the IRAS infrared observations with existing neutral atomic hydrogen observations. Before both sets of data are useful for comparison they have to be manipulated considerably.

The IRAS infrared measurements were obtained by observing through the zodiacal cloud of radiating particles that considerably contaminate the emission from material outside the solar system (Hauser, 1987). In collaboration with R.D. Wolstencroft I have developed a physical model describing the contribution of the zodiacal emission at all four IRAS wavelengths for all scans in the survey (Deul and Wolstencroft, 1987). After parameter determination on a subset of scans I have subtracted the modelled emission from all scans in the first two hours confirmation periods.

The neutral atomic hydrogen comprize the standard surveys: Burton (1986), Cleary *et al.*(1979), Heiles and Habing (1974), Kerr *et al.*(1987), and Weaver and Williams (1973). A thorough description of the creation of a 3–dimensional HI datacube is given by Deul and Burton (1987).

In section 2 I will present preliminary results for the large scale distribution of gas and dust as derived from the all–sky maps that were described above. Section 3 will deal with the intermediate scale structure ($< 10°$). I will show for two representative regions how the detailed correlation between infrared and 21–cm emission can be used to separate cirrus structures in velocity space and how we can interprete differences in the correlation between infrared cirrus features and corresponding HI material at kinematically separated velocities.

2. THE LARGE SCALE STRUCTURE

A straightforward comparison between the $100\,\mu m$ all–sky map corrected for zodiacal emission and the all–sky map of neutral atomic hydrogen column densities shows that the gas and dust emissivities correlate well at the higher galactic latitudes ($b > 10°$). To quantify the correlation we have taken strips in galactic latitude at constant galactic longitude. These strips were choosen at arbitrary positions in longitude and cover the entire latitude range ($-60° \leq b \leq 60°$) of the all–sky maps. Figure 1 shows scatter plots on a pixel–by–pixel basis for the strips with the $100\,\mu m$ flux on the horizontal and the 21–cm column densities on the vertical axis. The general correlation is evident from the scatter plot. There are, however, deviations from a one–to–one correlation which we will examine. A flattening of the correlation occurs at the higher flux levels . The galactic positions corresponding to these intensity levels are located near the galactic plane. We know that the neutral atomic hydrogen shows optical depth effects for all longitudes particularly for the inner Galaxy directions and between a few degrees latitude away from the galactic plane (Burton, 1976). Therefore the flattening can be attributed to the HI optical depths which cause a lowering of the integrated HI intensities in the manner shown in figure 1. A second characteristic of the correlation is the dependence on galactic longitude. This can be seen in figure 1 as the gradual increase of the correlation from the latitude strip at $l = 0°$ (filled triangles) to the strip at $l = 135°$ (filled stars). This is quite noticable at the higher intensity levels, but persists to higher galactic latitude positions where HI optical depths are low and therefore this must be caused by another effect. The interstellar radiation field responsible for heating the dust particles is known to decrease with galactocentric radius both in intensity and in relative amount of UV photons (Mathis *et al.*,1983). Consequently the $100\,\mu m$ emission per dust particles decreases. Assuming a constant dust to neutral gas ratio, appropriate for higher galactic latitudes where the contribution of molecular gas is negligible, the resulting $100\,\mu m/$HI ratio will decrease with galactocentric radius. This view is supported by examining the ratio in the galactic plane.

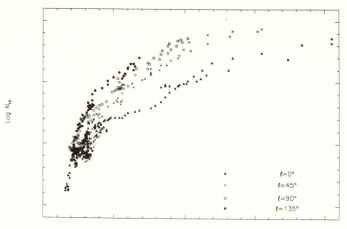

Figure 1. *Scatter plot of the HI column density as a function of the 100 μm brightnesses corrected for zodiacal emission. Scans in latitude averaged over 0°.25 at arbitrary longitudes were choosen. The correlation shows a flattening at the higher flux levels; the degree of flattening depends on galactic longitude.*

Having obtained all–sky infrared maps that are free of zodiacal emission the radial unfolding technique (Strong *et al.*, 1976) can now be applied to all four IRAS wavelengths. A solar galactocentric distance of $10\,kpc$ is assumed. The profiles presented in figure 2 show an enhancement of the emissivities near the molecular ring between $4 < R < 8\,kpc$ as well as a steep increase with decreasing galactocentric radius at $R < 2\,kpc$. Both these characteristics can be understood as the result of a non–uniform interstellar radiation field. The infrared emissivity per dust particle is directly related to the amount and spectral shape of the incident radiation. The number of population I stars, responsible for a large fraction of the interstellar radiation field increase in number density around the molecular ring and for $R < 2\,kpc$ the number density of population II stars strongly increases. Thus we expect the emissivities to increase near these positions. The profiles at 12 and $100\,\mu m$, resulting mainly from small ($a < 0.01\,\mu m$) and large ($a \simeq 1\,\mu m$) dust particles respectively, are strikingly similar. The region beyond $R \sim 2\,kpc$ shows a constant $100/12\,\mu m$ ratio. Inside $2\,kpc$ this ratio increases strongly, showing that the smaller particles are relatively scarce in the inner Galaxy. This could be the result of the destruction of small particles due to the increased particle density and interstellar radiation flux in this part of the Galaxy.

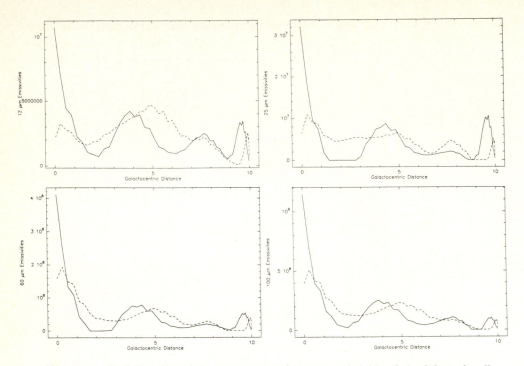

Figure 2. *Radial profiles of the 12, 25, 60, and 100 μm emissivities derived from the all-sky maps with zodiacal emission removed, using a radial unfolding technique appropriate for continuum data. The dashed line represents the profile for the fourth quadrant; the full drawn line that for the first quadrant. Note the close correspondance among the profiles. The molecular ring are around 4 < R < 8 kpc can be seen, as well as a steep increase of emissivities toward the galactic center.*

3. THE INTERMEDIATE SCALE STRUCTURE

In section 2 the apparently tight gas–to–dust ratio was discussed on galactic scales. In this section we examine the continuation of this correlation to the lower intensity levels and smaller spatial scales. A number of aspect of the gas–to–dust correlation have been discussed by *eg.* de Vries and Le Poole (1985), Tereby and Fich (1987), and Weiland *et al.*(1986), using starcounts, and HI or CO observations. Together with W.B. Burton, I have obtained with the NRAO 140–foot telescope in Green Bank, new HI observations of a number of large fields, at usefully high sensitivity and resolution, typically $10° \times 10°$ wide. The HI profiles at a resolution of $0.5\,km\,s^{-1}$ cover

$250\,km\,s^{-1}$ sufficient to span the velocity range characteristic of intermediate to high galactic latitudes. The fields were choosen by examining the infrared all–sky images and identifying regions that contain well–defined cirrus structure on a smooth background. Two examples from this set of fields will be shown here.

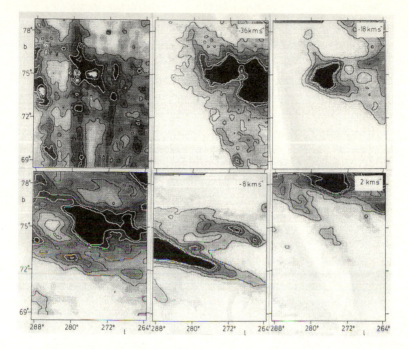

Figure 3. *Gray scale plot showing the 100 μm (top left–hand corner), the total column density of neutral atomic hydrogen (bottom left–hand corner), and four channel maps of the 21–cm line emission integrated over 2 km s⁻¹ centered at the indicated velocities. This region contains the scan observed by IRAS on June 23, 1983, and described by Low et al.(1984).*

The first region is centered at $l = 278°, b = +73°$. It is illustrated in figure 3. The two left–hand panels of this figure show the 100 μm (top) and HI column density (bottom) maps. This region is particularly interesting because it contains the scan observed by the IRAS telescope on June 23, 1983. Low *et al.*(1984) used this scan in their preliminary analyses of the IRAS data, in which they found that although most isolated infrared peaks correlate well with HI emission obtained from the Heiles and Habing (1974) survey, at least one feature was especially intersecting for lacking such a correlation. Two of the three features from that scan are contained in our first region. These features are labelled "B" and "X" by Low *et al.*, who found no detected HI counterpart for feature X. Feature B is the bright blob of emission at $l = 277°, b = 74°$ and feature X corresponds to the ridge of emission crossing $l = 277°, b = 72°$. Our new observations show that both cirrus features have HI counterparts at levels close to the noise level of the Heiles and Habing survey and that they are part of a larger cirrus complex extending over several tens of degrees over the sky. Although the infrared map shows considerable striping, notice the close correlation between the $100\,\mu m$ emission and the HI column density for this region. The small point–like sources in the infrared map all are associated with galaxies.

The four right–hand panels of figure 3 also show HI channel maps integrated over $2\,km\,s^{-1}$ and centered on the velocity specified in the top right–hand corner of each panel. They clearly illustrate the kinematic separation of cirrus features. The structure containing feature B is totally separated in velocity space from that containing feature X. The anomalous velocity at which feature B is most prominent indicates that there is dust emission associated with HI material moving at velocities well above the sound speed for that region. This phenomenon, which can be observed in the two fields presented in this paper, is a common characteristic among the other fields of our sample. The small negative velocity at which feature X peaks is expected for this high galactic latitude. Notice the slanted structure running from the top left–hand corner to the bottom right–hand corner persisting in all panels. This general morphology can be attributed to the northern extention of the magnetic field from the North Polar Spur.

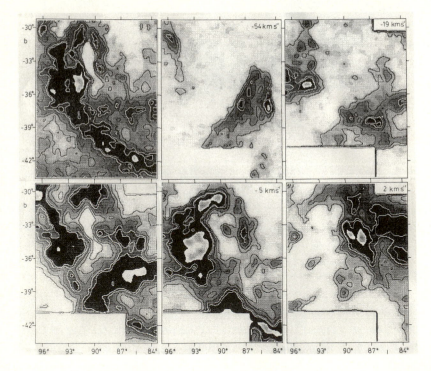

Figure 4. *Gray–scale plots of a well–defined cirrus structure on a smooth, low intensity background. The arrangement of the panels is like that of figure 3. This feature contains a cloud of HI emission at highly negative velocities. The cloud is associated with infrared dust that shows different properties than those of the surrounding material.*

The second region, which is closer to the galactic plane than the first one, is centered on $l = 90°, b = -37°$, and extends over 14 ° x 15 °. This field was choosen because it contains a cirrus feature that looks like a coherent structure on a smooth, low brightness background. The two left–hand panels of figure 4 show the 100 μm map (top) and the HI total column density (bottom). The areas blanked out in the HI map were not observed. Although both maps look remarkably alike, some minor but interesting deviations occur. The structure near $l = 86°, b = -38°$ shows a different 100 μm/HI ratio than the rest of the field. Not only is this ratio different, the 60/100 μm ratio also deviates to higher values. The central velocity of the structure is highly negative.

The four right–hand side panels of figure 4 illustrate the velocity structure of the HI in this region. The main body of emission occurs at slightly negative velocities appropriate for these galactic latitudes. Although there is some velocity structure in the main feature it must be seen as a filament that is virially unbounded. One outstanding feature, however, can be seen at $-54\,km\,s^{-1}$. It has a velocity profile width that is twice as large as the low velocity material. The corresponding 60 and 100 μm emission is quite different from that of the surrounding cirrus in that the 60/100 μm ratio is much higher. Because it is impossible to heat interstellar grains via such low velocity shocks another mechanism must cause the additional 60 μm flux.

One mechanism which could be responsible for this effect may be due to spectral line emission from collisionally exited trace atoms. A candidate is the fine structure line of [OI] at 63 μm (Petrosian, 1970). Assuming a cosmic abundance for Oxigen, the estimated transition probability from Petrosian, a line–of–sight path length of 1 pc, and using the HI resolution the calculated line strength could significantly influence the observed 60 μm flux. If this is the case, more features with additional 60 μm flux should be detectable.

It has been shown that the correlation between the IRAS 100 μm brightnesses and the neutral atomic hydrogen column densities does not result in a straight line but that the infrared flux is the result of multiple components. The nature of the large scale correlation is influenced by a changing interstellar radiation field, the HI optical depth effects, and the apparent variations in the small particle ($a < 0.01\,\mu m$) component. The intermediate scale correlations show characteristics which, assuming a constant interstellar radiation field, could be due to the effects of spectral line emission associated with shock exited atoms. Although the infrared emission is optically thin the observed flux is a complicated function of particle density, interstellar radiation field, and chemical composition of the radiating material.

References

Burton, W.B., 1976, *Ann. Rev. Astron. Astrophys.*, **14**, 275

Burton, W.B., 1986, *Astron. Astrophys. Suppl.*, **62**, 365

Deul, E.R., Burton, W.B., 1987, in *Mapping the Sky*, A.R. Upgren ed., Reidel, Dordrecht, in press

Deul, E.R., Wolstencroft, R.D., 1987, *Astron. Astrophys.*, in press.

Cleary, M.N., Heiles, C., Haslam,C.G.T., 1979, *Astron. Astrophys. Suppl.*, **36**, 95

Hauser, M.G., 1987, in *Comets to Cosmology.*, M. Rowan–Robinson ed., Springer Verlag, London, in press

Heiles,C., Habing, H.J., 1974, *Astron. Astrophys. Suppl.,* **14**, 1

Kerr, F.J., Bowers, P.F., Jackson, P.D., Kerr, M., 1986, *Astron. Astrophys. Suppl.,* **66**, 373

Low, F.J.,Beintema, D.A., Gautier, T.N., Gillet, F.C., Beichman, C.A., Neugebauer, G., Young, E., Aumann, H.H., Boggess, N., Emerson J.P., Habing, H.J., Hauser, M.G.,Houck, J.R., Rowan–Robinson, M., Soifer, B.T., Walker, R.G., Wesselius, P.R., 1984, *Astrophys. J. Lett.,* **278**, L19

Mathis, J.S., Mezger, P.G., Panagia, N., 1983, *Astron. Astrophys.,* **128**, 212

Petrosian, V., 1970, *Astrophys. J.,* **159**, 833

Strong, A.W., 1975, *J. Phys. A.,* **8**, 617

Tereby, S., Fich, M., 1987, *Astrophys. J. Lett.,* **309**, L73

Vries, C.P. de, Le Poole, R.S., 1985, *Astron. Astrophys. (Letters),* **145**, L7

Weaver, H.F., Williams, D.R.W., 1973, *Astron. Astrophys. Suppl.,* **8**, 1

Weiland, J.L., Blitz, L., Dwek., E., Hauser, M.G., Magnani, L., Rickard, L.J., 1987, *Astrophys. J. Lett.,* **306**, L101

MODEL FOR THE GALACTIC INFRARED EMISSION

P. Cox and P. G. Mezger

Max Planck Institut für Radioastronomie
Auf dem Hügel, 69
5300 Bonn 1, W-Germany

ABSTRACT

A model of the infrared/submillimeter emission from the galactic disk is presented on the basis of the IRAS results. The 60/100 μm brightness ratio is used as a very rigid observational constraint of the model. It is found that dust associated with atomic hydrogen is the dominant contribution to the total infrared luminosity. Correlations between warm and cold dust emission, HII regions and the distribution of the atomic gas are analyzed. Dust properties are briefly discussed.

1. INTRODUCTION

The basic radiation mechanism at the origin of the infrared/submillimeter emission of galaxies is simple . Dust grains absorb stellar radiation at optical/ultraviolet wavelengths and reradiate it in the infrared. On the average *one third/one fourth* of the total stellar emission is absorbed and reradiated by dust.

To interpret the infrared emission of galaxies, three main problems must be resolved:
  - the dust characteristics
  - the spatial distribution of both interstellar dust and stars
  - the geometrical association between stars and dust, which determines dust temperature
    and infrared luminosity

In our Galaxy, we believe to know with some accuracy the distribution of stellar populations, of the atomic and molecular gas components and hence of the interstellar dust which is thought to be uniformly mixed with the gas. Moreover during the past years our understanding of dust properties considerably increased. Using better model parameters and comparing model predictions with improved observations should lead to more realistic models for our Galaxy, which in turn can be used as basis for the modelling of the infrared/submillimeter emission of external galaxies.

The origin of the far-infrared radiation of our Galaxy has been investigated by us in a series of papers, mainly on the basis of balloon and rocket observations (see Cox et al. (1986) and references therein). The release of the IRAS data with its unprecedented sensitivity, resolution and completeness prompt a new and critical investigation. Recent interpretations of the infrared emission of galaxies, including our own Galaxy, which are based on the IRAS data base can be found in: Persson and Helou (1987), Crawford and Rowan-Robinson (1987) , Sodroski et al. (1987), Boulanger and Pérault (1987) and Pérault et al. (1987).

Within the past year, we have worked on an improved model for the infrared emission from the galactic disk, including the best to our knowledge of the IRAS data base. The following pages give a status report on our work in progress.

Figure 1 shows the longitudinal profiles of the observed 60 and 100 μm galactic plane emission, corrected for the zodiacal contribution. The profiles represent average intensities over the latitude interval |b| ≤ 1° in longitude increments of 0°5. The overall structure of the infrared emission of our Galaxy shows a number of strong peaks - all associated with bright Giant Molecular Clouds/HII complexes, spiral arms segments and the Galactic Center - which are superimposed on a continuous ridge, which we will define as the *diffuse emission component*. The exact definition of this diffuse emission is however somewhat arbitrary; it is defined in this paper as the lower envelope of the longitude profiles which passes at the nearest to the sources.

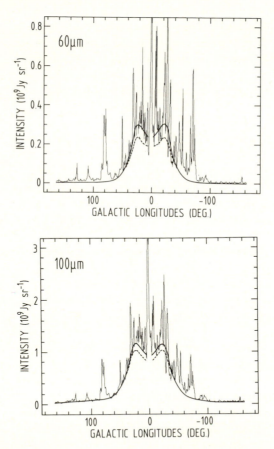

Figure 1. Galactic longitude profiles of the 100 and 60 μm galactic plane emission, averaged over the latitude interval |b|≤ 1°. The zodiacal contribution has been subtracted . The dashed curves represent the contribution of the cold dust associated with the atomic hydrogen; the solid curves are the sum of the contributions of the cold dust and the warm dust associated with the molecular cloud/HII complexes, and define the *diffuse emission component*.

The spectrum of the dust emission within the solar circle (but excluding the contribution of the Galactic Center regions) is shown in Figure 2, from 4 to 900 μm. This composite spectrum of the inner Galaxy includes both the diffuse component and the sources and has been compiled from all existing data available to date. References are given in the figure caption. The Galactic infrared spectrum has two distinct peaks , the most intense one at 100 μm and a weaker one at about 10 μm, which contains 30 % of the total infrared luminosity. This fact emphasizes the need of different dust components characterized by different temperatures in order to explain the infrared emission of normal spiral galaxies.

The following discussion will focus upon the infrared emission longward of 40 μm. The origin of the conspicuous mid-infrared emission is discussed in detail in the contributions by J.-L. Puget and F. Boulanger in this volume; the stellar contribution at these wavelengths is reviewed in the contribution to this conference by H. Habing.

The basic assumptions of our model are given in Cox et al. (1986). We emphasize the fact that we use the dust model originally introduced by Mathis et al. (1977) and modified by Draine and Lee (1984). This model consists of a mixture of silicate and graphite grains with a size distribution which *does not* include particles smaller than 50 Å. Comparison of the model predictions with the IRAS results requires , however, modifications of the dust model, which are discussed in Sect. 3.2.

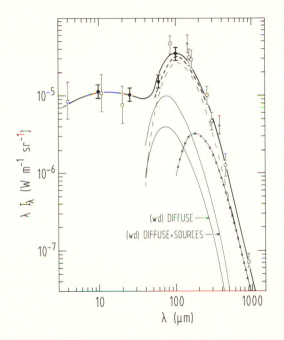

Figure 2. Spectrum of the dust emission between 4 and 900μm from the inner part (R ≤ 8 kpc) of our Galaxy, averaged over galactic longitude 3° - 35° and latitude |b|≤ 1°. This composite spectrum includes both the diffuse component and the sources. Filled circles are the IRAS observations (corrected for the zodiacal contribution; the points at 12 and 25 μm have been corrected for extinction, as well). Spectral points at 4, 10 and 20 μm are from Price (1981). 80 and 150 μm points are from Gispert et al. (1982), the 150, 250 and 300 μm points from Hauser et al. (1984), 380 μm from Caux et al. (1986), 450 μm from Owens et al. (1979) and 900 μm point is taken from Pajot et al. (private communication). Light curves represent the contributions of the individual components (crosses for the very cold dust, dotted line for the cold dust, solid line for the warm dust, diffuse component and sources as indicated). The heavy dotted line represents the total contribution of the diffuse components, whereas the heavy solid line represents the total contribution from both diffuse components and sources. The solid line shortward of 40 μm is a simple extrapolation through the observed points.

2. THE DIFFERENT DUST COMPONENTS

2.1 Cold Dust mixed with the Atomic Hydrogen

Dust grains associated with atomic hydrogen are heated by the general interstellar radiation field (ISRF). The temperatures of the graphite and silicate grains were evaluated using the ISRF and its dependence on the galactocentric distance as computed by Mathis et al. (1983). The variation of the ISRF with galactocentric distance and the associated dust temperatures are shown in Figure 3. Note that the graphite grains have a temperature range of 25 - 17 K.

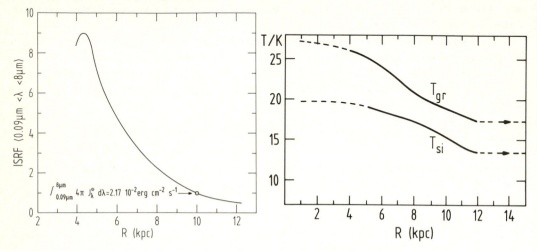

Figure 3. (a) Variation of the intensity of the ISRF, integrated between 0.09 and 8 μm with galactocentric distance. (b) Temperatures of graphite (T_{gr}) and silicates (T_{si}) associated with atomic hydrogen and heated by the ISRF.

The HI distribution is derived from the HI survey of Weaver and Williams (1973). The density is almost constant throughout the plane (0.35 - 0.45 cm^{-3}) and the scale height is an increasing function of the radius.

These parameters fix the contribution of cold dust to the total dust emission of the Galaxy. Results of the model calculations are shown as dotted lines in the averaged 60 and 100 μm longitudinal profiles (Fig. 1) . As can be seen from these figures, most of the diffuse infrared emission can be accounted for by this cold dust component.

2.2 Very Cold Dust mixed with the Molecular Hydrogen

Dust grains associated with the quiescent molecular clouds are also heated by the ISRF, which is, however, attenuated and partly converted into infrared radiation. Dust grains inside molecular are therefore colder than dust grains mixed with the atomic hydrogen and attain typical temperatures of 13 - 14 K (see Mathis et al., 1983 and Puget, 1985). We modelled the infrared/submillimeter emission of very cold dust associated with the molecular hydrogen using the H_2 distribution from Sanders et al. (1984) but modified according to Puget (1985). The very cold dust contribution is too small to be seen on the 60 and 100 μm longitudinal profiles. It accounts however for half of the emission longward of 300 μm .

2.3 Warm Dust Contribution

Subtraction of the contributions of cold dust and very cold dust - which are fixed by the model parameters which in turn are determined by observational constraints - from the observed Galactic emission leaves a warm dust component, whose spectrum is characterized by a temperature of 30 K. We attribute this emission to dust heated by O and B stars. This is supported by model calculations which show that only O and B stars have enough luminosity to heat required amount of dust to these temperatures. This warm dust associated with HII region/Molecular Cloud complexes traces the locations of massive star formation in our Galaxy. Since the intensity of dust emission increases like T_d^6, a relatively low mass of dust is needed. Our model fit, in fact, requires that only *one percent* of the interstellar molecular hydrogen is associated with warm dust.

Up to this point we have modelled the diffuse galactic infrared emission. The sources, which are superimposed on the diffuse components (see Fig. 1) have also typical colour temperatures of 30 K and contribute thus to the warm dust emission. To reproduce the spectrum of the dust emission from the inner part of our Galaxy, our model requires that discrete sources contribute about twice as much to the warm dust luminosity than the diffuse warm dust emission.

2.3 Comparison with Previous Work

The contributions of the different dust components to the Galactic infrared luminosity are summarized in Table I. Dust heated by the general ISRF and mixed with the atomic hydrogen dominates the far-infrared emission of the Galaxy and accounts for two-third of the total far-infrared luminosity. Warm dust (30 K) heated by O and B stars and tracing the recent star formation contributes one third to the far- infrared luminosity: one third of it is diffuse emission, and the remaining two third is associated with the compact HII regions. As shown in Table I, the Galactic infrared luminosity originates mainly inside the solar circle.

Table I. Characteristics of the the Galactic Disk for $R_O = 8.5$ kpc.

	inside the solar circle $1.7 \leq R(kpc) \leq 8.5$		integrated out R = 20 kpc Disk	
Dust associated with molecular gas	M_{H2} L_{vcd}	= 1.3 E9 M_O = 6.3 E8 L_O	M_{H2} /	= 1.4 E9 M_O
warm dust — diffuse	L_{wd}	= 7 E8 L_O	/	
warm dust — sources	L_{sou}	= 1.3 E9 L_O	/	
Dust associated with atomic hydrogen	M_{HI} L_{cd}	= 1 E9 M_O = 5 E9 L_O	M_{HI} L_{cd}	= 2.1 E9 M_O = 6 E9 L_O
Total far-infrared luminosity ($\lambda \geq 25$ μm)	L_{IR}	= 7.6 E9 L_O	L_{IR}	= 8.6 E9 L_O

As compared to our previous models, the contributions of the different components (warm and cold dust) have significantly changed. In particular the diffuse warm dust accounts now only for 10 % of the total infrared luminosity, as compared to our previous value of about 40 %. Two reasons can be invoked for this decrease. First, we used as basic constraints to the model the IRAS data and neglected the (less reliable) balloon observation at 80 μm: using this point would imply an increase of a factor of 3 in the contribution of the warm dust component. This illustrates how sensitive any infrared model is to and how much it is weighted by the exact knowledge of the data points between 50 and 100 μm. The carefully derived point at 60 μm gives us some confidence as to the actual results. Secondly, we used as a new constraint the dependence of the 60/100 μm brightness ratio on galactic longitude. The low dust temperature derived from this ratio (23 K) - see below - readily indicates that the contribution of diffuse warm dust cannot be very high.

The fact that the Extended Low Density (ELD) HII regions contribute heavily (85 %) to the Lyman Continuum production rate but not much to the warm dust emission may be understood as follows. To attain temperatures of 30 K dust must be fairly close to the OB association. Since the dust temperature decreases with electron density, dust associated with ionized gas of a few tenths cm^{-3} electron density (typical for ELD HII regions) tends to have low temperatures and will become indistinguishable from the dust mixed with the atomic hydrogen and heated by the general ISRF.

3. GENERAL COMMENTS

3.1 Star Formation

A direct implication of the high contribution of the cold dust to the galactic infrared emission is that the total far-infrared luminosity should not be interpreted as an indicator of recent (10^7 years) star formation activity, but rather as a measure of the *total* stellar luminosity. However, the bright infrared sources remain excellent tracers of recent star formation.

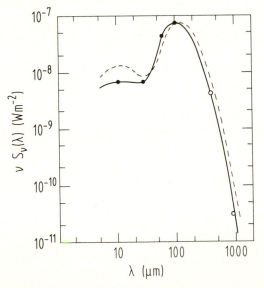

Figure 5. Spectrum of the dust emission associated with the Galactic Center within an area Δl x Δb~2° x 0°5 (full line). Filled circles are the IRAS observations. Spectral points at 380 μm and 900 μm are derived from Caux and Serra (1986) and Pajot et al. (1986), respectively. The dotted line represents the Galactic spectrum -taken from Fig. 2- normalized to the Galactic Center 100 μm flux density.

As shown before, the 50 - 100 µm wavelength region is the most sensitive region to disentangle the relative contributions of cold and warm dust. Hence, for galaxies in which most of the luminosity is derived from main sequence stars, the 60/100 µm brightness ratio is probably the best indicator of massive star formation activity.

To illustrate how regions known for star formation (and traced by the warm dust) do differ in their infrared characteristics from the Galactic Disk, Fig.4 shows the infrared spectrum from the Galactic Center . The quoted fluxes are intrinsic fluxes i.e the underlying emission from the Galactic plane has been subtracted. References to the observations are given in the figure caption. For comparison, the Galactic spectrum (taken from Fig. 2) is also plotted normalized to the Galactic Center 100 µm flux density. From the superposition, two differences between the Galactic Center and the Galactic Disk are apparent: the Galactic Center radiates more in the 60 µm band than does the Galactic Disk and the mid-infrared of the Galactic Center is definitely flatter than the Galactic Disk. The total luminosity associated with the Galactic Center is $2.8 \; 10^8 \; L_\odot$. For a total molecular mass of $1-4 \; 10^7 \; M_\odot$ (see Mezger et al., 1986 and Osborne et al., 1987), a typical luminosity over mass ratio of 10-25 is found, as compared to the value of 5 found in the Galactic Disk.

3.2 Dust Properties

Fig. 5 shows the 60/100 µm brightness ratio as observed by IRAS as a function of galactic longitude. This ratio remains almost constant for the diffuse emission in the Galactic Plane with a value of 0.22, implying a dust temperature of ~ 23 K. The peaks superimposed on this constant level trace the sources, which have typical temperatures of about 30 K. Shown as a solid line is the prediction made by the model. The strong decrease of the 60/100 µm brightness ratio is due to the decrease of the intensity of the ISRF (see Fig. 3). Whereas observed and computed ratio agree pretty well in the inner part of the Galaxy (l ≤ 30°), the disagreement is obvious in the outer Galaxy: the calculated ratio is systematically smaller than the observed one by factors up to 3.

The model is certainly at fault in the prediction of the 60 µm intensities. The constancy of the brightness ratio can be explained if a large fraction of the 60 µm emission is due to *temperatures fluctuations* in the grains smaller than 100 Å - this appears whenever the energy input associated with

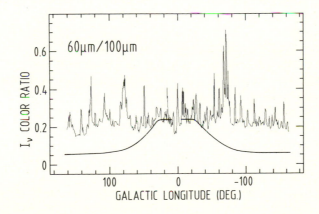

Figure 6. The 60/100 µm brightness ratio as observed by IRAS as a function of galactic longitude. Solid line represents the prediction based on the model.

a discrete heating event is comparable to or larger than the average heat content of the grains. The effects of temperature fluctuations on the infrared spectra have been investigated by Draine and Anderson (1985). Although they reproduce correctly the 60/100 μm ratio, they need 20 % of the carbon cosmic abundance in grains < 20 Å, which is at fault with the local extinction curve. Moreover, the nature of these grains is still unknown.

A complete model of the infrared of our Galaxy would incorporate such small grains, and should also incorporate even smaller grains down to polycyclic aromatic hydrocarbons to explain the secondary maximum in the galactic spectrum at about 10 μm. Since a population of such very small grains is responsible for 30 % or more of the total observed infrared luminosity (see Puget, this volume), the absorption cross sections as given by a classical grain model have to be revised in such a way that adding the absorption of the smallest grains to the absorption of a distribution of "normal" sized grains still accounts for the observed interstellar extinction curve.

Acknowledgments: We would like to thank J.-L. Puget and M. Pérault for useful discussions. S. Shigihara is kindly thanked for her help in the text editing. P. C. gratefully acknowledges the support of an Alexander von Humboldt fellowship.

REFERENCES

Boulanger, F. and Pérault M., 1987, submitted to *Astrophys. J.*

Cox, P., Krügel, E. and Mezger, P., G., 1986, *Astron. Astrophys.*, **155**, 380

Caux, E. and Serra, G., 1986, *Astr. Ap. Letters*, **165**, L5

Draine, B. T. and Anderson, N., 1985, *Astrophys. J.*, **292**, 494

Gispert, R.., Puget, J.-L., Serra, G., 1982, *Astron. Asrophys.*, **106**, 293

Hauser, M.G., Silverberg, R.F., Stier, M.T., Kelsall, T., Gezari, D. Y.,
        Dwelz, E., Walser, D., Mather, J. C., 1984, *Astrophys. J.*, **285**, 74

Mathis, J. S., Mezger, P. G., Panagia, N., 1983, *Astron. Astrophys.*, **128**, 212

Mezger, P. G., Chini, R., Kreysa, E., Gemünd, H.-P., 1986, *Astron. Astrophys.*, **160**, 324

Osborne, J.L., Parkinson, M., Richardson, K. M.,and Wolfendale, A. W., 1987,
        Astron. J., in press

Owens, D. K., Mühlner, D. K., Weiss, R., 1979, *Astrophys. J.*, **231**, 702

Pajot, F., et al., Gispert, R., Lamarre, J.-M., Peyturaux, R., Pomerantz, M.A.,
        Puget, J.-L-, Serra, G., 1986, Space-Borne Sub-Millimeter Astronomy Mission-
        ESA Workshop, Segovia, Spain. p.189

Pérault, M., Boulanger, F., Puget, J.-L-, Falgarone, E., 1987, preprint

Puget, J.-L., 1985, Proc. of XVI Les Houches meeting "Birth and Infancy of Stars"
        Lucas, Omont and Stora, eds., North-Holland, p.77

Sodroski, T. J., Dwek, E., Hauser, M.G., Kerr, F.J., 1987, *Astrophys. J.*, in press

Walterbos and Schwering, 1987, *Astron. Astrophys.*, **180**, 27

Weaver, H. and Williams, D. R. W., 1973, *Astron. Astrophys. Suppl. Ser.*, **8**, 1

THE NATURE OF THE GALACTIC BULGE

Robin Harmon and Gerard Gilmore

Institute of Astronomy
Madingley Road, Cambridge, CB3 0HA

ABSTRACT

The bulge of the Galaxy *as seen by IRAS* forms a spatially concentrated system of long period cool variable stars. From the period distribution we show that a significant fraction of the IRAS bulge is younger than the globular cluster system. Thus the metal rich stars in the central ~ 1 kpc of the Galaxy are not obviously part of the extended stellar spheroid.

1. INTRODUCTION

The stellar distribution of our galaxy is seen to comprise at least three components: a thin disc of vertical exponential scale about 300 pc; a thicker disc with a vertical exponential scale of order 1 kpc and radial scale ~ 4 kpc and a spheroidal distribution of stars which has a density rising towards the Galactic centre with a radial scale (r_e) of about 2.7 kpc. The high luminosity central ($\lesssim 2$ kpc) region of the Galaxy, often loosely termed the 'bulge', is presumably related to one or more of these components although this relation remains unclear. Available data show the bulge to have a velocity dispersion the same as that of the old disc near the Galactic centre (~ 110 kms^{-1}), a metallicity distribution which peaks at $\sim$ twice solar and a population of late M giants whose surface density falls by a factor of 10 in 200 pc. An important parameter is the age of the stars which can be derived from the period distribution of long period variables and it is this approach which we follow here for the long period variables discovered by IRAS.

2. THE BULGE AS SEEN BY IRAS

To determine the range of colour and flux in which the bulge is most readily apparent in the IRAS data we show the relative numbers of objects in two regions (one bulge-like $3° < |b| < 10°$, $|\ell| < 10°$ and one disc-like $3° < |b| < 10°$, $10° < |\ell| < 20°$) as a function of colour (f_{12}/f_{25}) and flux (f_{12}) in Figure 1. The bulge is dominant in the colour range $0.4 < (f_{12}/f_{25}) < 1.6$ and flux range 1 Jy $< f_{12} < 8$ Jy and, as selected by these parameter ranges, is shown in Figure 2. The effects of point source confusion are clearly visible towards its centre. The bulge is a compact object with a total vertical extent of less than 10°

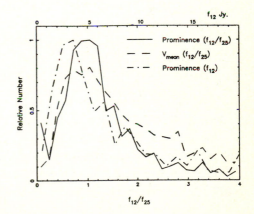

Figure 1.

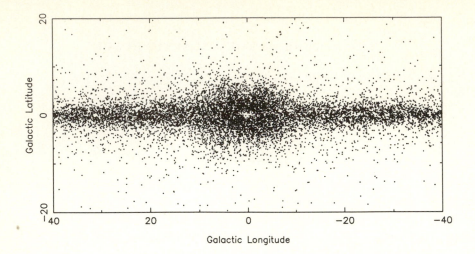

Figure 2. The Galactic Bulge as seen by IRAS.

and an axis ratio of about 0.6. Detailed fitting of the luminosity profile indicates an exponential scale height of $\lesssim$ 400 pc and corresponding radial scale of $\lesssim$ 650 pc, a less steep decline than that of the late M giants, but still more than an order of magnitude more concentrated (radially) than any of the three components mentioned above. The bulge sources have an usually high mean probability of being variable (v) (Figure 1). The colours of these objects (corresponding to 200 K $\lesssim$ T $\lesssim$ 450 K) their large luminosity ($\sim 10^4 \mathcal{L}_\odot$ and their high probability of being variable suggest that they are dust shrouded variable stars (Habing 1987). Indeed of the IRAS sources with this colour in Baade's Window 50% can be identified with the longest period sub-group (periods > 350 days) of the Mira variables seen there by Lloyd-Evans, (cf Glass 1986). The other IRAS sources in this field are redder in J-H and H-K and have a ratio of f_{12}/f_{25} closer to that of OHIR stars and hence are likely to have longer periods. IRAS missed objects with periods less than 350 days because they are neither red enough or luminous enough to be detected. They have not been excluded by our selection criteria.

3. MIRAS SEEN BY IRAS

We wish to use the parameter v quoted in the Point Source Catalogue to determine the period range of the variables seen in the bulge by IRAS. With only three observations (at t, t+14 days, t+180 days) we cannot hope to determine the period of any one variable but we are able to constrain the period range of a group of variables. In order to transform a distribution N(v) (the number of objects with a given v) to N'(T) (the number with a given period T) we need to know the 12 micron light curves of Mira variables and OHIR stars and the definition of the parameter v. The latter is explained in the IRAS Explanatory Supplement. The light curves of long period variables in the infra-red are approximately sinusoidal and independent of wavelength beyond $\sim 3\mu$m (Engles et al 1983) and have amplitudes A which increase with period:$A = \log(f_{max}/f_{min}) \approx 9.5(\pm 0.18) \times 10^{-4} \times T - 0.05(\pm 0.2)$. Simulation of the IRAS sampling process of the light curve of a variable source requires modelling a variety of instrumental and source parameters, in particular the standard deviation of the excursions of the sources in the Point Source Catalogue and the number of sources showing correlated and anticorrelated excursions of a given statistical significance. These are discussed in detail elsewhere (Harmon & Gilmore 1988). We have tested these simulations by successfully reproducing the IRAS variability distribution N(v) of 3 groups of sources from their known period distributions N'(T) ($\sim$1500 Miras drawn from the GCVS with periods in the range 100-600 days; 11 Miras from the $\ell = 0°, b = -4°$ window in the bulge and a set of about 30 OHIRS listed in Herman et al 1986).

4. APPLICATION TO THE BULGE

Figure 3 shows the simulated distributions of N(v) for a constant value of N'(T) in four period ranges. It can be seen that IRAS was particularly sensitive to periods in excess of 400 days. This is due both to its sampling interval and the increasing amplitude with period observed in long period variables. Comparing these four distributions with the observed N(v) for the bulge as defined in l, b, $f_{12}, f_{12}/f_{25}$ earlier (Figure 3 rightmost panel) it is apparent that objects with periods in excess of 400 days must comprise $\gtrsim$ 90% of this sample. Indeed a significant fraction of the sources could have periods of $\gtrsim$ 600 days.

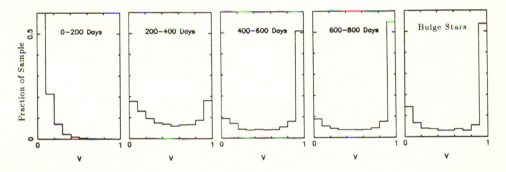

Figure 3. The sensitivity of IRAS to variables of a given period (first 4 panels) and the observed distribution of variability (last panel).

The determination of a main sequence mass corresponding to a given period is possible through pulsation theory and, via the period - luminosity relation, from stellar evolutionary models (with uncertainties due to mass loss) but are uncertain by more than a factor of 2 (Table 1a). The conversion of this mass to a stellar age is further uncertain due to the extreme metallicity-dependence of the stellar lifetimes (cf VandenBerg & Laskarides 1987, Mengel et al 1979 and Table 1b). For the shortest periods in the IRAS sample ($\sim$ 400 days) it is possible, by adopting the smallest allowed masses ($\sim 1.1 \mathcal{M}_\odot$), that these stars are as old as the globular clusters. For the longer period variables present the derived ages are significantly less than that of the globular cluster system however metal rich they may be.

Table 1a Initial Mass Estimates.

Author	Period	$L(\mathcal{L}_\odot)$	$M_{initial}(\mathcal{M}_\odot)$	Period	$L(\mathcal{L}_\odot)$	$M_{initial}(\mathcal{M}_\odot)$
Iben & Renzini	400	6500	1.1	600	9000	1.7
Habing	400	6500	2.4	600	9000	3.0
Wood & Bessel	400	7700	2.0			

Table 1b Age Estimates.

[Fe/H]	Y	$M(\mathcal{M}_\odot)$	Age(Gyr)	[Fe/H]	Y	$M(\mathcal{M}_\odot)$	Age(Gyr)
0.4	0.25	1.0	23	0.8	0.25	1.0	27
0.4	0.20	1.4	5	0.8	0.20	1.4	8.6
0.4	0.20	2.2	1	0.8	0.20	2.2	1.7

5. CONCLUSIONS

The majority of sources which delineate the "bulge" in the IRAS data are variable, with periods of at least 400 days, corresponding to an initial mass of between 1.1 $\mathcal{M}_\odot$ and 2.7 $\mathcal{M}_\odot$ and a maximum age of $\lesssim$ 15 Gyr for a $2.5\mathcal{Z}_\odot$ star. Longer period variables also exist in the bulge, showing that star formation continued there for at least several Gyr after the formation of the globular cluster system and the metal poor spheroid. The surface density of IRAS sources is much more concentrated towards the Galactic centre than would be expected if they were a core to the spheroidal component. The bulge *as seen by IRAS* is quite unlike the extended spheroidal component of the Milky Way.

REFERENCES

Engels, D., Kreysa, E., Schultz, G. V. & Sherwood, W. A., 1983. Astr. Astrophys., **124**, 123.

Feast, M. W., 1984. Mon. Not. R. astr. Soc., **211**, 51p.

Glass, I. S., 1986. Mon. Not. R. astr. Soc., **221**, 879.

Habing, H. J., 1987. in 'The Galaxy', p. 173, eds Gilmore, G. & Carswell, R. F., D. Reidel.

Harmon, R. T. & Gilmore, G., 1988. preprint.

Herman, J., Burger, J. H. & Penninx, W. H., 1986. Astr. Astrophys. J., **167**, 247.

Iben, I. & Renzini, A., 1983. Ann. Rev. Astr. Astrophys., **21**, 271.

Mengel, J. G., Sweigart, A. V., Demarque, P. & Gross, P., 1979. Astrophys. J. Suppl., **40**, 733.

Wood, P. R. & Bessel, M. S., 1983. Astrophys. J., **265**, 748.

VandenBerg, D. A. & Laskarides, P. G., 1987. Astrophys. J. Suppl., **64**, 103.

A PROCEDURE FOR DISTINGUISHING THERMAL AND SYNCHROTRON COMPONENTS OF THE RADIO CONTINUUM EMISSION OF THE GALACTIC DISC

A. Broadbent and J.L. Osborne

Department of Physics, University of Durham, South Road, Durham DH1 3LE, U.K.

C.G.T. Haslam

Max-Planck-Institute for Radioastronomy, Auf dem Hügel 69, D-5300 Bonn 1, F.R.G.

ABSTRACT

We point out the detailed correlation between the IRAS 60 μm band emission from the inner part of the Galactic disc and most of the features of the 11cm and 6cm radio continuum surveys made with the same angular resolution. As well as indicating that the 60 μm emission from this region comes in a large part from dust associated with the extended low density Hɪɪ regions, this provides a way of separating the thermal and synchrotron components of the radio emission.

1. INTRODUCTION

In a recent paper (Haslam and Osborne, 1987, 'paper I') we have remarked upon the detailed correlation between the maps of the 11 cm radio continuum emission within 1.5° of the galactic plane (Reich et al.,1986) and the IRAS 60 μm band emission. This correlation implies that a large part of the 60 μm emission from this part of the Galaxy comes from dust associated with ionised gas. This was illustrated in paper I by a grey-scale representation of the radio and infrared intensities in part of the first quadrant of galactic longitude. Figure 1 shows that there is an equally good correlation between the 60 μm and 6 cm radio emission of Haynes et al. (1978) in the fourth quadrant.

2. INFRARED EMISSION FROM THE GALACTIC PLANE

Such a conclusion had been drawn earlier from balloon flight infrared surveys but an equally strong claim could be made for correlation of the infrared emission with ^{12}CO column density, where the emission would be from dust permeating giant molecular clouds and being heated by O and B stars. It is our contention that, at least in the 60 μm band, with improved angular resolution (~4.5' for the 60 μm, 6 cm and 11 cm surveys) it is now clear that the primary correlation is with the radio continuum. This is shown in Figure 2. The undersampling of the CO survey (beamsize 45": sampling interval 3') does not account for the much less detailed agreement between profiles (a) and (c) than between (a) and (b). A quantitative comparison is given in Figure 3 for a 6° region of longitude. If the points due to the supernova remnant W44 are omitted from the plot the correlation coefficient becomes ~95%. The conclusion that a large part of the 60 μm emission

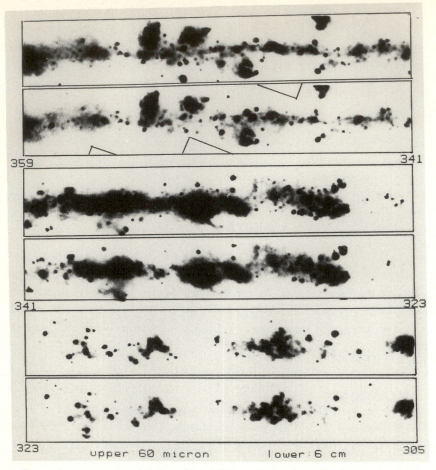

Figure 1. Grey-scale representation of the intensities of the 60 μm infrared emission (upper map of each pair) and the 6cm radio contiuum emission (lower map) of the galactic plane between longitudes 305° and 359°. Each map covers a range of 18° of galactic longitude and 3° of galactic latitude from -1.5^o to $+1.5^o$.

is associated with ionised gas gives support to the model of Cox *et al.* (1986) in which it arises mainly from dust asociated with extended low density HII regions.

The zodiacal light contribution has been removed from the 60 μm data of Figures 1, 2 and 3, using the simplest model in which the zodiacal light intensity varies only with ecliptic latitude. This is sufficiently accurate in this region of the galactic plane. From the 60 μm intensities of Figures 1 and 3 the estimated contribution of HI associated dust has been removed under the assumption that the empirical relationship between the 60 μm intensity and the HI column density determined for $|b| > 20^o$ (viz. 0.14 MJy sr^{-1}/10^{20} cm^{-2}) holds at lower latitudes. If this were true the HI contribution would be of the order of 10% of the total intensity. If the radial gradient in the Galaxy of the interstellar radiation field and the metallicity is allowed for the low latitude HI contribution towards the inner Galaxy will be higher but the observed radio-infrared correlation indicates that it should not be dominant.

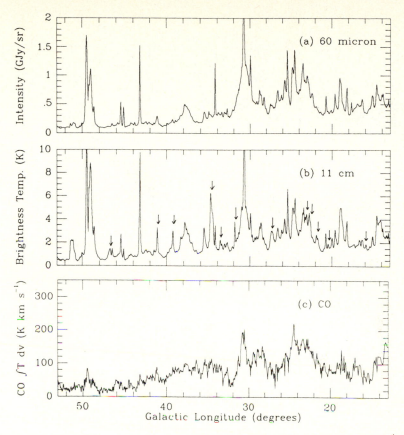

Figure 2. Profiles of (a) 60 μm emission, (b) 11cm emission and (c) CO column density (Sanders *et al.* 1986). The values are averaged over ±0.5° of latitude. The arrows on the radio profile show peaks due to catalogued supernova remnants.

3. SEPARATION OF THERMAL FROM NON-THERMAL EMISSION

An obvious difference between the profiles (a) and (b) is that the supernova remnants stand out as radio but not infrared sources. For example the ratio of 60 μm to 11 cm flux from the Crab Nebula is 0.3 compared with ratios ≥ 500 for individual HII regions. This gives a simple method of picking out supernova remnants close to the galactic plane. A search has been made of the 6 cm survey and a list of candidate remnants has been prepared. On a larger scale after the zodiacal light and HI associated components have been subtracted the 60 μm emission acts as a tracer of the thermal radio emission. We consider the points on the well-defined lower envelope to the plot in Figure 3 to correspond to directions where the non-thermal emission is negligible. The linear relationship between the thermal radio and 60 μm emission can then be used to identify the thermal component in directions where both are present. As the spectral index of the thermal component is fixed by the frequency dependence of the free-free absorption cross-section the thermal component at othe frequencies can be calculated. Figure 4 shows the galactic plane profile of the 408 MHz continuum emission of Haslam *et al.* (1982) and its division into thermal

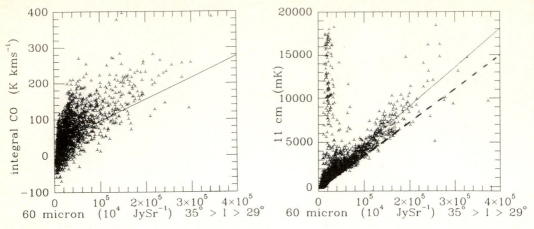

Figure 3. Point by point comparison of the 60μm intensities with (i) CO column densities (left-hand plot) and (ii) 11 cm intensities (right-hand plot). The correlation coefficient is 56% for (i) and 82% for (ii). Note the points due to the bright supernova remnant W44 lying close to the vertical axis of plot (ii). The dashed line is the adopted lower envelope of this plot.

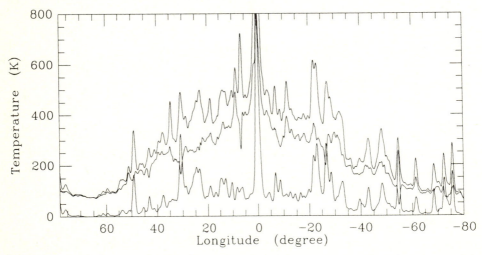

Figure 4. Galactic plane profile showing separation of radio emission into thermal and synchrotron components. Top line: observed total 408 MHz emission. Middle line: synchrotron component. Bottom line: thermal component.

and non-thermal components as outlined here. This allows a study of the Galactic magnetic field in the galactic plane.

References

Cox,P., Krügel & Mezger, P.G., 1986. *Astr.Astrophys.*, **155**, 380.

Haslam, C.G.T., Salter, C.J., Stoffel, H. & Wilson, W.E., 1982. *Astr.Astrophys.Suppl.Ser.*, **47**, 1.

Haslam, C.G.T. & Osborne, J.L., 1987. *Nature*, **327**, 211.

Haynes, R.F., Caswell, J.L. & Simons, L.W.J., 1978. *Aust.J.Phys.Astrophys.Suppl.*, **45**, 1.

Reich, W., Fürst, E., Steffen, P., Reif, K. & Haslam, C.G.T, 1986. *Astr.Astrophys.Supp.Ser.*, **58**, 197.

Sanders, D.B., Clemens, D.P., Scoville, N.Z. & Solomon, P.M., 1986. *Astrophys.J.Suppl.Ser.*, **60**, 1.

INFRARED CIRRUS

Jean-Loup PUGET

RAdioastronomie, Ecole Normale Supérieure
24 rue Lhomond,
75231 PARIS Cedex 05

ABSTRACT

The infrared cirrus are the filamentary emission seen at high galactic latitude in the IRAS data. This emission is associated with galactic interstellar clouds and was expected.

The average brightness at 100 μm is in rather good agreement with predictions based on classical interstellar dust models. Nevertheless the emission at shorter wavelengths is far in excess of the predictions of these models, and requires the presence of very small particles such as polycyclic aromatic molecules which are heated for short periods by single photons.

The structure and brightness of these cirrus clouds are such that they will be a limitation to the sensitivity of future space observatories in the far infrared.

1. INTRODUCTION

The high latitude infrared emission seen with IRAS has one component which exhibits a cloudy and filamentary structure and is commonly known as the infrared cirrus (Low et al 1984). There are several interesting questions to be studied in connection with this emission:

-we would like to separate the respective contributions of the various components of the diffuse high latitude emission (zodiacal, galactic, extragalactic);

-current models of interstellar dust and estimates of the interstellar radiation field (ISRF) can be best tested by observing the infrared emission of optically thin high latitude clouds;

- we can use the infrared emission as a probe of the structure of these clouds;

-an estimate of the fundamental limitations due to the cirrus for future infrared observations of weak extra galactic sources from space.

2. LOCAL CIRRUS

There have been many sudies of individual cirrus clouds since the first paper by Low et al (1984) which discussed this component of the IRAS data. Detailed comparison between 100 μm brightness, HI column density, extinction maps have shown that the infrared cirrus are made of the infrared emission by dust in high latitude interstellar clouds. Weiland et al (1986) have extended the comparison to high latitude CO emission showing an improved correlation when both atomic and molecular hydrogen are taken into account. They have demonstrated that some of these clouds are within 100 pc from the sun as expected for interstellar clouds making the HI disk. The 100μm brightness is proportional to the hydrogen column density up to Av≈1.5 (de Vries and Le Poole 1985) then it slowly saturates (Boulanger and Pérault 1987a and b). An extensive correlation with HI is also presented by Deul and Burton (1987) at this conference. The emissivity normalized to one magnitude of visual extinction :

$$L_{IR,H}(100\mu m) = I_v/A_v$$

observed in various individual clouds is shown in Table 1 for various studies of individual clouds.

Table 1

100 μm emissivity MJY/sr per magnitude	reference
7 to 30	Low et al (1984)
8 to 10	de Vries and Le Poole (1985)
8 to 23	Weiland et al (1986)
8.4	Terebey and Fich (1986)
28	Boulanger et al (1985)

The variations are not within the error bars and show intrinsic variations of the interstellar radiation field (ISRF) from cloud to cloud.

Low et al (1984) gave also the color ratio 60 μm/100 μm which was shown by Draine and Anderson (1985) to be too high when compared to the prediction of the classical dust models (Mathis et al 1983, Draine and Lee 1984). They argued nevertheless that keeping the same chemical composition but extending the grain size distribution down to 3 Angstroms could explain the observations. Puget et al (1985) made a model for the interstellar dust infrared emission including a component of polycyclic aromatic molecules identified by Leger and Puget (1984) as the possible carriers of the unidentified infrared bands. This model could account for the observations of reflection nebulae by Sellgren (1984); when applied to intersellar matter in the diffuse ISRF this model predicted that about 10% of the energy should be radiated in the 12 μm IRAS band. The cirrus emission in the 12 μm and 25 μm IRAS bands has been found with a level of that order by Boulanger et al (1985).

The conclusions to be drawn from these various studies of individual infrared cirrus can be summarized as follows:

1-The infrared cirrus are associated with interstellar clouds (mostly HI)

2-Locally as long as the optical depth does not exced 2, the infrared brightness is proportional to the column density measured by extinction or 21 cm emission although the proportionality factor varies from region to region by factors up to 4.

3-The spectrum of the infrared emission does not agree with the prediction of classical grain models: there is too much emission at 60 μm . Some of the brightest infrared cirrus have been shown to radiate also at 25μm and 12μm at a level which is orders of magnitude above the prediction of the classical grain models.

These qualitative features being well established, one is interested in quantitative and statistical questions:

1-Study the average properties of the cirrus emission over the whole sky outside of the main molecular clouds and the brightest part of the galactic disc.

2-find how much of the 100 μm brightness at the coldest areas at high latitude is due to interstellar emission

3-discuss dust models and the ISRF and see how they account for the intensity and spectrum of the emission

4-find out the properties of the infrared cirrus in other parts of the galaxy to use them as a pribe of the galactic structure

5-study the spatial structure of the emission and find out to what extent the infrared cirrus will be a limitation to future space infrared observations.

The best systematic study of this emission has been performed over the entire sky above latitude 10° by Boulanger and Pérault (1987a and 1987b) and is presented in these proceedings. We just recall here their main conclusions. After subtracting the zodiacal emission, the galactic emission can be identified by its cosecant dependance; this component is clearly visible at 100μm and 60 μm at all latitudes; at shorter wavelengths (12 μm and 25 μm) it is seen only at b<30° because the procedure used to remove the zodiacal component has also removed most of the extended galactic emission at b>30°. The average brightness of the galactic emission at the poles as deduced from the cosecant dependance is given for each of the four wavelengths in table 2 together with the corresponding HI 21 cm emission. The total column density estimated from extinction and reddening data on extragalactic sources estimated by de Vaucouleurs and Buta (1983) corresponds to an average extinction at the pole A_v=0.15 and is also given in Table 2.

Table 2

100 μm	2.6 MJy/sr
60 μm	0.55 MJy/sr
25 μm	0.1-0.17 MJy/sr
12 μm	0.12 MJy/sr
HI 21cm	167 K.Km/s
NH from extinction	$3\ 10^{20}$ cm^{-2}

This galactic emission together with the zodiacal component does not account for all the emission seen by IRAS at 100 µm. A residual which amounts to 2 MJy/sr is either due to uncertainties in the zero level of the IRAS data or to an isotropic (extragalactic?) background. An earlier analysis by Rowan-Robinson (1986) reached the same conclusion but overestimated this residual to be 5.7 MJy/sr. The galactic emission includes the emission of all components of the interstellar gas which contains dust. Boulanger and Pérault (1987a and b) investigate the respective contributions of the various components of interstellar gas by looking at the correlation of the deviations from the cosecant law in the distribution of the infrared brightness (at 100µm and 60 µm) with similar deviations in the neutral hydrogen column density as measured by the 21 cm emission. The correlation is good ; the slope of the best fit are

$1.65 \ 10^{-2}$ (MJy/sr)/(K.Km/s) for 100 µm and

$3.3 \ 10^{-3}$ (MJy/sr)/(K.Km/s) for 60 µm .

These values, which are about equal to those derived from the cosecant fits, imply that all the galactic infrared emission at high latitude is associated with HI gas seen in 21 cm emission. Molecular hydrogen is expected to give a negligible contribution but ionised gas seen through pulsar dispersion measures (Harding and Harding 1982) should contribute about 20 % of the flux.One possible explanation discussed by Boulanger and Pérault (1987a) is that neutral and ionised hydrogen are well correlated spatially, the alternative explanation being that the ionised gas is dust free.

The infrared luminosity of HI gas normalised per hydrogen atom is then

$$L_{IR,H}=6.3 \ 10^{-31} \ \text{W/H atom}$$

This value might be overestimated by a foctor up to 20% if most of the ionised gas is associated with HI clouds in form of ionised envelopes.

In an analysis of the large scale galactic emission , Pérault et al (1987) find that about two thirds of the infrared emission comes from a diffuse component when only one third is due to infrared sources defined as bright spots seen in the vicinity of young stars. In these regions the dust is heated by the radiation field associated with the young stars which is much larger than the average interstellar radiation field (ISRF) which heats the dust in the diffuse component; the emissivity and the spectrum of the diffuse infrared emission are the same as those of the cirrus and in fact the high latitude cirrus really makes the bulk of the infrared emission in the solar vicinity. The luminosity of this HI diffuse component in the solar neibourhood is thus $9 .5 \ L_0/pc^2$ for 4.8 M_0/pc^2 of atomic hydrogen. (to this should be added the radiation of molecular clouds and infrared sources associated with young stars to get the total infrared emission in the solar neigbourhood: $\approx 14 \ L_0/pc^2$).

3. CIRRUS IN THE INNER GALAXY

The diffuse component identified by Pérault et al (1987) in the inner galaxy is divided in two of comparable luminosity: one associated with molecular clouds and identified by its narrow latitude distribution, the second being associated with HI and thus being the equivalent of the local cirrus for the inner galaxy as discussed in the previous section.

The cirrus emission above the molecular ring (4-6 Kpc from the galactic center) can be directly observed by looking at the highest altitude HI clouds (≈500 pc above the mid plane, Lockman 1984). Figure 1 shows the longitude profile at latitude b=+/-3° for both infrared and HI column density. A conspicuous excess appears in the four bands for l<30° when there is no equivalent excess in HI column density. Pérault et al (1987) attribute this excess to HI clouds above the molecular ring which are heated by the stronger ISRF due to the active star formation going on in the molecular ring. We deduce a value for the total infrared emissivity per hydrogen atom for this region:

$$L_{IR,H} = 4. \ 10^{-30} \ W/H \ atom$$

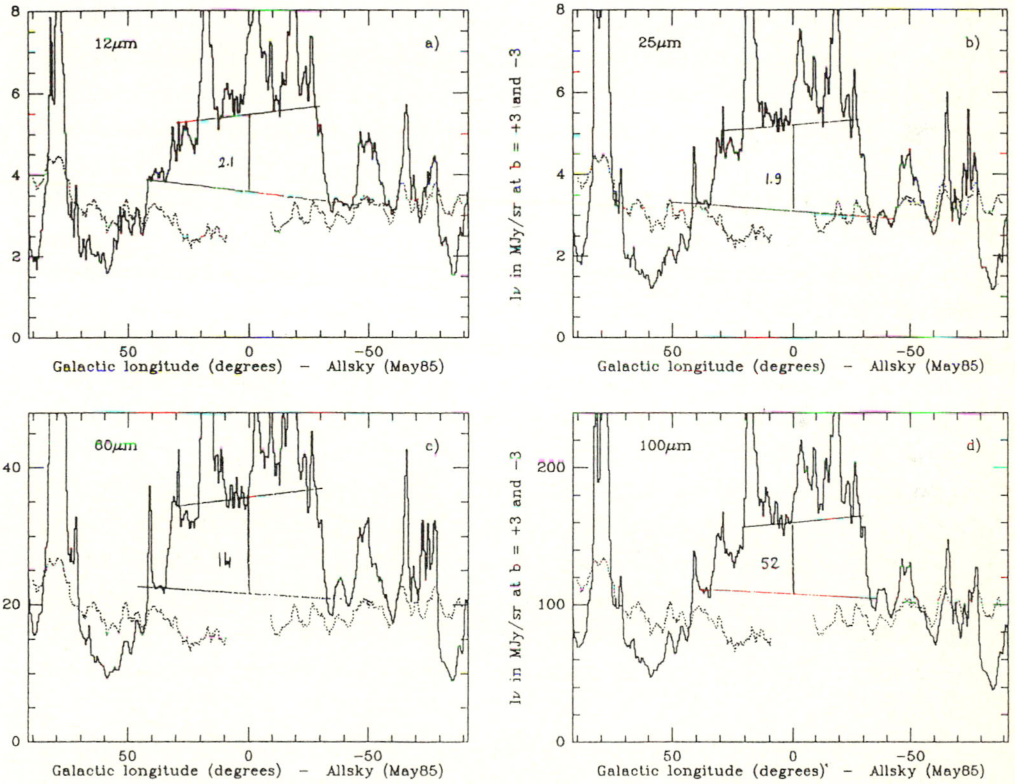

Figure 1: *Longitude profiles for infrared (full lines) and HI 21 cm emission (dotted lines) averaged for latitudes +3° and -3°. Note the excess in the infrared emission for longitudes between +30° and -30° due to high altitude cirrus clouds above the molecular ring.*

This value is about 6 times larger than the local one and is in good agreement with the value obtained for the 4-6Kpc ring in the unfolding of the radial distribution of the infrared production rate; although not different in principle, this determination is more staightforward and gives the best determinations of the emissivity and spectrum of the cirrus component in the molecular ring (the ISRF is almost constant with altitude as shown by Mattila 1980)

4. DUST MODELS

The absorption cross section as a function of wavelength for the diffuse interstellar medium is known empirically rather well in the visible the near infrared and near ultra violet. It is somewhat more uncertain in the far UV. We use the compilation of Draine and Lee (1984) for extinction ; the albedo in the ultra violet adoped agrees with the values deduced from various measurements by Lillie and Witt (1976), Joubert et al (1983). It should be noted that the albedo is not known empirically as well as the extinction; the values adopted here are somewhat smaller than those given by the dust model used by Mathis et al (1983) in the near UV.

The second ingredient needed to model the diffuse infrared emission is the ISRF. A thorough discussion of the relevant data is given by Mathis et al (1983); in the near infrared better measurements at high galactic latitude became available from rocket data (Matsumoto et al 1987) but do not lead to any significant revision of the ISRF. In the far infrared the IRAS data allow a better determination of the ISRF than the previous estimates (Mathis et al 1983, Puget 1985). This part of the spectrum plays a negligible part for the heating in the diffuse medium; the most uncertain part remains the far UV where various estimates differ by up to a factor of two. The values adopted here are obtained with a model where the ISRF is decomposed in several contributions from various stellar populations (Pérault et al 1987); they fit the existing data and are very close to those of Mathis et al (1983). $s_{abs,H}$ is the absorption cross section for the diffuse interstellar medium normalized per hydrogen atom. The total energy absorbed per hydrogen atom is given by:

$$L_{IR,H} = \int u_\nu \cdot \sigma_{abs,H}(\nu) \, d\nu = 5.7 \ 10^{-31} \ W/H$$

We notice that this value is larger than the estimate of Mathis et al (1983) :4.2 10^{-31} W/H, the reason being the smaller albedo adopted here. The value obtained here is close to the observed value when no correction is done for the ionised gas which might be associated to all HI clouds; if this correction is done the smaller value given by Mathis et al (1983) is in better agreement with the observed value.

The good agreement obtained (within the uncertainties discussed above) is an a posteriori check that the values adopted for the ISRF in the far UV are good within 30 % excluding the highest estimate of Witt and Johnson 1973.

The emission at long wavelengths $n.I_n(100 \ \mu m)$ is about one half of the total and close to earlier estimates. The average brightness at 100 μm at the galactic poles predicted by Stecker, Puget and Fazio (1977) : 3 MJy/sr is very close to the observed value given in Table 1: 2.6 MJy/sr. This illustrates that the discovery of the infrared cirrus at high latitude radiating at 100μm was not a surprise contrary to what has been written in some of the papers on the subject. It reflects the rather good empirical knowledge that we have had for more than 10 years of the ISRF and the absorption properties of interstellar dust which allowed a good estimate of the total energy to be reradiated in the infrared. The details of the dust models built to account for the extinction and absorption are almost irrelevant to the prediction of the 100 μm flux as long as the bulk of the energy is radiated by interstellar grains having a typical radius of 0.1 μm. There is much more to learn on the dust properties from the spectrum than from the integrated emission which is already well constrained by other observations as we have shown above.

As already discussed in the section on the local cirrus, the emission in the mid infrared requires a component of very small grains to be included in the interstellar dust model. Leger and Puget 1984 argued that one cannot use the optical properties of bulk graphite for particles as small as 3 Angstroms for which the infrared emissivity is dominated by lattice vibration modes and not by continuum associated with conduction electrons. The polycyclic aromatic molecules are so far the best candidate to account for the cirrus emission seen in the 12 μm band of IRAS (Puget et al 1985). More refined discussions for the physics of the emission have been proposed recently: Allamandola, Tielens and Barker 1987 Leger and d'Hendecourt 1987; a thermal model for the interstellar emission including both large particles and PAH's but taking into account correctly multi photon processes is presented by Desert (1987) at this conference.

Leger and d'Hendecourt (1987) have mesured the absorption cross-section of mixture of PAH's in the ultra-violet. Integrating the quantity

$$L_{PAH,C} = \int u_v \cdot \sigma_{PAH,C}(v) \, dv = 2.4 \; 10^{-27} \; W/C$$

it is then easy to evaluate the fraction of interstellar carbon which is locked in PAH's particles to account for the energy radiated both in the 12 μm band of IRAS and at shorter wavelengths using the spectrum of reflection nebulae observed by Sellgren (1986) for the extrapolation at shorter wavelengths. This fraction is 15%. Another 17% is needed to account for out of equilibrium emission at 25 μm and at 60 μm which is evaluated by Ryter et al (1987) to be about two thirds of the total for the average cirrus. This emission is probably associated with particles small enough to fluctuate in temperature but larger than the PAH's which radiates in the bands. The total amount of carbon locked in very small particles is thus 32% of all interstellar carbon.

5.STRUCTURE OF CIRRUS

The first statistical analysis of the stucture of cirrus clouds has been carried out by Gautier and Boulanger (1987) and presented at this conference. After taking out the detector noise from the power spectrum of the spatial distribution done along the scans in several high latitude clouds they find that the resulting power spectrum can be fitted by power laws around 2.5 at 100 μm. Their analysis gives flatter power spectra at shorter wavelengths.

A preliminary analysis of the consequences of such fluctuations for future space borne infrared observations shows that for photometry with small cryogenically cooled telescopes like ISO or SIRTF in the range 100 μm to 1 mm the sensitivity is limited to sources brighter than about 0.1 Jy in regions showing cirrus emission. For large uncooled space borne submillimeter antennas like the FIRST project of ESA or the LDR project of NASA, the better angular resolution makes the limitation to sensitivity due to the cirrus structure negligible when compared to the photon noise. It is likely that such uncooled telescopes will out perform the small cryogenically cooled ones for photometry at wavelengths larger than 100μm over most of the sky.

REFERENCES

Alamandola,L.J., Tielens,A.G.G.M. and Barker,J.R. 1987 Polycyclic Aromatic Hydrocarbons and Astrophysics, ed Leger,A., d'Hendecourt,L. , and Boccara,N.; Reidel, p.255

Boulanger,F., Baud,B., and van Albada,G.D., 1985, Astr. Ap. **144**, L9

Boulanger,F. and Pérault, M. 1987 a this conference

Boulanger,F. and Pérault,M. 1987 b Ap. J. sumitted

Désert,F.X., 1987, this conference

Deul,E.R. and Burton,B., 1987, this conference

Draine,B.T., and Anderson,N. 1985, Ap. J. Ap. J. **292**,494

Draine,B.T.,and Lee,H.M., 1984, Ap. J. **285**,89

Harding,D.S. and Harding,A.K., 1982 Ap. J. **257**,603

Joubert,M., Masnou,J.L., Lequeux,J., Deharveng,J.M., and Cruvellier,P., 1983, Astr. Ap. **128**,114

Leger,A., and Puget,J.L., 1984 Astr. Ap. **137**,L5

Leger,A. and d'Hendecourt,L. 1987 Polycyclic Aromatic Hydrocarbons and Astrophysics, ed Leger, A., d'Hendecourt,L., and Boccara,N.;Reidel,p.223

Lillie,C.F., and Witt,A.N. 1976, Ap. J. **208**,64

Lockman,F.J. 1984, Ap. J. **283**,90

Low,F.J., et al 1984, Ap. J. letters **278** L19

Mathis,J.S., Mezger,P.G., and Panagia,N., 1983,Astr. Ap. **128**,212

Mattila,K., 1980, Ap. J. **82**,373

Matsumoto,T., Akiba,M., and Murakami,H. 1987 preprint and this conference

Pérault,M., Boulanger,F., Puget,J.L., and Falgarone,E., 1987 in preparation

Puget,J.L. 1985, Birth and infancy of stars, ed Omont,A., Lucas,R. and Stora; North Holland Pub. Co.

Puget,J.L., Leger A., and Boulanger, F., 1985 Astr. Ap. **142**, L19

Rowan-Robinson,M.,1986, Lighgt on dark matter, ed Israel,F.P., (D. Reidel),p.499

Ryter,C., Puget,J.L., and Pérault,M. 1987, Astr. Ap in press

Sellgren,K., 1984, Ap. J. **277**,623

Sellgren,K., Allamandola,L.J., Bregman,J.D., Werner,M.W. and Wooden,D.H. 1985 Ap. J. **299**,416

Stecker,F.W., Puget,J.L., and Fazio,G.G. 1977 Ap. J. Letters **214**,L51

Terebey,S., and Fich, M., 1986, AP. J. Letters **309**, L73

de Vaucouleur,G.,and Buta ,R., 1983, Aston. J. **88**,939

de Vries,C.P., and Le Poole,R.S., 1985, Astr. Ap. **145**, L7

Weyland,J.L., Blitz,L., Dwek,E., Hauser,M.G., Magnani,L., and Rickard,L.J., 1986, Ap. J.(letters), **306**, L101

Witt,A.N., and Johnson,M.W. 1973 ,Ap. J. **181**,363

Infrared Emission from the Solar Neighborhood

F. Boulanger

IPAC, Caltech 100-22, Pasadena CA 91125, USA

M. Pérault

Ecole Normale Supérieure, 24 rue Lhomond, 75016 Paris, France

ABSTRACT

The main results of an extensive study of the infrared emission from the nearby interstellar medium are presented. These results are used to discuss the origin of the infrared emission from the solar neigbourhood. We show that altough most of the emission comes from interstellar matter not associated with current star formation, about half of the heating of grains is provided by ultraviolet photons radiated by stars younger than few 10^8 yrs.

1. INTRODUCTION

Within a few kpc from the sun the different components of the interstellar medium - neutral atomic gas, molecular clouds, and HII regions - can be observed and studied separately. For each of these components we can identify the heating source of the dust, and understand the origin of the infrared emission because we know the interstellar radiation field and the distribution of stars. In that sense the solar neighborhood is a unique study-case allowing us to get a *microscopic understanding* of the infrared emission from the interstellar medium, which will be the fundamental basis of a more general understanding of the infrared emission of galaxies.

An extensive study of the infrared emission from the solar neighborhood has been presented by Boulanger and Pérault (1988) (herinafter BP). Here, we summarize the main results of this study. These results are subsequently used to discuss the origin of the infrared emission on the scale of a few kpc around the sun, and a few 100 pc around the star-forming region of Orion.

2. INFRARED EMISSIOIN FROM THE NEARBY ISM

Galactic latitude profiles of the infrared and H I emission are presented in figure 1. The profiles were built by averaging the pixels of all-sky maps within 50 latitude bins with a width $\Delta(1/|sin(b)|)$ of 0.1; data in the direction of the Magellanic Clouds, the main nearby molecular clouds and OB associations, and few bright point sources were discarded when computing the profiles. The absence of emission above 30° in the 12 and $25\mu m$ profiles results directly from the assumption made in the subtraction of the zodiacal light (see BP). The latitude profiles of figure 1 prove the existence of Galactic emission outside the Galactic plane at all IRAS wavelengths. Using the IRAS catalog and the deep survey of Hacking and Houck (1987), BP have shown that photospheres and circumstellar dust-shells account for about 10% and 2.5% of the 12 and $25\mu m$

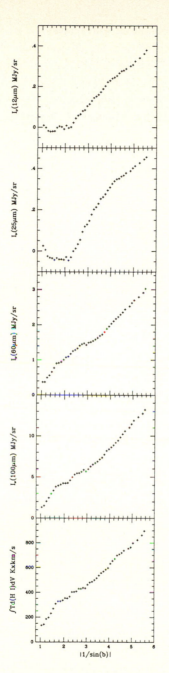

Fig. 1. Profiles of infrared and H I emission for absolute latitudes larger than 10°. Data in the direction of the Magellanic Clouds and in the vicinity of the main nearby molecular clouds and OB associations, and few point sources were discarded when computing the infrared and H I profiles.

emission seen at $|b| > 10°$. Therefore, at all wavelengths the emission measured by the latitude profiles originates from the interstellar medium. A spectrum of this emission is presented in figure 2. Since large grains heated by the local ISRF have an equilibrium temperature much too low to account for any significant emission at 12 and $25\mu m$ this result demonstrates that grains, small enough to be transiently heated to temperatures of several hundred degrees when they absorb a single photon, are a pervasive component of interstellar dust.

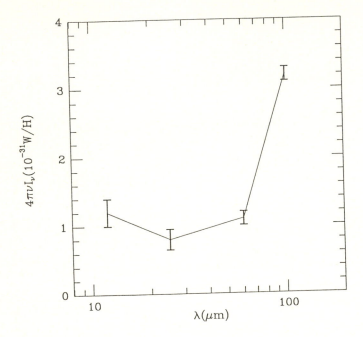

Fig. 2. Infrared spectrum of emission at $10° \leq |b| \leq 30°$. This spectrum is derived from the slopes of the latitude profiles presented in figure 1. The normalisation per hydrogen atom was done by dividing the infared slopes by the slope of the H I profile.

Away from heating sources, and molecular clouds the far-infrared emission of the nearby ISM is well correlated with the column density of H I gas (figure 3). On scales larger than few hundred parsecs the ratio between infrared emission and gas column density varies significantly from one region to another. The variations in the infrared emission per H atom are probably related to changes in the intensity of the ISRF. Most of the infrared emission measured away from OB associations comes from dust associated with H I gas; the average infrared luminosity to mass of gas ratio is 1.6 $L_\odot/M_\odot$ for this component of the interstellar medium. The high latitude molecular clouds surveyed by Magnani et al. (1985) contribute for a negligible fraction of the emission seen at $|b| \geq 20°$. The diffuse ionized gas observed outside discrete H II region (Reynolds 1984) which accounts for about 20% of the local density of atomic gas could contribute for part of the infrared emission. However, the direct correlation between $100\mu m$ and H I emission at high latitude leads to the same infrared emissivity per hydrogen atom than the one derived from the cosecant laws of figure 1. A possible explanation of this result is that the distribution of the diffuse ionized gas is correlated with the distribution of the H I gas.

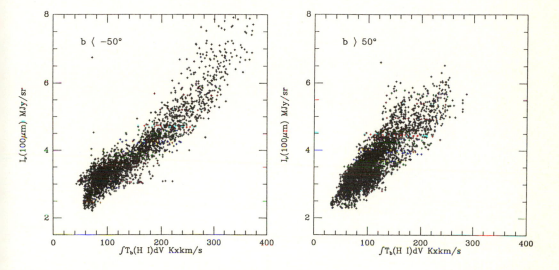

Fig. 3. Pixel-by-pixel comparison between $100\mu m$ and H I emission in the northern and southern polar caps. To avoid contamination by stray-radiation we used for this compariosn the H I survey made with the horn-reflector of the Bell Laboratories (Stark et al. 1987). Each cross represents a $1.5°x1.5°$ pixel

Embedded stars have a negligible contribution to the infrared emission of molecular clouds not associated with HII regions. Those clouds which account for most of the molecular mass in the Galaxy are mainly heated by the interstellar radiation field (ISRF). In the solar neighborhood, the infrared luminosity per mass of gas of molecular clouds not associated with H II regions is of the order of 0.8 $L_\odot/M_\odot$, half of the value derived for the atomic gas. The difference between atomic and molecular gas is due to the optical depth of molecular clouds to UV and optical photons from the ISRF. Figure 4 presents a pixel-by-pixel comparison of the $100\mu m$ and CO emission of the Orion A and Chamaeleon II molecular clouds chosen as example of clouds of high and low star formation activity (Boulanger et al. 1988). For clouds like Chamaeleon II that are largely devoid of star formation, the infrared emission which is dominated by the external heating correlates with the CO emission. In more active clouds, like Orion A, the infrared and CO emission are correlated only outside star-forming regions. At the bottom of the $100\mu m$-CO diagram, a lower envelope shows this correlation. Points representing lines of sight in the direction of star-forming regions appear scattered above the line of correlation; the radiation absorbed from embedded stars is for these points comparable to or much greater than the emission associated with the external heating. The correlation between CO and $100\mu m$ emission seen outside star forming regions indicate that for clouds not associated with H II regions the far-infrared and CO emission are equally good tracers of the molecular gas.

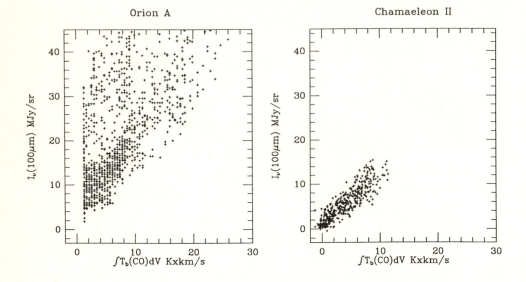

Fig. 4. Pixel-by-pixel comparison between $100\mu m$ brightness and CO integrated emission for the Orion A and Chamaeleon II clouds. The pixel size is 8'x8'.

The good correlation between infrared emission and gas column density implies that IRAS data can be used to study the distribution of molecular and atomic gas away from the Galactic plane, and to measure the extinction at high Galactic latitude. For relative measurements over angular scales smaller than typically $10°$ the sensitivity limit at $100\mu m$ corresponds to emission from a column density of gas of few $10^{19} H\,cm^{-2}$ or A_V of 0.01 mag; uncertainties on absolute measurements of A_V are of the order of 0.05 mag. The $100\mu m$ maps of the northern and southern polar caps presented in figure 5 give an overview of the extinction at high Galactic latitude.

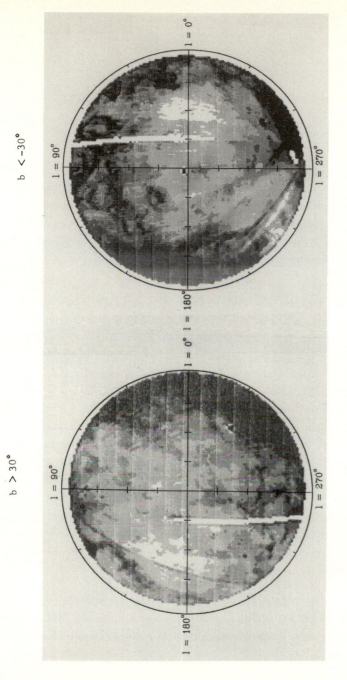

Fig. 5. Grey-scale maps of the $100\mu m$ emission around the northern and southern Galactic poles ($|b| \geq 30°$); tick marks on the axis are spaced by $15°$. The resolution of the maps is $1.5°$. To fix the grey-levels of the plot we converted the $100\mu m$ brightness into visual extinction using the slope and the intercept of the $100\mu m$-H I correlation presented in figure 3. The grey-scale maps are white for $Av < 0.04$ mag; the three levels of grey correspond to the ranges of Av: 0.04 to 0.14 mag, 0.14 to 0.20 mag, and 0.20 to 0.30 mag; regions in black have an $Av > 0.30$ mag. The positive error-bar on these measurements is 0.07 mag; the negative error-bar is 0.03 mag. The bright source near the south Galactic pole is the spiral galaxy NGC 253.

3. ORIGIN OF THE EMISSION

3.1 Solar Neighborhood

To interpret the infrared emission of galaxies in terms of star formation properties, one must answer two related questions:

(1) What are the relative contributions of stars of different type and age to the heating of dust?

(2) What are the respective contributions of molecular clouds, atomic gas and HII regions to the infrared emission?

Attempts to answer these questions from observations of the Galactic plane are hampered by confusion because the infrared emission arises in different components of the interstellar medium integrated along the line of sight, and by extinction which obscures the distribution of star light across the Galaxy. In the solar neighborhood, the question of the origin of the infrared emission can be addressed more directly, because the different components of the interstellar medium are observed separately, and also because the distribution of stars is known.

The IRAS, H I and CO surveys enabled BP to estimate the overall emission from the different components of the interstellar medium. The infrared luminosity per unit surface is 14.5 $L_\odot/pc^2$. Dust associated with atomic and ionized gas away from star-forming regions accounts for 70% of this emission; HII regions and molecular clouds contribute for about 20% and 10%, respectively (see BP). These numbers show that 80% of the infrared emission from the solar neighborhood comes from dust heated by the ISRF. Using the extinction curve, albedos and the radiation field tabulated by Mathis et al. (1983), we computed that ultraviolet ($0.0912\mu m \leq \lambda \leq 0.346\mu m$), optical ($0.346\mu m \leq \lambda \leq 0.8\mu m$), and near-infrared photons ($\lambda \geq 0.8\mu m$) account for 50%, 30%, and 20%, respectively, of the heating of dust by the ISRF. Ultraviolet photons of the ISRF originate mainly from B stars (Viallefond 1987); therefore, we find that although most of the emission of the solar neighborhood is coming from interstellar matter not associated with star formation, about half of the heating of interstellar grains comes from stars younger than few $10^8 yrs$. The other half comes from the older stars which are responsible for the optical and near-infrared emission of the Galaxy. This result probably applies to most galaxies with a low ratio between $100\mu m$ and optical emission like the solar neighborhood where $\nu I_\nu(100\mu m)/\nu I_\nu(0.44\mu m)$ is 0.3. For other galaxies we expect the fraction of the infrared emission coming from young stars to increase with the ratio between infrared and optical emission.

3.2 Star Forming Regions

To investigate how the luminosity of OB associations is spread over the ISM, and to what extent the luminosity of young stars is converted into infrared radiation, we studied in detail the infrared emission of the Orion region.

Data gathered by BP lead to the following picture of the Orion region. The OB association with a luminosity of 5 $10^6 L_\odot$ is the main source of radiation over a sphere of 300 pc diameter containing 2.4 $10^5 M_\odot$ of molecular gas and 1.6 $10^5 M\odot$ of atomic hydrogen. With a total infrared luminosity of 1 $10^6 L_\odot$ molecular clouds and associated HII regions are the dominant sources of the emission; with a luminosity of $4 \, 10^5 L_\odot$ the H I gas accounts for only 30emission. Therefore in the 300 pc sphere surrounding the OB association the origin of the infrared emission is the opposite of what we found for the whole solar neighborhood.

The total infrared emission of the Orion region, about $1.4 \, 10^6 L_\odot$, represents only 30% of the luminosity of the OB association; thus, we find that a large fraction of the radiation of the association gets spread over an extended volume of the interstellar medium; a study of six OB associations in the outer Galaxy led Leisawitz(1987) to a similar conclusion. Close to the Galactic plane, the average density of the interstellar medium (atomic + molecular gas) is of the order of 0.6 Hcm^{-3} (Lockman 1984, Dame et al 1987). For this density, in the plane the opacity in the ultraviolet is 2.2 mag/kpc; however, as the ISM is highly inhomogenuous the effective opacity is certainly lower. Thus, we estimate that photons can spread as far as 1kpc from the OB association. As the size of this distance is larger than the scale height of the disk a large fraction of the radiation of the OB association leaves the Galaxy. On one hand, by emphasizing the fact that the radiation of an association of young stars spreads over a wide volume of the Galaxy, the study of Orion strengthens the idea that throughout the interstellar medium a large fraction of the heating of interstellar dust is coming from young stars. On the other hand, the fact that a significant fraction of the luminosity of OB associations leaves the Galaxy questions the use of infrared observations to measure the overall luminosity of star forming regions in the Galaxy, and the luminosity of young stars in external galaxies.

References

Boulanger, F., Cohen, R.S., Gaida, M., Grenier, I., Koprucu, M., Maddalena, R.J., Thaddeus, P., and Ungerechts, H. 1988, In preparation

Dame, T.M., Ungerechts, R.S., Cohen, R.S., E. de Geus, Grenier, I.A., May, J., Murphy, D.C., Nyman, L.A., and Thaddeus, P. 1987, Ap. J., In press

Hacking, P., and Houck, J.R., 1987, Ap. J. Suppl. **63**, 311.

Leisawitz, D. 1987, Star Formation in Galaxies, ed. C.J. Lonsdale Persson

Lockman, F. J. 1984, Ap. J. **283**, 90.

Magnani , Blitz, L., Mundy, L., 1985, Ap. J. **295**, 402.

Mathis, J.S., Mezger, P.G., and Panagia, N., 1983, Astr. Ap. **128**, 212.

Reynolds, R.J., 1984, Ap. J. **282**, 191.

Stark, A.A., Bally, J., Linke, R.A., and Heiles, C., 1987, In preparation.

Viallefond, F., 1987, Thèse d'Etat, Univeristy Paris VII

A POST-IRAS INTERSTELLAR DUST MODEL

Francois-Xavier Désert
NASA Goddard Space Flight Center, Code 685, Greenbelt, MD, 20771, USA

Francois Boulanger
IPAC, Caltech 100-22, Pasadena, CA, 91125, USA

Michel Pérault
Groupe Radioastronomie, E.N.S., 24 rue Lhomond, 75005 Paris, France

ABSTRACT. A composite interstellar dust model is built in order to understand the infrared colors of various astronomical objects including the low-density interstellar medium (HI, H_2, and HII) in cirrus, reflection nebulae, and HII regions. Taking into account the various constraints on the different dust components, in particular coming from IRAS data, we show that a bimodal rather than continuous size distribution between small and big grains is the most favorable solution.

1. INTRODUCTION

The wealth of data coming from the Infrared Astronomical Satellite has implied what we think is a profound revision of the composition of dust in the interstellar medium. Since the discovery of small grains emitting between 2 and 13 μmin in reflection nebulae (Sellgren 1984), identified as PAHs (Léger and Puget 1984), the IRAS data has shown their presence even at wavelengths as long as 60 or 80 μm. The two basic characteristics of the infrared emission of small grains (radius less than about 10 nm) are due to their low heat capacities and therefore their temperature fluctuations (see e.g. Désert, Boulanger and Shore 1986) : 1) the emission wavelengths are smaller than for big grains, emitting at their equilibrium temperature, at the same distance to the illuminating object, 2) the shape of the emitted spectrum does not depend on the distance to the illuminating star. Owing to the IRAS data, these two characteristics can be recognized in many different places in The Galaxy and in external galaxies. In the following, we describe a composite dust model and try to confront it to some obervations. More details can be found in Désert and Boulanger (1987).

2. A DUST MODEL

   Table 1 summarizes the adopted optical properties of the different components of grains and Table 2 gives their size distributions and mass abundances relative to hydrogen.

Table 1

Grain type	UV and Visible	Infrared optical properties
PAHs	$\alpha\ N_C/\lambda$	Léger and d'Hendecourt (1986)
Amorphous Carb.	$\alpha\ m/\lambda$	Draine (1981)
'Bump' grains	Graphite-like bump	" (Hecht's(1986) model)
Graphite	Draine and Lee (1984)	
Silicate	"	

Table 2

Grain type	Mass/M_H	a_{min}(nm)	a_{max}(nm)	Density
PAHs	$4.0\ 10^{-4}$	0.4	1.5	$2.3\ 10^{-7}$ g cm^{-2}
Amorphous Carb.	$7.3\ 10^{-4}$	4.0	4.5	1.0 g cm^{-3}
'Bump' grains	$4.7\ 10^{-4}$	2.0	5.0	2.3 g cm^{-3}
Graphite	$2.5\ 10^{-3}$	20	250	2.3 g cm^{-3}
Silicate	$6.0\ 10^{-3}$	100	250	3.4 g cm^{-3}

The size distribution is taken as a power-law with an exponent 3.5 as in the model by Mathis, Rumpl and Nordsieck (1977). The exact exponent is not very important for small grains due to their small size ranges. Table 2 has been deduced by trying to explain simultaneously the absorption curve in the UV and visible spectrum and the infrared emission of cirrus clouds : namely the cosecant law deduced by Boulanger and Pérault (1987, see also this conference).

3. RESULTS AND DISCUSSION

Figures 1, 2, 3, and 4 show the different infrared colors of the dust illuminated by the solar neighborhood interstellar radiation field (Mathis, Mezger and Panagia 1983) multiplied by a factor X (the crosses correspond to X = 0.3, 0.5, 1, 1.5, 2, 3, 5, 10, 30, 50, 100, 300, 10^3, 3 10^3). The dashed curves show the colors of big grains only (graphite and silicate). The continuous curves (complete model) show a plateau for the colors due to the presence of small grains : $I_\nu(12\mu m)/I_\nu(100\mu m) \simeq 0.04$, $I_\nu(25\mu m)/I_\nu(100\mu m) \simeq 0.06$ for X between 1 and 50 and $I_\nu(60\mu m)/I_\nu(100\mu m) \simeq 0.21$ for X between 0.3 and 2. The cosecant law is fitted with X = 1.5. Figure 3 shows the far-infrared color variations with the radiation field energy density. The open circles represent cirrus clouds at 12, 10 (cosecant law), and 5 kpc and the full circles are IRAS observations of ρ Oph region analysed by Ryter et al. (1987). The agreement is relatively satisfactory over more than 4 orders of magnitude of the energy density.

In a minimalist sense, the model could be reduced to two types of grains : the small grains with a PAH like structure from about 20 to 6 10^4 carbon atoms and the big grains which could be silicate cores with various hydrocarbon mantles (between 20 to 250 nm). However, the color-color diagrams should not be affected too much by the different underlying hypotheses of the proposed model.

Acknowledgements This work was done while one of us (F.X.D.) held a National Research Council-NASA Research Associateship. We wish to thank A. Léger, J.L. Puget and C. Ryter for helpful discussions.

References
Boulanger, F., and Pérault, M. 1987, Ap. J., submitted.
Désert, F. X., Boulanger, F., and Shore, S. 1986, As. Ap., 160, 295.
Désert, F. X., and Boulanger, F. 1987, As. Ap., to be submitted.
Draine, B. T. 1981, Ap. J., 245, 880.
Draine, B. T., and Lee, H. M. 1984, Ap. J., 285, 89.
Léger, A., and Puget, J. L. 1984, As. Ap., 137, L5.
Léger, A., d'Hendecourt, L. 1986, in PAHs and Astrophysics, Kluwer.
Mathis, J. S., Rumpl, W., and Nordsieck, K. H. 1977, Ap. J., 217, 425.
Mathis, J. S., Mezger, P. G., and Panagia, N. 1983, As. Ap., 128, 212.
Sellgren, K., 1984, Ap. J., 277, 623.
Ryter, C., Puget, J. L., and Pérault, M. 1987, As. Ap., in press.

133

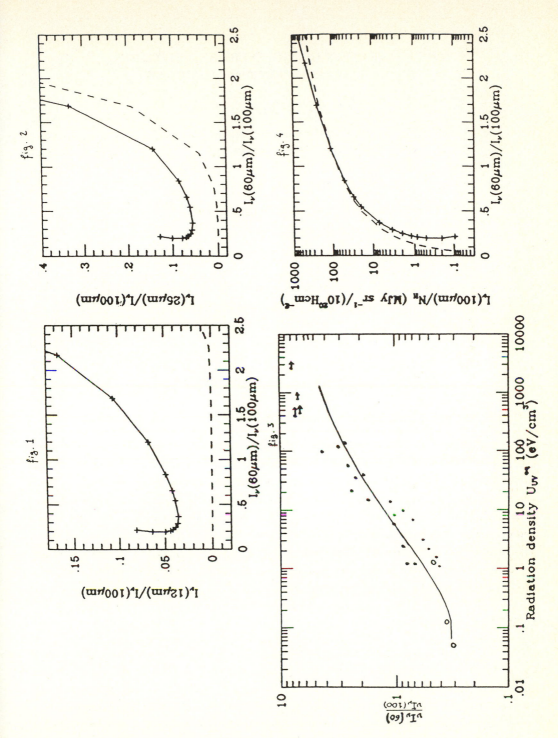

THE EXCITATION OF THE INFRARED EMISSION
FROM VISUAL REFLECTION NEBULAE

K. Sellgren

Institute for Astronomy, University of Hawaii
Honolulu, Hawaii 96822 USA

M. W. Castelaz

Allegheny Observatory, Observatory Station
Pittsburgh, Pennsylvania 15214 USA

M. W. Werner

Space Science Division, M/S 245-6, NASA Ames Research Center
Moffett Field, California 94035 USA

L. Luan

Lick Observatory, University of California at Santa Cruz
Santa Cruz, California 95064 USA

ABSTRACT

We present current results on the infrared emission from visual reflection nebulae. A detailed study of the Pleiades reflection nebulosity shows that the 12 and 25 μm emission from this nebula is primarily due to nonequilibrium emission from very small particles, while the 60 and 100 μm emission is primarily due to thermal emission from dust in equilibrium with the stellar radiation field. We discuss briefly a comparison of the Pleiades emission to that of infrared cirrus clouds and discuss in more detail preliminary results of a survey of the IRAS emission from reflection nebulae illuminated by stars with temperatures between 3,000 K and 21,000 K. The goal of this survey is a determination of the excitation of the small particles believed to be responsible for the 12 μm emission of these reflection nebulae.

1. INTRODUCTION

Visual reflection nebulae provide an ideal environment for studying the excitation of the small particles believed to be responsible for the diffuse 12 μm emission seen in reflection nebulae, infrared cirrus clouds, and other galaxies. The discovery of anomalous extended near-infrared emission in visual reflection nebulae (Sellgren, Werner, and Dinerstein 1983) originally led to a model in which the near-infrared continuum emission (1–5 μm) was attributed to nonequilibrium thermal emission by very small grains (radius 10 Å) transiently heated to high temperatures by single UV photons (Sellgren 1984). This model, coupled with observations of the 3.3 μm unidentified emission feature in these reflection nebulae, in turn led to an identification (Léger and Puget 1984; Allamandola, Tielens, and Barker 1985) of the emission features at 3.3, 6.2, 7.7, 8.6, and 11.3 μm with fundamental vibrational transitions in a specific class of large molecules (radius 5–10 Å) known as polycyclic aromatic hydrocarbons (PAHs). Spectrophotometry of these reflection nebulae from 1 to 13 μm (Sellgren et al. 1985) showed that the emission from reflection nebulae in the IRAS 12 μm band is probably dominated by nonequilibrium emission, with contributions both from the features and from the associated continuum. Subsequent observations of the infrared cirrus (Boulanger, Baud, and van Albada 1985; Weiland et al. 1986), of the diffuse infrared radiation from our Galaxy (Boulanger and Perault 1987), and of normal spiral galaxies (Helou 1986), have found that nonequilibrium 12 μm emission is widespread in the interstellar medium of our own and other galaxies and accounts for a large fraction (20–40%) of the total infrared emission from dust. Visual reflection nebulae offer a unique opportunity to study this important component of the interstellar medium because they provide an environment where the response of the small particles to differing radiation fields can be studied by choosing illuminating stars of different effective temperatures. The outstanding sensitivity of IRAS to low surface brightness emission makes its data base the ideal tool for surveying a large number of reflection nebulae illuminated by stars of varying temperature in order to characterize the excitation of the 12 μm emission.

2. IRAS OBSERVATIONS OF THE PLEIADES

As the first step in this survey, we studied in detail the IRAS emission from the Pleiades (Castelaz, Sellgren, and Werner 1987). This is a nearby group of four reflection nebulae, illuminated by B stars, whose infrared emission is very bright and very extended. We derived 12/25 μm and 60/100 μm color temperatures as a function of distance from one star, 23 Tau, which are shown in Figure 1. These temperature profiles show that the 12/25 μm color temperature is independent of distance from the star, while the 60/100 μm color temperature decreases with distance from the star. This demonstrates that the 12 and 25 μm emission is nonequilibrium emission, while the 60 and 100 μm emission is due to equilibrium thermal emission from dust.

3. COMPARISON OF THE PLEIADES EMISSION
WITH THE INFRARED CIRRUS

Our next step in understanding the emission from small particles has been to quantitatively compare the infrared emission from the Pleiades with that from the infrared cirrus (Werner et al. 1987). Our approach has been to adopt the two-component grain model used by Weiland et al. (1986) to model the infrared emission from high-latitude cirrus clouds and to apply the same grain model to the Pleiades to determine whether it can fit the emission from both reflection nebulae and high-latitude clouds. Our specific grain model combines a Mathis, Rumpl, and Nordsieck (1977) grain size distribution, with its small grain size cutoff extended to 3 Å, and an enhanced small-grain component that has a power-law size distribution with a power-law index of −5 and whose sizes extend from 3 to 10 Å. The larger grains include both silicate and graphite

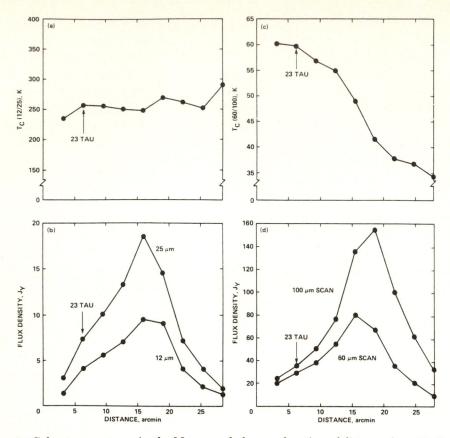

Figure 1. Color temperatures in the Merope nebula as a function of distance from 23 Tau. (a) Color temperatures derived from the 12 and 25 μm data along a path from the northeast to the southwest through the Merope nebula. (b) Color temperatures derived from the 60 and 100 μm data along the same path. (c) and (d) The 12, 60, 25, and 100 μm flux densities along the same paths. These flux densities are the observed values rather than the color-corrected flux densities (from Castelaz, Sellgren, and Werner 1987).

grains, while the enhanced small-grain component consists of graphite grains. These grains are heated by the radiation field appropriate to the high-latitude clouds (the interstellar radiation field) and to the Pleiades (the diluted stellar energy distribution), including both equilibrium and nonequilibrium heating. The only free parameter is the relative amounts of large and small grains needed to fit the 12–100 μm energy distribution observed. We find that the relative amounts of large and small grains required are the same in the cirrus clouds and in the Pleiades, within a factor of two. This indicates that the grains in reflection nebulae are similar to those in more diffuse regions, such as the cirrus clouds. This gives us confidence that the results we hope to achieve on the excitation of 12 μm emission in reflection nebulae will apply to the interstellar medium elsewhere in our galaxy.

4. SURVEY OF 12 μm EMISSION IN REFLECTION NEBULAE

We are currently working on a survey of reflection nebulae to determine the excitation of the 12 μm emission. The results of this survey will be presented in greater depth by Sellgren et al. (1987). We have selected 45 reflection nebulae from the van den Bergh (1966) catalog of visual reflection nebulae. The illuminating stars of these nebulae range in temperature between 3,000 K and 33,000 K. We used the IPAC facility to coadd images in all four IRAS bands for all of these nebulae. These images were used to measure the total nebular flux at each wavelength for each nebula. This approach gave us the highest sensitivity for detecting the very faint nebulae illuminated by late-type stars. We also selected a nebular position offset from the illuminating star, typically 3′ away in declination, and then obtained nebular surface brightnesses in a 3′ by 3′ box at all four IRAS wavelengths. Our motivation for this second measurement, which is less sensitive than the total flux approach, is to avoid correcting for the direct radiation from the central star, which dominates the total flux (star plus nebula) at 12 and 25 μm for nebulae illuminated by K and M stars. This offset nebular measurement additionally has the advantage of providing surface brightnesses at a specific distance from the star, so that observed dust temperatures can be easily compared to model calculations. Positions north and south of the star were averaged after discarding positions contaminated by other stars in the field. Analysis of our entire data sample using this nebular offset technique is not yet complete; to date we have data for 12 nebulae, whose illuminating stars range in temperature from 5,000 K to 21,000 K. We caution that further analysis of our entire sample may modify some of the tentative conclusions we present below.

We have used the offset nebular surface brightnesses to determine the relative amounts of infrared luminosity that emerge in the IRAS 12 μm band and IRAS 100 μm band. We find that the ratio of $\nu F_\nu (12~\mu m)/\nu F_\nu (100~\mu m)$ has values between 0.1 and 0.9, with most values around 0.4. In the sample to date, this ratio does not seem to depend on the temperature of the illuminating star for temperatures between 5,000 K and 21,000 K. The constancy of this ratio, which should reflect the relative amounts of nonequilibrium emission from small particles and equilibrium thermal emission from large grains, is puzzling in view of current models invoking UV excitation of the small particle emission.

One possible explanation of the constancy of the 12 μm to 100 μm luminosity ratio with temperature of the exciting star is that both the 12 and 100 μm emission come from nonequilibrium processes. We therefore examined whether the dust temperature derived from the 60 and 100 μm surface brightnesses was consistent with equilibrium dust temperatures predicted for ordinary dust grains heated by the stellar radiation field. Our technique was to calculate the ratio of the average absorption cross section at the peak of the stellar energy distribution, Q_*, to the absorption cross section at 100 μm, Q_{100}. The ratio Q_*/Q_{100}, derived from a standard energy balance equation for equilibrium radiative grain heating, depends only on observed quantities such as the 60/100 μm dust temperature, the observed stellar flux, and the angular offset between the star and the observed nebular position. We find that the derived values of Q_*/Q_{100} are in very good agreement with those expected if the 60 and 100 μm emission from reflection nebulae is due to equilibrium thermal emission. In particular, Q_*/Q_{100} for B stars agrees with the values found by other more detailed studies where the spatial distribution of the 60/100 μm dust temperature shows clearly that the far-infrared emission is due to thermal emission from dust in equilibrium with the B-star radiation field (Whitcomb et al. 1981; Harvey, Thronson, and Gatley 1980; Castelaz, Sellgren, and Werner 1987; Fig. 1). The values of Q_*/Q_{100} derived for cooler stars in our sample clearly follow the expected trend with stellar temperature, in that Q_* decreases for cooler stars as their stellar energy distributions peak at longer wavelengths. This indicates that the 60 and 100 μm emission for all of the reflection nebulae measured, independent of stellar temperature, is primarily due to equilibrium thermal emission.

Our results for the 12/100 μm luminosity ratio, which we have shown represents the relative amounts of nonequilibrium emission from small particles and equilibrium thermal emission from large grains, have several consequences. The constancy of this ratio over a range of exciting star temperatures implies that the excitation of the small particles must be possible over a wide range of wavelengths, not merely confined to UV wavelengths as originally thought. This suggests that the absorption of the small-particles at UV and visual wavelengths must be broadband in nature, rather than due to a single strong absorption feature such as the 2200 Å absorption feature. Also, because small-particle emission at 12 μm accounts for a large fraction of the total stellar radiation reradiated at infrared wavelengths, the small particles must therefore account for a similarly large fraction of the UV and visual absorption. This also suggests that the small particle absorption must be over a broad range of wavelengths, since there are no single absorption features or group of features in the observed interstellar extinction curve to which such a large fraction of the total absorption can be easily attributed. Our observations therefore may lead not only to a revised understanding of the excitation mechanism for the small particle emission, but also to further constraints on the composition of the small particles based on their inferred absorption characteristics. Finally, the important role the small particles play in the total energetics of interstellar dust indicates that the small particles are critical contributors not only to the infrared emission of the interstellar medium, but also to the UV and visual interstellar extinction curve.

5. ACKNOWLEDGMENTS

The work discussed here on the comparison of the Pleiades and the infrared cirrus is being done in collaboration with E. Dwek and M. Hauser. We thank the IRAS General Investigator program for generous support of our observations. K. S. gratefully acknowledges NASA Ames University Consortium Agreement NCA2-75, the American Astronomical Society travel grant program, and the conference organizers for travel support. K. S. is supported by NASA contract NASW 3159. L. L. and M. C. acknowledge the support of NASA Ames University Consortium Agreements NCA2-129 and NCA2-194, respectively.

References

Allamandola, L. J., Tielens, A. G. G. M., and Barker, J. R. 1985, *Astrophys. J. Lett.*, **290**, L25.
Boulanger, F., Baud, B., and van Albada, G. D. 1985, *Astron. Astrophys.*, **144**, L9.
Boulanger, F., and Perault, M. 1987, preprint.
Castelaz, M. W., Sellgren, K., and Werner, M. W. 1987, *Astrophys. J.*, **313**, 853.
Harvey, P. M., Thronson, H. A., and Gatley, I. 1980, *Astrophys. J.*, **235**, 894.

Helou, G. 1986, *Astrophys. J. Lett.*, **311**, L33.

Léger, A., and Puget, J. L. 1984, *Astron. Astrophys.*, **137**, L5.

Mathis, J. S., Rumpl, W., and Nordsieck, K. H. 1977, *Astrophys. J.*, **217**, 425.

Sellgren, K. 1984, *Astrophys. J.*, **277**, 623.

Sellgren, K., Allamandola, L. J., Bregman, J. D., Werner, M. W., and Wooden, D. H 1985, *Astrophys. J.*, **299**, 416.

Sellgren, K., Castelaz, M. W., Werner, M. W., and Luan, L. 1987, in preparation.

Sellgren, K., Werner, M. W., and Dinerstein, H. L. 1983, *Astrophys. J. Lett.*, **271**, L13.

van den Bergh, S. 1966, *Astron. J.*, **71**, 990.

Weiland, J. L., Blitz, L., Dwek, E., Hauser, M. G., Magnani, L., and Rickard, L. J. 1986, *Astrophys. J. Lett.*, **306**, L101.

Werner, M. W., Castelaz, M. W., Dwek, E., Hauser, M. G., Luan, L., and Sellgren, K. 1987, in preparation.

Whitcomb, S. E., Gatley, I., Hildebrand, R. H., Keene, J., Sellgren, K., and Werner, M. W. 1981, *Astrophys. J.*, **246**, 416.

Extended Infrared Emission near Stars

F.O. Clark, R.J. Laureijs, C.Y. Zhang, G. Chlewicki, P.R. Wesselius

Laboratory for Space Research Groningen, Postbus 800
9700 AV Groningen, The Netherlands

FOC also: University of Kentucky, Lexington, KY, 40506 USA
CYZ also: Purple Mountain Obs., Academica Sinica, Nanjing, China

Abstract

Extended infrared emission detected near many young stars can be used to divine
the luminosity, apparent color temperature, and thermal emitting mass of the surrounding regions. The dominant dust heating mechanism appears to be ultraviolet
radiation, either from stellar wind shocks or direct stellar radiation from hotter
stars. The infrared luminosities from these extended emitting regions are often an
appreciable fraction of that of the central star.

1. Extended Infrared Emission

Extended infrared emission is often detectable around young stars (Clark et al.
1986, Zhang et al. 1987), and can reveal morphology of dust color temperature, mass
of warm dust, and stellar wind luminosity if shock heated (Clark et al. 1986).
Molecular spectral lines offer diagnostics which can confirm the presence of winds
and shocks (Snell et al. 1980, Clark and Turner 1987).

L1551 veils a young pre-main-sequence star (IRS-5) with accompanying extended
infrared emission from dust surrounding a bi-polar flow (Clark and Laureijs 1986).
The dust in this case is heated by shock produced ultraviolet radiation (Edwards et
al. 1986, Clark et al. 1986). The luminosity of the central star is ~38 $L_\odot$, while
the extended infrared bolometric luminosity is L > 19-28 $L_\odot$. The dust apparent color
temperature is 21-24 K (emissivity ~λ^{-1} to $^{-2}$). The mechanical luminosity in the
flow must be even higher, and these enormous wind luminosities strongly suggest
that the wind derives it's energy from the process of star formation itself, i.e.
from the gravitational energy of the cloud, and not by stellar radiation pressure.

CED 110 exhibits extended infrared emission detected by the IRAS Chopped Photometric Channel (a high spatial resolution instrument on board IRAS) Clark et al.
1987). The visible central star lies well above the main sequence, and is one of
the younger visible objects in the Chameleon Cloud. We estimate an L(bol) = 21 $L_\odot$
for the star (.337 to 100 μm). The extremities of the double peaked extended 50 μm

emission exhibit apparent color temperature peaks of 34 K. The surrounding ambient density is 10 times that of L1551, ~3 10^3 cm^{-3} (Toriseva and Mattila 1985). The OH data also reveal anomalously strong OH lines and high velocity wings at CED 110, which speak for a stellar wind (Clark & Turner 1987), presumably also responsible for the bipolar infrared emission. The extended infrared luminosity is 2.2 $L_\odot$, ~10 % of that of the star. The dense surrounding cloud results in a short ultraviolet path length from the working surface of the conjectured shock.

A wedge shaped region of infrared emission lies adjacent to the bipolar flow source IRAS 05553+1631 (Figure 1). CO data taken with the University of Cologne 3^m telescope show broad CO along the edges of the infrared emission, suggesting a limb brightened mostly hollow structure, and strong narrow ambient CO next to each infrared

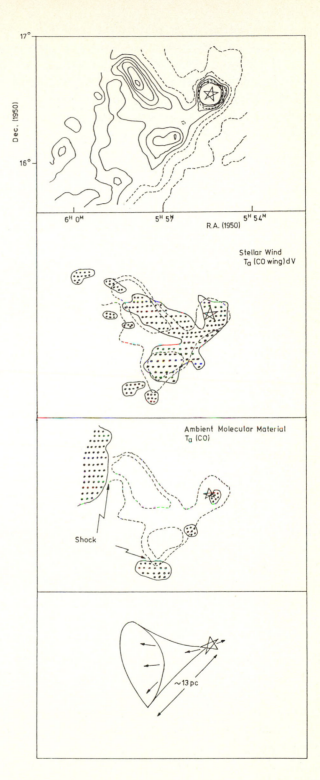

Figure 1. a- 100 μm emission , b- CO wings, c-CO T(peak), d- sketch

peak, on the side of the peak facing away from the star. The peak infrared color temperature is high, ~40 K, and broad OH and CO emission lines (~50 km/s) are detected, suggesting shock heating. These data are interpreted as a large wind cavity, with attendant local heating by shock produced ultraviolet at both indicated locations of substantial ambient material.

Infrared emission near the young open cluster NGC 2169 apparently exhibits the final step in the destruction of the parent molecular cloud (Clark et al. 1987). NGC 2169 is younger than the Pleiades in consequence of the presence of main sequence B1 stars. The infrared luminosity of this object is large, ~10,000 $L_\odot$, nearly half that of the cluster. Such an enormous fractional luminosity suggests that a significant fraction of the dust heating may derive from the diffuse galactic radiation field, as was also found for extended emission in Serpens (Zhang et al. 1987). The distance of the cloud is established by extinction in foreground stars, and stars within the cloud which produce extended dust heating.

Broad OH emission lines indicate shocks in NGC 2169. The average of the stellar velocities is comparable to that of the gas, indicating intimate association. Direct radiative heating is probably the dominant dust heating mechanism in NGC 2169 for two reasons: (1) the infrared luminosity is orders of magnitude greater than that of measured winds from comparable stars, and (2) the dust temperature is observed to peak on stars of early B type.

2. Results

Table 1 enumerates the results for extended infrared emission from the three recognized bipolar flows to date: CED 110, L1551, and IRAS 05553+1631, and for the apparent radiatively heated extended emission from the open cluster NGC 2169.

Table 1

Source	spectral class	$L_{*,bol}$ $L_\odot$	L_{IR} $L_\odot$	$V_{heating}$ km/s	V_{OH} km/s	[1]$<T_{dust}>$ K	E_{tot} ergs	10^4 yrs
CED 110	G2	21	2.2	56	>20*	32	$2\ 10^{45}$	
L1551	F-G	38	19	32	>30#	23	$1\ 10^{46}$	
05553+1631	B3?	>900	~450	43	~32	26	$9\ 10^{46}$	
NGC 2169	B1	~25,000	~10,000	--	41	27	(rad. heating)	

* maximum extent not determined by observations.

\# broad component may be greater.

[1]average dust temperature assuming a dust emissivity of λ^{-1}.

A signature of radiative heating of dust by local stars or the galactic diffuse radiation field is strong emission in all four IRAS bands, 12, 25, 60, and 100 μm, with similar morphology. This apparition is displayed by "normal" diffuse clouds with no internal heating sources. In contrast, the three bipolar flows recognized to date are only clearly detected in 60 and 100 μm emission, perhaps with accompanying very weak 12 μm emission.

The mechanical luminosities in these flows must be even higher. Clark et al. (1986) made a simple shock heating model which indicated that the L1551 mechanical luminosities were 2-7 times the infrared luminosities. Such enormous mechanical luminosities presumably could not be produced by the central star (Table 1). A plausible energy source is the small scale gravitational energy of the cloud from regions as small as 1 au.

3. Summary

Very luminous extended infrared emission is sometimes detected near stars. Supporting spectroscopic observations are required to understand the physical nature of this emission. Shock produced ultraviolet is the inferred heating mechanism for three stars, while direct radiative heating is implicated for a young cluster. Perhaps the change occurs when the star ceases to be shrouded by dust, and the local stellar UV exceeds that of the wind. A diagnostic indicator is the morphology of color temperature. If hottest at the star, the mechanism is presumably direct radiative heating, and if hottest away from the star, shock heating prevails. Extended infrared emission provides a new probe of physical processes near stars.

REFERENCES

Clark, F.O. Laureijs, R.J. 1986 A.&A. Letters 154, L26.

Clark, F.O., et al. 1986a "Space-Bourne Sub-Millimetre Astronomy Mission" ESA SP-260, 173

Clark, F.O., et al. 1986b A.&A. Letters 168, L1.

Clark and Turner 1987 A.&A. 176, 114.

Clark, F.O., Laureijs, R.J., Zhang, C.Y., Chlewicki, G., and Wesselius, P.R. 1987 Mass outflows from Stars, 2nd Torino Workshop (in press).

Edwards, et al. 1986 Ap.J. Letters 307, L65.

Snell, R.L., Loren, R. Plambeck, R. 1980 Ap.J. Letters 239, L17.

Toriseva and Mattila 1985 A&A 153, 207.

Zhang, C.Y., et al. 1987 A.& A. "IRAS Study of the Serpens Molecular Cloud: I. Large Scale (submitted)

BEYOND THE ASYMPTOTIC GIANT BRANCH

Sun Kwok and Kevin Volk
Department of Physics, University of Calgary
Calgary, Alberta, Canada T2N 1N4

ABSTRACT

The evolution of the IR spectra of asymptotic giant branch (AGB) stars
has been obtained by time-dependent radiative transfer models. These
models are compared with IRAS photometric and spectroscopic observations. Extrapolation of the models to beyond the AGB lead to predictions of the spectra of proto-planetary nebulae. Candidates of proto-
planetary nebulae have been successfully found from these predictions.

1. INTRODUCTION

We now know that asymptotic giant branch (AGB) stars suffer from
increasing loss of mass as they evolve. Conventional spectral classification schemes work to about spectral type of M10, beyond which the
stellar photosphere is completely obscured by circumstellar dust
created by the mass loss process. It would therefore be desirable to
find a circumstellar feature which can be used to replace the photospheric lines as a diagnostic of the state of evolution of the underlying star. The IRAS Low Resolution Spectrometer (LRS) has observed over
two thousand stars with the silicate dust feature with the strength of
the feature ranging from emission to absorption (Volk and Kwok 1987).
This suggests a range of optical thickness in the circumstellar
envelope from τ(9.7 μm) = 0.1 to 100, corresponding to a change of mass
loss rate of approximately three orders of magnitude. Most of the
stars with the silicate feature in emission have been found to be Mira
Variables whereas those with the feature in absorption are often OH/IR
stars with no optical counterparts. Plotting of the emission and
absorption sources in a [25μ/12μ]-[60μ-25μ] colour-colour diagram shows
that the emission and absorption sources are located in different parts
of the diagram but lie on a continuous band which can be interpreted as
an evolutionary sequence (Kwok, Hrivnak and Boreiko 1987). In this
paper, we report our efforts in modelling the evolution of AGB stars by
calculating circumstellar spectra and compare the results with IRAS
photometric and spectroscopic observations.

2. THE MODEL

Since stars evolve through the AGB in less than 10^7 years, a time-dependent model is required. Fortunately, the core and envelope of an AGB star are virtually decoupled, and as a result, mass loss and nuclear burning can be treated separately. The Paczynski core mass - luminosity relationship suggests that the luminosity of AGB stars increases exponentially with time. If one assumes a mass loss formula $\dot{M}(L_*, M_*, R_*, ..)$, where L_*, R_* and M_* are respectively the luminosity, radius, and current mass of the star, then the circumstellar spectrum can be calculated by a continuum dust radiative transfer code. The AGB evolution is carried through until the hydrogen envelope is completely

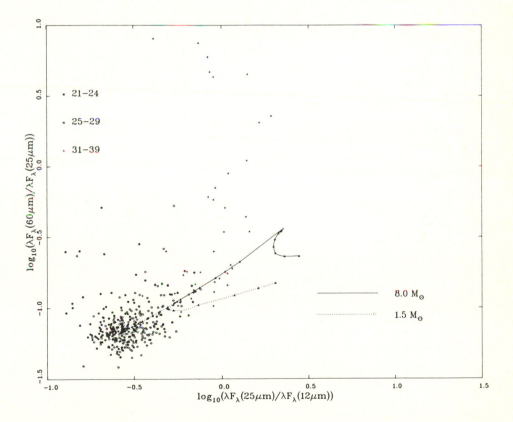

Fig. 1. The evolutionary tracks on the colour-colour diagram for AGB stars with initial masses 1.5 and 8 $M_\odot$. Numbers in legend are LRS classes.

removed by mass loss. We first derive the dust opacity function from the LRS spectrum of optically thin sources. The density profile $\rho(r,t)$ in the circumstellar envelope is related to the mass loss formula by $M(t-r/V)=4\pi r^2 V\rho(r,t)$, where V is the dust velocity. A number of mass loss formulae were tried but few were able to reproduce the extreme strengths of the silicate feature as observed by LRS. An example of a mass loss formula with which reasonable agreements were obtained is

$$\dot{M} = 1.8 \ 10^{-12} \ (M_*(0)/8M_\odot) \ L_* R_* M_*^{-1} \ M_\odot \ yr^{-1}$$

where $M_*(0)$ is the main-sequence mass of the star. The model emergent spectra are also convolved with the IRAS instrumental profile to obtain simulated photometric measurements at the four IRAS bands. The evolutionary tracks on the colour-colour diagram for $M_*(0)=1.5$ and 8 $M_\odot$ stars are shown in Figure 1.

3. PROTO-PLANETARY NEBULAE

The above models can be extrapolated to beyond the AGB where mass loss has stopped. It is found that the spectrum takes on a peculiar shape with a sharp decline in flux at wavelengths shortward of the silicate feature. This is due to the cooling of the dust as it disperses and the resulting shift of the peak of the dust continuum to longer wavelengths. A search of the LRS catalogue for objects with such spectral behaviour results in a number of candidates, including IRC+10420, V1027 Cygni, 18095+2704 and 10215-5916. 18095+2704 was observed by Hrivnak, Kwok and Volk (1988) at the Canada-France-Hawaii Telescope and was identified with a star of V=10.6 magnitude. Spectroscopic observations at the Dominion Astrophysical Observatory suggest that it has the spectral type of F2II/Ib, similar to a number of high galactic latitude F supergiants (e.g. 89 Her) which have been suggested as possible proto-planetary nebulae (Parthasarathy and Pottasch 1986). The association of the cool IRAS source with a bright visible object can be explained by the light of the central star emerging from the dispersing circumstellar envelope.

While the evolution from the AGB to planetary nebulae cannot be followed in the visible because of circumstellar extinction, the colour temperature of the star appears to obey a monotonically decreasing sequence from >600K for Mira Variables to <50K for planetary nebulae. It is interesting to note that while the colour temperature is continuously decreasing, the physical reason responsible is different on and beyond the AGB. On the AGB, the decreasing colour temperature

TABLE 1
EVOLUTION FROM AGB TO PLANETARY NEBULAE

evolutionary phase	example	optical image	period (days)	colour tempera- ture (K)	silicate dust	OH
AGB	Mira Variables	bright	300-600	>600K	emission	yes
LAGB	OH/IR stars	no optical counterpart	600-2000	250-600K	absorption	strong
post-AGB	19454+2920	no	non-variable	150-250	?	weak
proto-PN	18095+2704	yes	non-variable	150-250	emission	weak
young PN	Vy2-2 Hb 12	bright	non-variable	100-200	emission	single peak
PN	many	bright	non-variable	<100	no	no

is the result of increasing optical thickness of the circumstellar envelope, whereas beyond the AGB it is due to geometric dilution as the dust shell disperses into the interstellar medium. A proposed scenario for the evolution from AGB to planetary nebulae is summarized in Table 1.

4. CONCLUSIONS

IRAS spectroscopic and photometric observations have provided important data for the understanding of the evolution from AGB to planetary nebulae, a transition phase in stellar evolution which cannot be effectively studied in the visible. The infrared sequence proposed here can be tested by the identification and observations of proto-planetary nebulae.

References
Hrivnak, B.J., Kwok, S., and Volk, K. 1988, submitted to Astrophys. J.
Kwok, S., Hrivnak, B.J., and Boreiko, R.T. 1987, Astrophys. J., in press.
Parthasarathy, M. and Pottasch, S.R. 1986, Astron. Astrophys., 154, L16.
Volk, K. and Kwok, S. 1987, Astrophys. J., 315, 654.
Volk, K. and Kwok, S. 1988, submitted to Astrophys. J.

The Study of Star Formation with IRAS

Charles A. Beichman

Infrared Processing and Analysis Center
California Institute of Technology
Jet Propulsion Laboratory
Pasadena, CA 91125

"[In] the survey...one source...was found which could not be identified...We feel that the object is in the nebula and has a gravitationally associated cool shell..It is well known that the Orion Nebula is a very young association and that the probability of finding a star in the process of forming should be relatively high. Thus an attractive interpretation of the observations is that the infrared object is a protostar."

E. E. Becklin and G. Neugebauer (1967)

"The discovery of the infrared nebula in Orion means that the early stages of star and star cluster formation can be observed in the far infrared...fundamental problems concerning the earliest stages of stellar evolution can now be answered by direct observation."

D. E. Kleinmann and F. J. Low (1967)

ABSTRACT

A brief review is made of how IRAS has advanced our understanding of star formation. The IRAS survey has revealed the presence of embedded infrared sources within small, dense cores of molecular gas. Some of these objects may be protostars of roughly solar mass deriving a substantial fraction of their energy from infalling material. Circumstellar disks appear to surround many young stars, both in the embedded phase and, later, as T Tauri stars. IRAS also cataloged regions of high mass star formation across the galaxy. These data may lead to an understanding of what triggers the formation of such stars. Finally, while high mass star formation obviously contributes strongly to the infrared emission from galaxies, observations within the Galaxy show that a substantial fraction of infrared emission comes from dust heated by the ambient radiation field and not directly by young stars.

1. INTRODUCTION

It was almost exactly two decades ago that the Becklin- Neugebauer object (BN) and the Kleinmann-Low Nebula (KL) were discovered by pioneering infrared surveys of the central part of the Orion Nebula (Becklin and Neugebauer 1967; Kleinmann and Low 1967). Thus began a steady course of surveys and follow up observations that demonstrated that the formation of stars is best studied in the infrared; IRAS represents the culmination of that effort.

IRAS has contributed to our understanding of star formation in three ways: first, by providing large scale surveys of regions of nearby star formation and locating true low mass *protostars*, i.e., stars of roughly stellar mass in the earliest possible stages of evolution; second, by mapping star formation complexes throughout the Galaxy in a way that reveals the interaction between giant molecular clouds and their environment; and third, by detecting more than 20,000 galaxies for which a major source of infrared emission is the formation of high mass stars under a large variety of conditions.

In this review I will concentrate on IRAS results within the Galaxy; other articles in this volume deal with extra- galactic aspects of star formation. Covering all IRAS results on star formation is impossible in a short review like this; interested readers are referred to other papers in this volume and to recent reviews in the *Annual Review of Astronomy and Astrophysics* (Soifer, Neugebauer and Houck 1987; Beichman 1987).

2. LOW MASS STAR FORMATION

IRAS provided the areal coverage and sensitivity at long wavelengths to locate the youngest, most deeply embedded objects in nearby star forming regions. For example, IRAS had a sensitivity to embedded sources with luminosities as low as $0.1L_\odot$ in the Taurus and Ophiuchus molecular clouds. Since the evolutionary tracks for protostellar objects more massive than a few tenths of a $M_\odot$ achieve this luminosity in just a few thousand years, the IRAS survey is likely to be complete in star forming regions closer than 150 pc.

The combination of IRAS results, millimeter and ground based infrared observations and theoretical advances has led to the conclusions that: 1) small, quiescent clumps of dense molecular gas are favored sites for the formation of solar type stars; 2) many of the infrared sources embedded within molecular clouds are probably true protostars, defined here as objects still deriving a significant amount of luminosity from the accretion of nebular material; 3) disks of roughly solar system size surround many young stars, including optically visible T Tauri stars; and 4) an evolutionary scenario starting with the collapse of a rotating molecular cloud can explain many of the observations, including the infrared energy distributions and the existence of bi-polar outflows.

2.1 Sources Associated with Dense Cores

Beichman *et al.* (1986) examined the IRAS data for 95 small, dense condensations of molecular gas found within larger cloud complexes, notably Taurus and Ophiuchus. More than half of these clumps of gas, called "cores" (Table 1; Myers and Benson 1983), harbor IRAS sources of predominantly two types: optically invisible sources with energy distributions rising steeply to long wavelengths and found close to the densest parts of the cores; or optically visible objects with warmer energy distributions that resemble previously classified T Tauri stars. Other authors, including Clark (1987), have found a similar propensity for IRAS sources of these types to be located within or in close proximity to small, dense cores. Clemens and Barvainis (1987) cataloged 248 small molecular clouds identified on the Palomar Observatory Sky Survey (POSS) plates. Over 300 IRAS sources were found in close proximity to more than half of the clouds in their catalog.

Table 1. Properties of The Dense Cores

Gas Temperature	10-15	K
Density	$10^4 - 10^5$	cm^{-3}
Radius	0.03-0.3	pc
Mass	1-50	$M_\odot$
Δv	0.2-0.4	km s^{-1}
Free fall time	10^5	yr

An important property of the cores themselves is that the CO and NH_3 line widths are typically thermal in breadth, implying that little or no supersonic turbulence is present. The unsupported cores have free-fall times of approximately 10^5 yr, close to the ages inferred for the infrared sources found within them. Thus it appears that a gas cloud with temperature, density and mass similar to those of Table 1 will collapse into a star in little more than a free fall time, unless, perhaps, supported by a strong magnetic field.

2.2 Embedded Sources as Protostars

What are the properties of the embedded sources found within the nearby cores? The combination of ground based and IRAS data has helped lead to an understanding of these objects. Table 2 lists some of the properties of these objects. In the table, the color temperature depends on the wavelengths used to derive the temperature, a result normal for a centrally heated dust cloud. R_d is the characteristic radius at which emitting dust is found; the mass and age of the underlying protostar are inferred from evolutionary models, and A_V is an estimate of the optical extinction to the source (Myers *et al.* 1987a). Temperatures as high as 3000-5000 K are derived from near infrared observations and imply that underlying most of the IRAS sources are objects with photospheric temperatures.

Table 2. Properties of The Embedded Sources

Color Temperature	30-5000	K
Luminosity	0.5 - 50	$L_\odot$
R_d	50-5000	AU
Stellar Mass	0.25-1.5	$M_\odot$
Age	10^5	yr
A_V	10-30	mag

The spectral energy distribution from 1 to 100 μm demonstrates, however, that these embedded objects are not simply normal stars enshrouded within a molecular cloud. Figure 1 gives a typical energy spectrum for the embedded objects from the near infrared out to 100 μm. A detailed analysis of the emission (Myers *et al.* 1987) demonstrates that the observed radiation arises from three components: a hot, central star whose appearance is modified by extinction, a disk of material approximately 50 AU in diameter and dust in the molecular cloud core heated by the central star (Table 3).

To understand why the three component model is required by the observations, consider first a model wherein a normal star lies embedded within a molecular cloud with a density law that peaks toward the center of the cloud. The appearance of the source in the visual and near infrared will be that of a hot blackbody as modified by the effects of extinction. At longer wavelengths the appearance of the source will be determined by the temperature and density distribution of the dust grains heated by the central star. Thus, the infrared observations can constrain the temperature of star as well as the distribution of the dust surrounding it.

Table 3. Components of Radiation from Embedded Sources

Wavelength range (μm)	Radiation Source	Temperature (K)
0.7- 3	Star+Dust Abs.	3000-5000
3- 30	Star + Disk	100-1000
30-300	Core	10- 100

The shaded area in Figure 1 shows that a two component model fails to account for the strength of the mid-infrared emission, 3-30 μm. Approximately 25% of the bolometric luminosity of the embedded sources is emitted at these wavelengths, but to emit over this band, dust must have temperatures between 100-1000 K and must, therefore, lie within 1-100 AU of a star of

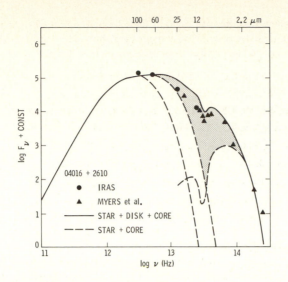

Figure 1. The typical energy of one of the embedded sources associated with a small, dense core of gas. The data are from Myers *et al. 1987.* The dashed curves represent the two component model described in the text (a star attenuated by dust that heats its surroundings). The shaded area shows the deficit between the observed emission and the prediction of the two component model, approximately 25% of the total. The solid line represents a fit to the data of the model given by Adams, Lada and Shu (1987).

roughly stellar luminosity. However, the presence of significant amounts of material that close to the star is inconsistent with the optical depth of dust expected from a spherically symmetric distribution of dust in the surrounding core. If the inner boundary of the core extends to within a few AU of the central star, then the A_V inferred from integrating the dust density law from the outer to inner edges of the cloud, for any reasonable power law dependence, is hundreds of magnitudes and would render the star invisible in the near-infrared. Alternatively, if the star is to be detectable in the near infrared, there should not be strong mid-infrared emission, if the dust distribution is spherically symmetric. This 3-30 μm excess is a general problem for the embedded sources (Beichman *et al.* 1986; Myers *et al.* 1987a).

A solution to the problem invokes the presence of a disk of material with a radius of approximately 50 AU, located within a hole of approximately the same size in the prevailing distribution of dust within the core (Myers *et al.* 1987a). An disk inclined to the line of sight can have dust close enough to the star to emit appreciably at 3-30 μm, yet will not occult the star in the near infrared. It is of interest that Adams and Shu (1986) have demonstrated on very general grounds that a passive disk that merely absorbs stellar radiation will absorb and re-emit 25% of the stellar luminosity. This fraction is of the same order as that seen in the 3-30 μm band for the embedded sources.

Although the above arguments are qualitative, they are borne out by more sophisticated calculations. A detailed physical theory by Adams and Shu (1986; see also Adams, Lada and Shu, 1987a; and Shu, Adams and Lizano, 1987, and references therein) specifies the rotational and infall velocities of the cloud and predicts in a natural manner the formation of a disk around a central star. Their models result in an excellent fit to the 1-100 μm spectral energy distributions of a variety of embedded sources and T Tauri stars (Figure 1).

An example of a source where the disk may, in fact, lie in the line of sight is the remarkable source found in Ophiuchus (Walker *et al.* 1986); suggestive, but inconclusive, evidence of infall in the spectral lines of some molecules implies that this source may be one of the youngest known protostars.

2.3 Disks Around T Tauri Stars

IRAS detected approximately 50-75% of the T Tauri stars known to abound in the Taurus-Auriga complex (Rucinski 1985; Emerson, Cohen and Beichman 1987; Harris, Clegg and Hughes 1987). Energy emitted longward of 12 μm is not a major contributor to a T Tauri star's luminosity; only 10-20% of the bolometric luminosity comes out in IRAS wavelengths. However, the colors of known T Tauri stars are quite well defined with $0.03 < log[f_\nu(25)/f_\nu(12)] < 0.58$, -$0.26 < log[f_\nu(60)/f_\nu(25)] < 0.41$ and $-0.61 < log[f_\nu(100)/f_\nu(60)] < 0.69$ (Harris *et al.* 1987); these ratios are for flux densities that have not been corrected for the effects of the broad IRAS passbands. Harris *et al.* used these colors to find new candidate T Tauri stars in the Taurus region. The new objects follow closely the spatial distribution of the previously known stars. The overall properties like luminosity and inferred mass of the T Tauri stars are similar to those of the embedded sources and a number of the above mentioned authors have asserted that the T Tauri stars are older siblings of the embedded stars that have already erupted from their chrysalids.

Arguments like those presented above for the embedded sources imply that the infrared emission from the T Tauri stars arises in three different components (star, disk and cool residual dust), albeit in different ratios than for the embedded stars. The photosphere and disk contribute $\sim$ 90% of the total luminosity; the bulk of the radiation seen by IRAS ($\lambda > 25$ μm) comes from a small residual dust envelope. (Rucinski 1985; Adams, Lada and Shu 1987a,b; Emerson, Cohen and Beichman 1987; Adams and Shu, this conference).

One of the most important questions to be resolved by future observational and theoretical work is whether the disks seen around the embedded sources and the T Tauri stars are passive, i.e. heated by radiation from a central star, or active, i.e. heated by accretion or other processes in the disk itself. A first attempt at understanding this can be made by dissecting the spectral energy distributions of T Tauri stars. A number of authors have used visible and near infrared observations to constrain the photospheric component, which can then be subtracted out to leave an excess that is attributable to the disk (Rydgren and Zak 1987; Emerson, Cohen and Beichman 1987). The spectrum of the 1 to $\sim$ 12 μm excess over the photosphere can be directly related to the temperature gradient in the disk. Spectral power laws, $\lambda F_\lambda \propto \lambda^{-0.75} to \lambda^{-1}$ are seen which are significantly shallower than the $\lambda^{-1.33}$ expected for either a passive disk or a classical Keplerian accretion disk (Adams *et al.* 1986; Lynden-Bell and Pringle 1974). Adams, Shu and Lada (1987b) have asserted that T Tauri stars with relatively flat infrared energy distributions must have disks that are actively heated, perhaps by non-viscous effects such as local gravitational heating.

In this discussion of disks it is important to note that high spatial resolution infrared imaging and millimeter interferometry have resolved disk-like structures around a number of young T Tauri stars (Beckwith *et al.* 1984; Grasdalen *et al.* 1984; Sargent and Beckwith 1987). Observations like these offer the prospect of characterizing directly the mass and other physical properties of the disks.

Another vital question for future work concerns the relation between these disks, the formation of planets and the disks seen around main sequence stars such as Vega and β Pictoris. It should not be forgotten that these nearby protostars are representative of what the primitive solar nebula was like some four billion years ago.

2.4 Outflows and HH Objects

The discovery of energetic, bi-polar outflows of molecular material from young stellar objects revealed an important new aspect of how stars form (Snell, Loren and Plambeck 1980) . CO observations of IRAS sources have added important new clues to the role of outflows in shaping the evolution of young stars. Myers *et al.* (1987b) have shown that outflows are often associated with IRAS sources found in the dense cores described above. Between 25%-50% of IRAS cores show outflows, although some are very weak, 10^{-8} $M_\odot$ yr^{-1}. The observations suggest that sources with outflows are in a different evolutionary state than those without flows (Figure 2). For example, IRAS sources with outflows are 3 times more luminous than those without (on average 7 $L_\odot$ vs. 2.4 $L_\odot$), although selection effects may play a role in producing this difference (Myers *et al.*). IRAS data also show that the exciting stars of Herbig-Haro objects, i.e. those embedded sources showing optical evidence for outflowing material, are more luminous than those objects without HH objects (Cohen and Schwartz 1986).

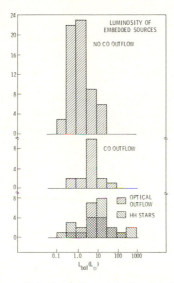

Figure 2. Histograms show that the bolometric luminosities of IRAS sources associated with bipolar outflows observed in CO (Myers *et al.* 1987b) or associated with optical outflows and/or Herbig-Haro objects are higher than those without activity of this kind. (Myers *et al.* 1987b; Cohen and Schwartz 1987).

An important result deduced by Myers *et al.* (1987b) is that the outflows probably carry enough momentum and kinetic energy to disperse a small core. Thus, outflows are probably the main agent of cloud dispersal, at least for small clouds. Evidence for this process comes from the observations of the Barnard 5 cloud (Beichman *et al.* 1984), where the source IRS 4 has cleared out a cavity in the surrounding cloud (Goldsmith, Langer and Wilson 1986). A molecular cloud which shows a remarkable collection of these phenomena is the cloud NGC 1333 which has 9 IRAS sources, 7 outflows and 15 HH objects. (Jennings *et al.* 1987; Knee *et al.* 1987).

The ubiquity of outflows is an important problem. Do all sources pass through an outflow stage? Are sources with outflows younger or older than those without? An important first step in answering these questions comes from the comparison of complete IRAS and CO surveys of the Mon OB1 molecular cloud by Margulis and Lada (1986). These authors found IRAS sources associated with 6 of the 9 CO outflows in the cloud. The limiting luminosity of the IRAS data, approximately 5 $L_\odot$, is consistent with the remainder of the outflows having slightly fainter infrared counterparts. More surprising, however, is the result that 24 of 30 IRAS sources *did not show outflows*, implying that outflow phase lasts only 20% of a star's lifetime as bright infrared source, or occurs for only 20% of all infrared bright stars. Margulis and Lada suggest that the objects without flows are very young protostars. While this explanation is possible, so too is the opposite; the objects without flows might be older than the outflow sources, possibly T Tauri stars that have evolved beyond the outflow phase, but which are still trapped within the large Mon OB1 cloud.

Finally, there remain uncertainties about the physics of outflows. Is the outflow a collimated stellar wind driven by the central star or is the outflow a hydrodynamic consequence of the infall process (Pudritz 1985)? In the former case, the start of outflow might mark the onset of a specific physical process within the star, such as deuterium burning (Shu 1985). In the latter case, infall and outflow may occur simultaneously. The kinetic energy in the outflow from L1551, as inferred from the infrared luminosity of dust in the lobes (Edwards *et al.* 1986; Clark and Laurejis 1985; Clark *et al.* 1986), is such a large fraction of the luminosity of the central star, from 18- 50%, that it is likely that the outflow is driven by some other agency than the central star itself.

3. HIGH MASS STAR FORMATION

IRAS scanned the entire Galaxy and these data, when combined with CO and HI surveys, give important new insights into how high mass star form. Conclusions based on the IRAS data include the possibility that star formation may not be triggered by internal events such as sequential star formation and the fact some, *but not all*, of the infrared emission from a galaxy comes from star formation.

3.1 Sites of High Mass Star Formation

Persson and Campbell (1987) found some 400 isolated, unconfused objects in the IRAS Point Source Catalog with colors similar to those of known high-mass, young stellar objects such as S140 and S255 (Wynn-Williams 1982). These authors examined a subset of 113 sources in the Southern Galactic plane using near infrared photometry, spectroscopy and optical imaging. Many of these sources appear to be younger than compact HII regions and to be associated with optical jets or nebulosity. Braz and Epchtein (1987) used a combination of near infrared photometry and OH/H_2O maser emission to identify new young stars. Terebey and Fich (1987) examined a region in the outer Galaxy. The evolutionary status of these various sources can, perhaps, be determined by the comparison of their spectral energy distribution with models such as those of Crawford and Rowan- Robinson (1986).

An important reason for the study of these objects is to elucidate the role of stellar winds and bi-polar outflows in the disruption of the clouds around these luminous, $\geq 10^3$ $L_\odot$, objects. The Serpens molecular cloud is an example of a region where newly formed B stars appear to have created cavities around themselves through the action of a stellar wind (Zhang, Laurejis and Clark 1987).

3.2 Triggers of Star Formation

A number of authors have investigated the star formation rate in giant molecular clouds (measured by the infrared luminosity, L_{IR}) as function of cloud mass (M_{H_2}, measured by CO emission and the virial theorem). Unfortunately, while the consequences of these studies are important, the results to date are ambiguous. If we let L_{IR}/M_{H_2} be proportional to $M_{H_2}^N$, then we can compare the results from these different groups. Scoville and Good derive N< 0, Solomon *et al.* and Rengarajan (1984) get N ~ 0 and Thronson and Mozurkewich get N=0.8 (Figure 3). The choice of samples and the measurement techniques appear to play a much greater role than is desirable, so that determinations of the slope of the relation seem suspect. At best it appears that the star formation rate is not strongly dependent on cloud mass, i.e. N~ 0.

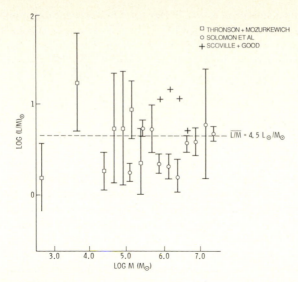

Figure 3. A plot of the luminosity to mass ratio, L/M in solar units, versus cloud mass as determined by three different groups. The uncertainties in the results of Solomon *al.* (1987) and Thronson and Mozurkewich (1987) were determined by averaging the data for individual clouds within logarithmic mass bins of width 0.2.

Scoville and Good (1987) as well as Solomon *et al.* (1987) argue that if the star formation rate is either constant or decreasing with increasing cloud mass (values of $N \leq 0$), then it is unlikely that nonlinear mechanisms, such as sequential or supernova induced star formation, which make the star formation rate a function of number of recently formed stars, are operating. Elmegreen (private communication) has countered this argument by noting that in most Galactic molecular cloud complexes there is only one OB cluster forming at a time so that the L/M ratio averaged over the entire cloud will appear to be constant. Only when the mean time between the formation of clusters is smaller than the lifetime of a cluster ($t \geq 10^6$ yr) will the non-linear effects cited by Scoville and Good and Solomon *et al.* become important.

3.3 Infrared as Tracer of Star Formation

Boulanger and Perault (1987) investigated the infrared emission from the solar neighborhood and found that the origin of the infrared varies drastically with position in the Galaxy. By examining the Galactic latitude dependence of the 60 and 100 μm intensity, these authors were able to show that in the solar neighborhood the infrared emission is dominated by emission from dust associated with diffuse HI gas and heated by the diffuse interstellar radiation field (ISRF); the amount of energy coming from molecular clouds or HII regions is small. On the other hand, emission from HII regions and from molecular clouds heated by the ISRF or by OB stars dominates the production of infrared energy in a region like Orion (Table 4).

However, despite the importance of star formation in Orion, it can still be misleading to use the infrared as direct tracer of **current** star formation rate in all but most extreme galaxies because of two competing effects: first, the interstellar medium is transparent on scale of OB association, a few hundred pc, since $\tau(UV) \sim 2.2$ mag kpc^{-1}. As a result, ionizing radiation from an OB association can leave the confines of the association; and second, dust can be heated by both OB stars and by the diffuse ISRF.

Table 4. Infrared From Different Galactic Components

Component	Infrared Emissivity[1] $(L_\odot/M_\odot)$	Fraction of Total IR Solar Neighborhood[2]	Orion[3]
Cold H_2/Diffuse HI	0.8-1.6	80%	30%
HII regions	5-25	20%	70%
Avg. Galactic Plane	2.5	-	-

Notes:

[1] Adapted from Boulanger and Perault.

[2] Total of 14.5 $L_\odot$ pc^{-2}.

[3] Total $10^6 L_\odot$ in 300 pc diameter.

Boulanger and Perault found that in Orion only about 1/3 of total luminosity of the OB association is trapped and radiated in the infrared by dust; the rest of the energy either diffuses throughout the plane or escapes from the Galaxy altogether. Leisawitz (1987) found a similar result for outer galaxy HII regions. In dense regions, like the molecular ring, however, the stellar luminosity is trapped and star formation accounts for more than 50% of total infrared output (Solomon *et al.* 1987).

The fact that dust near star forming regions is only partially heated by young stars can be illustrated by examining the Orion region. Integrating over the central 1°of Orion shows that 90% of infrared emission comes from young OB stars, while integrating over a 20°region implies that only 60% of the luminosity can be is attributed to star formation. The remainder of the heating comes from the diffuse ISRF. As Helou (1986) has pointed out, this effect means that IRAS data alone are not a good measure of the star formation rate in external galaxies. Emission from dust by the ISRF, i.e. cirrus emission, plays an important role in all but the most dusty, luminous galaxies.

Models of the infrared emission from the entire Galaxy confirm the importance of infrared from non-star forming regions (Cox and Mezger 1987; Cox, this conference). These authors found that HII regions and hot core molecular clouds account for only about 30% of total infrared output of the Galaxy; most of remainder comes from dust associated with HI heated by the ISRF, i.e. cirrus.

4. CONCLUSIONS

IRAS has added greatly to our understanding of how stars form. By examining thousands of square degrees instead of a few square arcminutes like the surveys that discovered BN and KL, IRAS was able to give a complete and unbiased view of where and under what conditions stars form. The major conclusions to be drawn from the studies discussed above include:

1) Low mass stars form out of small dense cores of molecular gas on time-scales of 10^5 yr. Some of these objects appear to be accreting material from the surrounding gas clouds.

2) The spectral energy distribution of the embedded IRAS sources and of T Tauri stars suggest that disks are an important structure throughout much of the pre-main sequence lifetime of stars (and beyond, cf. Vega and β Pic).

3) Outflows and accretion may co-exist throughout the infancy of a star until the protostar breaks out from its cocoon of dust and gas by means of the outflow.

4) The efficiency of star formation appears to be constant over a large range of cloud masses, but the extent to which this result can be used to rule out certain triggers of star formation remains ambiguous.

5) Infrared emission within the Galaxy originates in dust found in two distinctly different realms, either in massive star forming regions or in quiescent HI gas. This combination of mecha-

nisms makes it difficult to determine global properties, such as the star formation rate, in external galaxies.

5. ACKNOWLEDGMENTS

Part of this work was funded by the Infrared Processing and Analysis Center at the California Institute of Technology and the Jet Propulsion Laboratory as part of NASA's Extended Mission program. JPL is operated by the California Institute of Technology under contract to the National Aeronautics and Space Administration. I would like to thank M. Rowan-Robinson and the scientific organizing committee for providing travel support.

6. REFERENCES

Adams, F.C., Lada, C.J. and Shu, F.H. 1987a, *Ap.J.*, **312**, 788.

Adams, F. C., Lada, C. J., and Shu, F. H., 1987b, in preparation..

Adams, F. C. and Shu, F. H. 1986, *Ap. J.*, **308**, 836.

Becklin, E.E. and Neugebauer, G. 1967, *Ap. J.*, **147**, 799.

Beckwith, S., Zuckerman, B., Skrutskie, M. F. and Dyck, H.M. 1984, *Ap. J.*, **287**, 793.

Beichman, C. A., *et al.* 1984, *Ap. J. (Let.)*, **278**, L45.

Beichman, C. A., Myers, P.C., Emerson, J.P., Harris, S., Mathieu, R., Benson, P.J., and Jennings, R. E. 1986, *Ap. J.*, **307**, 337.

Beichman, C. A., 1987. in *Ann. Rev. Astron and Astrop.*, in press..

Braz, M.A. and Epchtein, N. 1987, *Astron and Astrop.*, **176**, 245.

Boulanger, F. and Perault, M. 1987, *Ap. J.*, in press.

Clark, F. O. 1987, *Astron. and Astrop*, in press..

Clark, F. O. and Laurejis, R.J. 1985, *Astron. and Astrop*, **154**, L26.

Clark, F. O., Laurejis, R.J., Chlewicki, Zhang, C.Y., van Oosterom, W. and Kester, D. 1986, *Astron. and Astrop*, **168**, L1.

Clemens, D. and Barvainis 1987, *Ap. J.*, in preparation..

Cohen, M. and Schwartz, R.D. 1986, *Ap. J.*, **316**, 311.

Cox, P. and Mezger, P.G. 1987 in *Star Formation in Galaxies*, ed. C. Persson (NASA Printing Office: Washington), p.23.

Crawford, J. and Rowan-Robinson, M. 1986, *Mon. Not. R. Astr. Soc.*, **221**, 923.

Edwards, S., Strom, S.E., Snell, R. L., Jarrett, T. H., Beichman, C. A., and Strom, K.M. 1986, *Ap. J. (Let.)*, **307**, L65.

Emerson, J.P. Cohen, M. and Beichman, C. A., 1987 in preparation..

Goldsmith, P.F., Langer, W. D. and Wilson, R.W. 1986, *Ap. J. (Let.)*, **301**, L11.

Grasdalen, G. L. Strom, S.E., Strom, K.M.,Capps, R.W., Thompson, D. and Castelaz, M. 1984, *Ap.J. (Let.)*, **283**, L57.

Harris, S., Clegg, P. and Hughes, J. 1987, in preparation..

Helou 1986, *Ap. J. (Let.)*, **311**, L33..

Jennings, R. E., Cameron, D.H.M., Cudlip, W. and Hirst, C.J. 1987, *Mon. Not. R. Astr. Soc.*, **226**, 461.

Knee, L.B.G., Liseau, R., Sandell, G. and Zealey, W.J. 1987, *Astron. and Astrop*, in press..

Kleinmann, D. E. and Low, F. J. 1967, *Ap. J. (Let.)*, **149**, L1.

Leisawitz,D. 1987 in *Star Formation in Galaxies*, ed. C. Persson (NASA Printing Office: Washington), P. 75.

Lynden-Bell, D. and Pringle, J.E. 1974, *Mon. Not. R. Astr. Soc.*, **168**, 803.

Margulis, M. and Lada, C. J. 1986, *Ap. J.*, **307**, L87.

Myers, P. C., and Benson, P.J. 1983, *Ap. J.*, **266**, 309.

Myers, P. C., Fuller, G.A., Mathieu, R.D., Beichman, C.A., Benson, P.J., Schild, R.E. 1987a, *Ap. J.*, in press..

Myers, P. C., Heyer, M., Snell, R.L. and Goldsmith, P.F. 1987b, *Ap. J.*, in press..

Persson, S.E. and Campbell, B. 1987, *Ap. J.*, in press.

Pudritz, R.E. 1985, *Ap. J.*, **293**, 216.

Rengarajan, T. 1984, *Ap. J.*, **287**, 671.

Rucinski, S.M. 1985, *Astr. J.*, **90**, 2321.

Rydgren, A.E. and Zak, D.S. 1987, *Publ. Astr. Soc. Pac.*, **99**, 141.

Sargent, A. I. and Beckwith, S. 1987, *Ap. J.*.., in press..

Scoville, N.Z. and Good, J. in *Star Formation in Galaxies*, ed. C. Persson (NASA Printing Office: Washington), p. 3..

Shu, F. H., Adams, F. C. and S. Lizano, 1987, in *Ann. Rev. Astron and Astrop.*, in press..

Shu, F. H., 1985 in IAU Symposium 106, *The Milky Way*, ed. H. van Woorden, W.B. Burton and R. J. Allen (Dordrecht:Reidel), p.561..

Snell, R. L., Loren, R.B. and Plambeck, R.L. 1980, *Ap. J. (Let.)*, **239**, L17.

Solomon, P.M., Rivolo, A.R., Mooney, T.J., Barrett, J.W. and Sage, L.J. in *Star Formation in Galaxies*, ed. C. Persson (NASA Printing Office: Washington), p. 37.

Soifer, B. T., Neugebauer, G. and Houck, J.R. 1987, in *Ann. Rev. Astron and Astrop.*, in press..

Terebey, S. and Fich, M. 1987, preprint.

Thronson, H.A., Jr. and Mozurkewich, D. *Ap. J.*, in press.

Walker, C.K., Lada, C.J., Young, E.T.,Maloney,P.R. and Wilking, B.A. 1986, *Ap.J.(Let.)*, **309**, L47.

Wynn-Williams, C.G. 1982, *Ann. Rev. Astron and Astrop.*, **20**, 58.

Zhang, C.Y., Laurejis, R.J. and Clark, F.O. 1987 *Astron. and Astrop.*, in press..

THE LUMINOSITY FUNCTIONS OF TAURUS AND CHAMAELEON

J.D.Hughes and J.P.Emerson
Department of Physics,
Queen Mary College,
Mile End Road,
London E1 4NS.

SUMMARY

The Infrared Astronomical Satellite (IRAS) surveyed the sky at 12, 25, 60, and 100 μm providing an invaluable, unbiassed far-infrared database for the study of regions of active star formation. We obtain the far-infrared luminosity functions of Taurus and Chamaeleon, within which the stellar populations appear to be at a similar evolutionary stage (within statistical errors). We then convert the far-infrared luminosity functions to equivalent bolometric luminosity functions, and find a shortfall in the number of intermediate and high mass objects when we compared to the initial luminosity function for field stars (Miller and Scalo 1979).

1 INTRODUCTION

Considerable progress has been made in the understanding of star formation over recent years (eg. Black and Matthews 1985, Peimbert and Jugaku 1986, Shu, Adams, and Lizano 1987), but much of this work has been severely biassed by selection effects. Past investigations have concentrated on known regions of interest, and it is unclear whether these regions are representative of the star formation process as a whole. This is a progress report of a program in which we aim to characterise the global (statistical) properties of many regions of low mass star formation on the basis of IRAS data, and then compare the IRAS derived properties of these regions with characteristics of the gas, and with theoretical expectations to better understand the star formation process.

We concentrate on the molecular cloud complexes in our galaxy which are only forming low mass stars since these regions are less complex than those with high mass star formation, and since current theory for these objects is comparatively simple and complete (Shu, Adams, and Lizano 1987). It has also been suggested (Herbig 1962, Mezger and Smith 1977) that star formation occurs by different processes for high and low mass stars, which may cause difficulties in the interpretation of the luminosity function in a region where both low and high mass stars are forming.

2 METHOD

We select dark cloud complexes with known distances and no high mass star formation. We chose Taurus and Chamaeleon as the first two regions to be investigated both because of their proximity (160pc (Cohen and Kuhi 1979), and 140pc and 200pc for Chamaeleon T1 and T2 (Whittet *et al.* 1987), respectively) and because they are well studied at other wavelengths (Elias 1978, Cohen and Kuhi 1979, Jones and Herbig 1979, Rydgren 1980), allowing us to check the validity of our approach by establishing whether conclusions drawn from far-infrared data are consistent with those obtained at other wavelengths.

TABLE 1.

IRAS colour definitions of the various population types given by Emerson (1986).

OBJECT TYPE[1]	[25-12][2]	[60-25]	[100-60]
1 Stars	-0.7 TO -0.2	-0.9 TO -0.4	-0.2 TO -0.6
2 Bulge Stars	-0.2 TO +0.3	-0.8 TO -0.2	-
3 Planetary Nebulae	+0.8 TO +1.2	0.0 TO +0.4	-0.4 TO 0.0
4 T Tauri Stars	0.0 TO +0.5	-0.2 TO +0.4	0.0 TO +0.4
5 Cores	+0.4 TO +1.0	+0.4 TO +1.3	+0.1 TO +0.7
6 Galaxies	0.0 TO +0.4	+0.6 TO +1.2	+0.1 TO +0.5

[1] *Molecular cloud hot spots, HII regions, reflection nebulae and star formation regions all occupy an area similar to that of molecular cloud cores (5) and galaxies (6).*

[2] $Log_{10}[\frac{S_\nu\{25\}}{S_\nu\{12\}}]$

Within the RA and Dec boundaries of the regions will lie fore- and background objects. We only consider IRAS sources with flux detections in three adjacent bands to enable us to classify objects by their IRAS colours alone (see Table 1, from Emerson 1986), and hence differentiate between true members of the star formation region (types 4,5 and 6 of Table 1) and field objects. Figs. 1a and 1b show IRAS colour-colour plots of all the sources in Taurus (a) and Chamaeleon (b) which have flux detections at 12, 25, and 60μm.

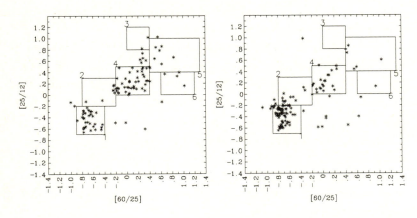

Fig.1a   **Fig.1b**

IRAS colour-colour plots of sources in Taurus (a) and Chamaeleon (b) with flux detections at 12, 25, and 60 µm. The boxes correspond to the object types given in Table 1.

3 THE FAR-INFRARED LUMINOSITY FUNCTION

Having thus selected a sample that is part of the star formation region, we calculate the total in-band flux in Wm^{-2} by multiplying the IRAS catalogue flux densities by appropriate bandwidths in Hz and adding together the fluxes in each band (not including upper limits). We then extrapolated the fluxes out to infinity by first calculating a colour temperature from the ratio of the flux detections at the two longest detected wavelengths, and then using a blackbody of that temperature scaled to the detection at the longest wavelength. The total flux (catalogue data plus extrapolation) was then used to find far-infrared luminosities (L_{FIR} in $L_{\odot}$).

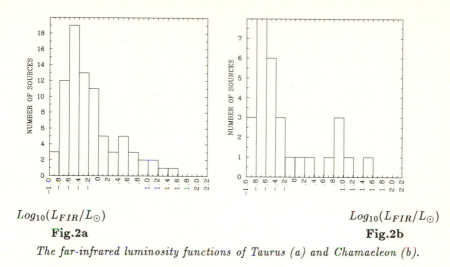

$Log_{10}(L_{FIR}/L_{\odot})$

Fig.2a

$Log_{10}(L_{FIR}/L_{\odot})$

Fig.2b

The far-infrared luminosity functions of Taurus (a) and Chamaeleon (b).

Figs. 2a and 2b show the far-infrared luminosity functions of Taurus and Chamaeleon, the profiles of the functions are similar, as are the maximum far-infrared luminosities which we observe (L_{max}, see Table 2).

Assuming that the luminosity function is a simple power law of the form:

$\phi(L) = BL^{-\gamma}$, where $B = \frac{N}{[\frac{L^{1-\gamma}}{1-\gamma}]_{L_{min}}^{L_{max}}}$, and N is the number of sources in the sample we can use the maximum likelihood method to estimate the parameters B and γ, the results are given in Table 2. The power law exponents are similar and the ratio of the B's is consistent with that expected from the ratio of cloud volumes (indicating similar source densities).

TABLE 2.

CLOUD	$L_{max}(L_{\odot})$	N	B	γ
Taurus	28	80	$8.22^{+0.71}_{-0.78}$	1.18 ± 0.06
Chamaeleon	26	29	$3.22^{+0.36}_{-0.45}$	1.13 ± 0.1

4 THE BOLOMETRIC LUMINOSITY FUNCTION

It is of interest to see how our instantaneous luminosity functions compare with the initial luminosity function (ILF) of Miller and Scalo (1979), so we estimated the bolometric luminosity functions of the regions. We do not know the entire spectral energy distribution of many of these young stellar objects, so we are unable to calculate their bolometric luminosities directly.

Within each star-forming region, we therefore divided the sources into three groups: T Tauri-type sources (including Herbig and Rao's (1972) Orion population objects), cores of molecular clouds, and "others" which were A- and B-type stars (probably ZAMS) which had considerable far-infrared-excesses, causing them to appear on the colour-colour plots in the "galaxies" area. Myers *et al.* (1987) calculate bolometric luminosities from observational data for a sample of T Tauri stars and young stellar objects associated with molecular cloud cores (Beichman *et al.* 1986): using their data we calculate the average value of $\frac{L_{FIR}}{L_{bol}} = 0.77 \pm 0.01$, for "cores", and $\frac{L_{FIR}}{L_{bol}} = 0.47 \pm 0.03$ for T Tauri stars (range: 0.13, 0.51, respectively). For "others", A0V stars with associated reflection nebulosity, $\frac{L_{FIR}}{L_{bol}}$ has a value of about 0.14 (HD97300), and 0.002 for B7–8 stars (standard M_V vs. L_{bol}). The error quoted is the standard deviation of the mean: the large range in T Tauri star values is due to the presence of objects with the IRAS colours of T Tauri stars (Harris, Clegg, and Hughes 1987) which are optically invisible and so radiate most of their energy in the infrared and beyond. We then use our FIR luminosities and these factors to estimate the bolometric luminosities for all our objects.

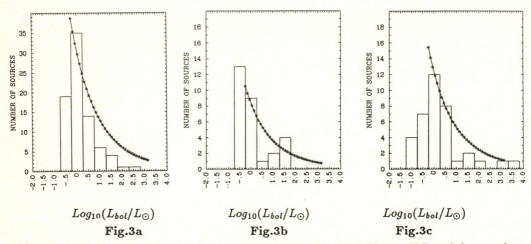

$Log_{10}(L_{bol}/L_{\odot})$

Fig.3a

$Log_{10}(L_{bol}/L_{\odot})$

Fig.3b

$Log_{10}(L_{bol}/L_{\odot})$

Fig.3c

The calculated bolometric luminosity functions shown with the field star ILF scaled to produce the observed number of sources between −0.25 and 0.75 in (a) for Taurus. The ILF in (b) for Chamaeleon is scaled to the bins between −0.75 and 0.25. (c). The ILF for ρ Ophiuchus (from Lada and Wilking 1984) is shown for comparison.

5 CONCLUSIONS

While we stress that the method employed to convert the far-infrared luminosity functions to bolometric luminosity functions is an approximation, several important results should be noted. In Fig.3a, scaling the ILF of Miller and Scalo (1979) to produce the observed number of IRAS sources in Taurus with log_{10} luminosity between −0.25 and 0.75 shows a discrepancy between the ILF and the number of sources observed at intermediate to high luminosities. In Chamaeleon, scaling the ILF predicts the presence of about two objects in the range $1.25 \leq Log_{10}(L_{bol}/L_{\odot}) \leq 1.75$, but here we see four objects, the A0V star HD97300, the very luminous T Tauri-type source 12496-7650, HD97048 and a "core", the latter two objects have less than half the luminosities of the former. Due to the small number of objects in our sample, it is difficult to say whether these results are significant, however, Fig.3b shows a distinct discrepancy between the ILF and the observed number of objects with intermediate luminosities. Lada and Wilking (1984) note the apparent lack of stars in ρ Oph with luminosities corresponding to the mass range 2–4$M_\odot$ (see Fig.3c), citing this as possible direct evidence for bimodal star formation (although they note statistical uncertainties prevent positive conclusions being drawn). Scalo (1986) shows that statistics alone predict that gaps might appear in previously continuous mass distributions as one moves to higher masses.

The IRAS properties of the young stellar objects in Taurus and Chamaeleon indicate that the two regions are at a similar evolutionary stage (within statistical uncertainties), forming low mass stars within quiescent clouds. Both regions exhibit a lack of intermediate- to high-luminosity objects when compared with the ILF, and to the ρ Oph population (Lada and Wilking 1984). Our conclusions agree with our preconceptions about these two well studied regions, and provide support for the hypothesis that ρ Ophiuchus is an intermediate stage between regions of low and high mass star formation, and give us confidence that this method can be used to characterise other star formation regions and investigate their origins.

REFERENCES

Beichman, C.A., *et al.*, 1986, *Astrophys. J.*, **307**, 337.

Black, D.C., & Matthews, M.S., Ed., 1985, *Protostars and Planets II,* University of Arizona Press.

Cohen, M., & Kuhi, L.V., 1979, *Astrophys. J. Suppl.*, **41**, 743.

Elias, J.H., 1978, *Astrophys. J.*, **224**, 857.

Emerson, J.P.,1986, I.A.U. Symposium **115** in *"Star Forming Regions,"* Ed. M. Peimbert, & J. Jugaku, D. Rcidel, p.16.

Harris, S., Clegg, P.E., & Hughes, J.D., 1987, in preparation.

Herbig, G.H., 1962, *Adv. Astr. Astrophys.*, **1**, 47.

Herbig, G.H., & Rao, N.K., 1972, *Astrophys. J.*, **174**, 401.

Jones, B.F., & Herbig, G.H., 1979, *Astr. J.*, **84**, 1872.

Lada, C.J., & Wilking, B.A., 1984, *Astrophys. J.*, **287**, 610.

Mezger, P.G., & Smith, L.F., 1977, in IAU Symposium **75**, *Star Formation,* ed. T. de Jong, & A. Maeder, (Dordrecht:Reidel) p.133.

Miller, G.E., & Scalo, J.M., 1979, *Astrophys. J. Suppl.*, **41**, 513.

Myers, P.C., *et al.*, 1987, *Astrophys. J.*, **319**, 340.

Peimbert, M., & Jugaku, J., 1986, Ed., I.A.U. Symposium **115** on *"Star Forming Regions,"* D. Reidel.

Rydgren, A.E., 1980, *Astr. J.*, **85**, 444.

Scalo, J.M., 1986, *Fundamentals of Cosmic Physics*, **11**, 1.

Shu, F.H., Adams, F.C., & Lizano, S., 1987, *Ann. Rev. Astr. & Astrophys.*, in press.

Whittet, D.C.B., *et al.*, 1987, *Mon. Not. Royal Astr. Soc.*, **244**, 497

INFRARED SPECTRA OF YOUNG STELLAR OBJECTS

FRED C. ADAMS AND FRANK H. SHU

Astronomy Department, University of California, Berkeley, CA 94720, USA

ABSTRACT

We present models of the spectral energy distributions for an evolutionary sequence of young stellar objects, from protostars to pre-main-sequence stars. The *protostellar* theory, characterized by a central star and disk embedded within an infalling envelope of dust and gas, can explain the observed infrared spectra of embedded *IRAS* sources associated with molecular cloud cores. Next, we find that T Tauri stars with near- and mid-infrared excesses can be understood as young stars surrounded by nebular disks. The disks in T Tauri systems are found in two varieties: *passive* disks which have no intrinsic luminosity and merely intercept and re-radiate stellar photons, and *active* disks which have appreciable intrinsic luminosity. In addition, there is another class of objects in which a residual dust shell still surrounds the star and disk.

1. INTRODUCTION: AN OVERVIEW OF STAR FORMATION

We begin by reviewing the basic picture of star formation that has emerged in recent years (see also Beichman, this volume). The genesis of stars takes place in molecular cloud cores, small centrally condensed regions within molecular clouds. When a cloud core collapses, it forms a central hydrostatic object surrounded by an infalling envelope of gas and dust; i.e., the object enters the protostellar stage of evolution. As the protostar evolves, the central object continuously gains mass at a constant rate. Eventually, a stellar wind develops and breaks through the infalling envelope at the rotational poles of the system. The object then enters the next stage of evolution, the bipolar outflow stage. As the outflow widens and reverses the infall, the newly formed star and disk become separated from the parent molecular cloud and the star becomes optically visible. In the earliest stages, the star and disk may be accompanied by a residual dust shell. However, this shell gradually disperses and the star/disk system is left behind. The disks in these systems may be either *passive* disks, which have no intrinsic luminosity but intercept and re-radiate stellar photons, or *active* disks, which have appreciable intrinsic luminosity. Here we show that the observed spectra of embedded *IRAS* sources and T Tauri stars with infrared excesses can be understood in terms of the theoretical evolutionary sequence outlined above. In particular, the embedded *IRAS* sources associated with molecular cloud cores correspond to *protostars*, whereas T Tauri stars with infrared excesses are examples of stars with circumstellar disks.

2. PROTOSTARS

In this section we discuss the emergent spectra of protostars, objects still gaining mass through infall. We limit our discussion to low-mass protostars (i.e., $M \leq 2M_\odot$), where the radiation field is not strong enough to affect the infalling envelope and the dynamical collapse is decoupled from the radiation. Hence, we can determine the basic structure of protostars from the dynamical collapse alone (see below); these results can then be used as a basis for performing the radiative transfer calculation required in order to determine the emergent spectra of these objects.

Since protostars form within molecular cloud cores (see Myers *et al.* 1987), the observed core properties represent the initial conditions for protostellar collapse. In nearby molecular cloud complexes (which are forming stars of low mass), these cores are observed to be nearly isothermal, with temperatures in the range $10 - 35$ K (e.g., Myers 1985) and corresponding sound speeds $a = 0.20 - 0.35$ km/s. In addition, the cloud cores are observed to be rotating slowly with angular velocities $\Omega = 10^{-14} - 10^{-13}$ rad/s (e.g., Myers, Goodman, and Benson 1987). At spatial scales much smaller than the radius of rotational support, $R = a/\Omega \sim 0.1 - 1.0$ pc, isothermal cloud cores are expected to have density profiles of the form $\rho = (a^2/2\pi G)r^{-2}$ (see Shu 1977; Terebey, Shu, and Cassen 1984, hereafter TSC).

In the idealized case, an isothermal cloud core will collapse from inside-out (Shu 1977); i.e., the interior of the core will collapse first, and the successive outer layers will follow. Thus, the collapse naturally produces a core/envelope structure, with a central hydrostatic object (i.e., the forming star) surrounded by an infalling envelope of dust and gas. This inside-out collapse progresses as an expansion wave propagates outward at the sound speed. The head of the wave defines the boundary of the collapse region at $r_H = at$, where t is the time since the beginning of the collapse. Outside this radius, the cloud is static; inside this radius, the flow quickly approaches free-fall velocities. The flow remains nearly spherical outside a centrifugal radius, $R_C = G^3 M^3 \Omega^2/16a^8$, the position where the infalling material with the highest specific angular momentum will encounter a centrifugal barrier (see TSC). Inside this radius, the flow becomes highly non-spherical as particles spiral inward on nearly ballistic trajectories. In the region immediately surrounding the star, the temperature will be too high for dust grains to exist and an opacity-free zone will result (see Stahler, Shu, and Taam 1980).

Notice that there is no mass scale in the collapse scenario outlined above. Instead, the collapse flow feeds material onto the central star and disk at a well defined mass infall *rate* $\dot{M} = 0.975a^3/G$ (Shu 1977 and TSC). The absence of a mass scale suggests the possibility that the origin of stellar masses is determined by *stellar* processes rather than by the interstellar medium (see Lada 1985 and Shu, Adams, and Lizano 1987a for recent reviews).

In order to calculate the spectra of protostellar objects, we first adopt the quasi-static approximation; i.e., we assume that the physical structure of the object (density distribution, etc.) is that given by the collapse scenario outlined above at a particular instant in time. We then perform the radiative transfer calculation using this quasi-static structure. The radiation field is divided into three separate components: the star, the disk, and the dust envelope. The stellar component is taken to be a blackbody with a (single) temperature T_*. The disk radiates like a series of blackbodies with a power-law temperature distribution $T_D \sim \varpi^{-q}$, where q is usually taken to be $3/4$ in accord with the theory of Keplerian accretion disks (Lynden-Bell and Pringle 1974). The disk radius is defined by the centrifugal barrier (see above). The effects of mutual heating and shadowing of the disk and star by each other are self-consistently taken into account (see Adams and Shu 1986; hereafter AS). Since both the stellar and disk components are attenuated by the total extinction of the infalling dust envelope, most of the luminosity of the system is absorbed and re-radiated at far-infrared wavelengths.

In order to calculate the third component, the diffuse radiation field of the dust envelope, we must specify the opacity and the density profile and then self-consistently determine the temperature distribution in the envelope (see AS, Adams and Shu 1985). Here, interstellar dust grains (a mixture of graphite, silicate, and ices) provide the dominant contribution to the opacity. At infrared wavelengths, the scattering cross section of these grains is much smaller than that of absorption; hence, the effects of scattering are neglected in this treatment. In addition, we adopt the technique of an "equivalent spherical envelope" by taking the spherical average of the non-spherical density distribution produced by the rotating infall solution (see AS and TSC). This approach allows us to include the main effects of rotation – a less centrally concentrated infalling envelope which leads to a lower column density, a lower luminosity for a given infall rate due to the storage of energy in the form of rotation, and the production of a circumstellar disk – while retaining a spherical radiative transfer calculation.

In our formulation, the luminosity of the system has six separate contributions. We first include the shock luminosity, the energy liberated as the perpendicular component of the infall velocity is dissipated at the shock fronts on the stellar and disk surfaces. Also, in both the star and the disk there is a mixing luminosity which arises from the further dissipation of energy as the infalling interstellar material becomes adjusted to stellar and disk conditions. The above components of the luminosity can be calculated directly (and analytically) from the rotating infall solution (see AS, and Shu, Adams, and Lizano 1987b). In addition, we take into account the further dissipation of energy in the disk arising from accretion; this luminosity is parameterized in terms of the fraction η_D of the material that initially falls onto the disk that is accreted onto the star. Similarly, we allow for the dissipation of rotational energy in the star itself.

We find that the theoretical spectral energy distributions calculated from this protostellar model are in good agreement with the observed spectra of embedded *IRAS* sources (see Fig. 1a; Adams, Lada, and Shu 1987a, hereafter ALSa). The inferred mass infall rates for these objects are generally consistent with the measured gas temperatures of ~35 K in Ophiuchus and ~10 K in Taurus.

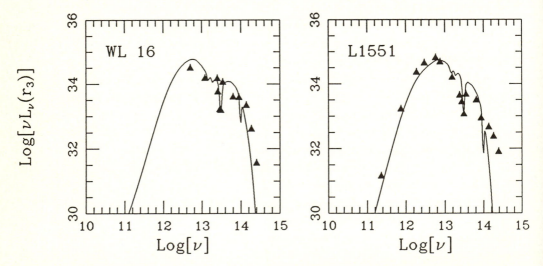

Figure 1. Theoretical and observed spectral energy distributions of protostellar candidates. (a) Infall source WL 16. Data taken from Wilking and Lada (1983), Lada and Wilking (1984), Young, Lada, and Wilking (1986), and from *IRAS*; theoretical model assumes $a = 0.35$ km/s, $M = 0.5 M_\odot$, $\Omega = 5 \times 10^{-13}$ rad/s, and $(\eta_*, \eta_D) = (0.5, 1.0)$. (b) Bipolar outflow source IRS5 L1551. Data taken from Cohen and Schwartz (1983), Cohen *et al.* (1984), Davidson and Jaffe (1984), and from *IRAS*; solid curve shows theoretical model with $a = 0.35$ km/s, $M = 1.0 M_\odot$, $\Omega = 1 \times 10^{-13}$ rad/s, and $(\eta_*, \eta_D) = (1.0, 0.5)$.

In order to fit the observed spectra, we must adopt cloud rotation rates in the range $10^{-14} - 10^{-13}$ rad/s, values which are consistent with the currently available observations. For these sources, the total visual extinction of the dust envelope is fairly large, $A_V = 40 - 200$, so that most of the luminosity of the system is reprocessed by the infalling envelope. The spectral energy distributions generally have maxima at wavelengths of $60 - 100\mu$m and have absorption features at 10 μm (from silicates) and at 3.1 μm (from water ice). As expected for extended atmospheres, the resulting protostellar spectra are always much broader than that of a blackbody. The spectra of sources in the next stage of evolution, the bipolar outflow stage, can also be explained by protostellar infall models (see Fig. 1b); this result suggests that both infall and outflow are taking place simultaneously in such objects.

3. T TAURI STARS WITH DISKS

When the infall is eventually turned off, the remaining star and disk will become optically visible. In this stage, the disk can be either active or passive. We first discuss the spectra of passive disks. All of the luminosity is intrinsic to the star, which is taken to be a blackbody at a temperature T_* given by the stellar spectral classification. The disk is assumed to be spatially thin and optically thick. In the limit that the disk radius is large compared to the stellar radius, the disk will intercept and re-radiate 25% of the stellar luminosity. The resulting surface temperature profile will approach the form $T_D \sim \varpi^{-3/4}$ at large radii (see AS); thus, the disk will radiate like a classical Keplerian accretion disk with an effective luminosity of $L_*/4$ and will add an infrared excess to the spectral energy distribution of the system. These passive disk models have no free parameters and produce the correct infrared excess for some observed T Tauri stars (see Fig. 2a).

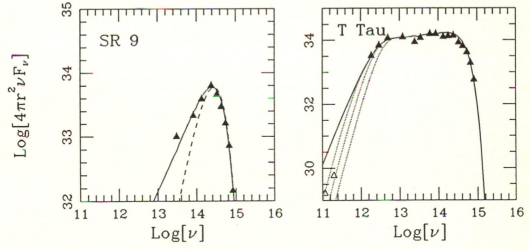

Figure 2. Theoretical and observed spectral energy distributions of T Tauri stars with circumstellar disks. (a) SR 9, a passive disk system. Data taken from Chini (1981) and from Lada and Wilking (1984); the solid curve shows a theoretical star/disk model with $L = 3.0L_\odot$, $A_V = 1.0$, and $T_* = 4000$ K, whereas the dashed curve shows the spectrum of a reddened blackbody. (b) T Tauri, an active disk system. Solid curve shows the spectrum in the optically thick $(M_D \to \infty)$ limit, whereas the dotted curves show spectra for finite disk masses of 1.0, 0.1, and 0.01 $M_\odot$. The theoretical model assumes $L_* = 5.0L_\odot$, $L_D = 12.0L_\odot$, $A_V = 1.44$, $R_D = 120$ AU, and $T_* = 5100$ K. Observational data taken from Cohen and Schwartz (1976), Rydgren, Strom, and Strom (1976), and from *IRAS*; the open symbols represent upper limits/marginal detections from Weintraub, Masson, and Zuckerman (1987) and Walker, Adams, and Lada (1987).

Next we consider another class of star/disk systems, those with active disks which have appreciable intrinsic luminosity in addition to the energy intercepted from the star. The star is taken to have a luminosity L_* and the spectrum of a blackbody, with the surface temperature T_* determined by the spectral classification. However, these systems often have spectra that are much flatter than that of a classical Keplerian accretion disk; hence, the disk temperature distributions must be flatter than $T_D \sim \varpi^{-3/4}$ (ALSa). In addition to the luminosity $(L_*/4)$ intercepted from the star, the disk must have an intrinsic luminosity L_D; this intrinsic energy source produces a *radial* temperature gradient of the form $T_D \sim \varpi^{-q}$, where q is determined from the slope of the spectrum. Again, we assume that the disk is spatially thin (with scale height $H \ll \varpi$) and flat, and has an isothermal *vertical* structure (see Adams, Lada, and Shu 1987b; hereafter ALSb).

We allow for the disk to be partially optically thin by introducing a surface density profile of the form $\Sigma \sim \varpi^{-p}$. Since the resulting spectra are insensitive to the value of p, we adopt $p = 7/4$, the result for a disk built up from the rotating infall solution of the protostellar theory (Cassen and Moosman 1981). Specifying the coefficient of the surface density profile is equivalent to specifying the mass of the disk M_D, which is left as a free parameter of the theory. In the limit that the disk becomes optically thick at all wavelengths (i.e., $M_D \to \infty$), the outer boundary of the disk defines a minimum disk temperature and hence a "turnover" wavelength in the spectrum. By fitting this turnover in the spectrum, we obtain a *lower limit* to the disk radius (see ALSb). By using models of this type, we can fit the observed spectral energy distributions of virtually all T Tauri stars with infrared excesses, from passive disks (in the limit $L_D \to 0$) to the extreme case of flat spectrum objects (see Fig. 2b and ALSb).

Since the stars in these systems do not produce all of the observed luminosity, stars with active disks have been placed incorrectly in the H-R diagram (e.g., Cohen and Kuhi 1979). By repositioning these stars in the H-R diagram, we can obtain new estimates for the masses and ages of the stars; however, these new estimates are also uncertain because the theoretical evolutionary tracks have been calculated without taking into account the presence of a disk.

Through fitting the observed spectral energy distributions, we find that the disks associated with the flat spectrum sources must have unorthodox radial gradients of temperature and contain intrinsic luminosity in addition to the energy intercepted and reprocessed from the central star; i.e., the disks must be *active*. The minimum values for the radii of these disks (see above and ALSb) are approximately 100 AU. Maximum values for the disk masses can be obtained if measurements of the dust continuum emission are available at low frequencies, where the disks are likely to be optically thin. These mass estimates, which depend on the dust opacity at submillimeter wavelengths, are somewhat tentative and lie in the range $0.1 - 1.0$ $M_\odot$. It is significant that the estimated disk properties (i.e., disk sizes, masses, and hence disk angular momenta) are in agreement with the disk properties predicted by the *protostellar* theory (see §2 and ALSa).

For a given star/disk system, we can use the derived disk temperature profile and stellar mass estimate (which determines the rotation curve of the disk) to calculate the maximum disk mass that is stable against self-gravity (ALSb). We find that the spectral limits and estimates of the disk masses derived for three systems (T Tau, DG Tau, and HL Tau) are close to the theoretical values that would make the self-gravity of the disks dynamically important; thus, gravitational instabilities may provide one possible source for the apparent disk activity.

There is another class of young stellar objects which probably represents an intermediate stage between the protostellar and the star/disk phases of evolution. These objects have two peaks in their emergent spectral energy distributions and can be explained as cases in which optically thin dust shells still surround the stars and disks (ALSa), perhaps because of residual infall. Again, the disks in these systems can be either passive (e.g., see ALSa) or active (e.g., see ALSb).

4. CONCLUSIONS

We have shown that the observed spectra of embedded *IRAS* sources associated with molecular cloud cores are in good agreement with the theoretical spectra of objects in the protostellar collapse phase. The spectra of the protostellar theory fit both the pure infall and (well-collimated) bipolar outflow phases of evolution; hence, infall is probably still taking place in the latter. In addition, we find that the spectral energy distributions of T Tauri stars with infrared excesses can be explained as young stars surrounded by circumstellar disks. Some sources can be understood in terms of passive disks that reprocess stellar radiation but have no intrinsic luminosity; other systems require appreciable intrinsic disk luminosity. For the extreme cases in the latter category, we obtain estimates for the disk masses $(0.1 - 1.0 \ M_\odot)$ and disk radii (~ 100 AU) through spectral modelling.

The picture of protostellar evolution espoused above is incomplete, especially our current understanding of disk physics. In the rotating protostellar collapse, most of the infalling material initially falls onto the disk rather than the star. However, a significant fraction of this disk material must eventually accrete onto the star. Otherwise, the masses of the forming stars would be unrealistically small, the luminosity produced by the system would be much smaller than that of observed sources, and the disk would become unstable (in principle, this instability could drive the accretion). Thus, there must be some disk accretion mechanism at work, but the detailed manner in which it manifests itself is not yet known. A similar situation holds for disks surrounding T Tauri stars: we find that such disks must have considerable intrinsic luminosity and unorthodox (i.e., shallow) radial gradients of temperature. At the present time, a detailed and complete theoretical explanation for this result is lacking. Various mechanisms have been invoked, including flaring disks (Kenyon and Hartmann 1987), radiative transfer effects, non-Keplerian rotation curves in the disk, and wave phenomena (ALSb), but the question remains open. Thus, the problem of understanding disk physics at a fundamental level should be the main focus of future work.

References

Adams, F. C., and Shu, F. H. 1985, *Ap. J.*, **296**, 655.

Adams, F. C., and Shu, F. H. 1986, *Ap. J.*, **308**, 836 (AS).

Adams, F. C., Lada, C. J., and Shu, F. H. 1987a, *Ap. J.*, **213**, 788 (ALSa).

Adams, F. C., Lada, C. J., and Shu, F. H. 1987b, submitted to *Ap. J.* (ALSb).

Cassen, P., and Moosman, A. 1981, *Icarus*, **48**, 353.

Chini, R. 1981, *Astr. Ap.*, **99**, 346.

Cohen, M., and Kuhi, L. V. 1979, *Ap. J. Suppl.*, **41**, 743.

Cohen, M., and Schwartz, R. D. 1976, *M. N. R. A. S.*, **174**, 137.

Cohen, M., Harvey, P. M., Schwartz, R. D., and Wilking, B. A. 1984, *Ap. J.*, **278**, 671.

Cohen, M., and Schwartz, R. D. 1983, *Ap. J.*, **265**, 877.

Davidson and Jaffe, D. T. 1984, *Ap. J. (Letters)*, **277**, L13.

Kenyon, S. J., and Hartmann, L. 1987, *Ap. J.*, in press.

Lada, C. J. 1985, *Ann. Rev. Astr. Ap.*, **23**, 267.

Lada, C. J., and Wilking, B. A. 1984, *Ap. J.*, **287**, 610.

Lynden-Bell, D., and Pringle, J. E. 1974, *M. N. R. A. S.*, **168**, 603.

Myers, P. C. 1985, in *Protostars and Planets II*, ed. D. C. Black and M. S. Mathews, p. 81, Tucson: Univ. Ariz. Press.

Myers, P. C., Fuller, G. A., Mathieu, R. D., Beichman, C. A., Benson, P. J., Schild, R. E., and Emerson, J. P. 1987, *Ap. J.*, in press.

Myers, P. C., Goodman, A., and Benson, P. J. 1987, submitted.

Rydgren, A. E., Strom, S. E., and Strom, K. M. 1976, *Ap. J. Suppl.*, **30**, 307.

Shu, F. H. 1977, *Ap. J.*, **214**, 488.

Shu, F. H., Adams, F. C., and Lizano, S. 1987a, *Ann. Rev. Astr. Ap.*, **25**, 23.

Shu, F. H., Adams, F. C., and Lizano, S. 1987b, in *Interstellar Dust and Related Topics, Fermi School Lectures*, ed. S. Aiello, New York: Academic Press, (in press).

Stahler, S. W., Shu, F. H., and Taam, R. E., 1980, *Ap. J.*, **241**, 637.

Terebey, S., Shu, F. H., and Cassen, P. 1984, *Ap. J.*, **286**, 529 (TSC).

Walker, C. K., Adams, F. C., and Lada, C. J. 1987, in preparation.

Weintraub, D. A., Masson, C. R., and Zuckerman, B. 1987, *Ap. J.*, **320**, in press.

Wilking, B. A., and Lada, C. J. 1983, *Ap. J.*, **274**, 698.

Young, E. T., Lada, C. J., and Wilking, B. A. 1986, *Ap. J. (Letters)*, **304**, L45.

Infrared Galaxies

HIGH LUMINOSITY GALAXIES IN THE IRAS SURVEY

D.B. Sanders, B.T. Soifer, and G. Neugebauer

Division of Physics, Mathematics and Astronomy
California Institute of Technology
Pasadena, California, 91125 USA

ABSTRACT

The IRAS survey has demonstrated the existence of a previously unrecognized class of ultraluminous objects. These ultraluminous infrared galaxies represent the dominant population of extragalactic emitters at luminosities where the only other objects known are quasars. Observations suggest that these objects are all found in merging galaxies, and appear to be dust enshrouded quasars. We suggest that these IR loud quasars represent the formation stage of quasars, with the latter stages in the evolution of such systems being found in the UV excess quasars.

1.0 OBSERVATIONS OF ULTRALUMINOUS INFRARED GALAXIES

Analysis of the galaxies detected in the IRAS survey has shown that for bolometric luminosities $L > 3 \times 10^{11}$ $L_\odot$, infrared bright galaxies are the dominant population in the local universe. Soifer, et $al.$ (1986) have found that for $L > 3 \times 10^{11}$ $L_\odot$, the space density of the infrared luminous galaxies equals the space density of Seyfert galaxies, while for luminosities above 10^{12} $L_\odot$, ultraluminous infrared galaxies exceed quasars in space density by a factor of 5.

We have studied in detail (Sanders, et $al.$ 1987) 10 ultraluminous infrared galaxies found in the IRAS Bright Galaxy Sample. We define an ultraluminous infrared galaxy to be a galaxy having an 8-1000 μm luminosity that is equal to the bolometric luminosity of a quasar, i.e. $L \gtrsim 10^{12}$ $L_\odot$. These 10 objects are most amenable for detailed studies, being the closest examples known of the ultraluminous infrared galaxies. The distances of the objects range from 77 Mpc for Arp 220 to 325 Mpc for IRAS 14348-1447. The most luminous object in the sample is the well known Seyfert galaxy Mrk 231.

Perhaps the most striking observational feature of these ultraluminous objects is that they are ALL found in apparently interacting systems. Optical imaging of these galaxies shows that only one of them, IRAS 05189-2524, has a predominantly stellar appearance. All the others are clearly extended, and so are unquestionably galaxies. Furthermore deep CCD images show evidence of large scale "peculiar" structures, i.e. tidal tails, or rings in all of the systems. Four of these object show double nuclei.

As infrared selected objects, it is natural that these systems should have the peak of their energy distributions in the infrared. However, it is the magnitude of this "infrared loud" property that is rather surprising. The average ratio of $\nu L_\nu(80\ \mu m)/\nu L_\nu(0.44\ \mu m)$ for the ultraluminous systems is ~ 40, as compared with ~ 2 for the Bright Galaxy sample as a whole.

Optical spectroscopy shows that the nuclei of all these objects have strong emission lines. The most fully studied of these systems, Mrk 231, has long been classified as a Seyfert 1 galaxy based on the very broad permitted lines. IRAS 05189-2524 and UCG 5101 also show line widths of > 2000 Km s^{-1} (FWZI) and as such are classified as Seyfert 1.5 and 1.8 systems respectively. Six of the 7 remaining objects from this sample show Hα linewidths in the range 1000 - 2000 Km s^{-1} and line ratios characteristic of Seyfert 2 or LINER systems.

The near infrared colors of these systems are not at all typical of the colors of normal galaxies. The J-H and H-K colors of these systems have a large dispersion, 0.5 mag in J-H and 1 mag in H-K, with the "centroid" of J-H ~ 0.9 mag, H-K ~ 0.7 mag. For comparison normal spiral galaxies have J-H $\sim 0.7\pm0.1$ mag, H-K $\sim 0.25\pm0.1$ mag, while quasars have J-H $\sim 0.9\pm0.2$ mag, H-K $\sim 1.0\pm0.2$ mag.

All of these ultraluminous systems are gas rich, and most of this gas is in molecular form. Of the 9 systems we have been able to observe to date in the 2.6mm line of CO, all are detected, with an average inferred H$_2$ mass of 2.6×10^{10} M$_\odot$, or 10 times that of the Milky Way. The infrared luminosity per unit mass of gas is also much higher in these systems than in molecular clouds or "starburst" galaxies. The average L$_{ir}$/M(H$_2$) for these systems is 60, with a range of 27 to 135. In comparison, the dense HII region cores of Giant Molecular Clouds in our galaxy show a L$_{ir}$/M(H$_2$) ratio of ~ 20, which is also found in "starburst" galaxies such as M82 and NGC 253.

Observations suggest that the infrared emission and molecular gas is concentrated to the nuclei of these systems. In Mrk 231 high resolution observations (Matthews, *et al.* 1987) have localized the 10 μm emission to within ~ 400 pc of the nucleus. In Arp 220 the 20 μm emission has been localized to r < 400 pc (Becklin & Wynn-Williams, 1987), while 10^{10} M$_\odot$ of molecular gas has been observed to be contained within r < 700 pc of the nucleus (Scoville, *et al.* 1986).

2.0 A MODEL OF ULTRALUMINOUS INFRARED GALAXIES

The observations described above suggest that the dominant power source in these systems is a central "quasar" so heavily enshrouded in dust that it is not readily seen through observations in the visible, but is primarily identified through the enormous bolometric luminosity that is emerging in the infrared. The fact that all of these systems are found in interacting/merging galaxies suggests that interactions are related in a fundamental way to the formation of the ultraluminous infrared galaxies. We suggest that what is occurring is the formation of quasars as a result of the collisions of gas rich galaxies.

In this model, the collision of two gas rich galaxies triggers the funneling of the available gas to the nucleus of the merger system. Presumably, this stage is accompanied by a vigorous period of star formation in the galaxies. The accumulation of $\sim10^{10}$ M$_\odot$ of molecular gas into the environment of a pre-existing (e.g. Fillipenko and Sargent 1985) or contemperaneously forming massive black hole would seem to be the ideal environment for the formation of a quasar. The physical processes involved in this merging of such gas rich systems are outlined in the work of Norman (this volume).

Initially, we would expect such a quasar to be totally enshrouded in gas and dust and would appear like Arp 220; i.e. the central source is nearly invisible except for its bolometric luminosity. In this regime the surrounding galaxy is plainly visible, indeed it is really all that is seen optically. As the quasar disrupts the enshrouding gas cloud through radiation pressure and outflows, the central source begins to become visible at shorter wavelengths. Mrk 231 might represent this stage where the surrounding galaxy is still quite visible but the central source is prominent. Ultimately, the central source sufficiently disrupts the enshrouding gas cloud to become visible as a "classical" UV excess quasar. At this stage the surrounding galaxy is very difficult to detect, because of the large luminosity of the central quasar. As the quasar evolves through the stages from dust enshrouded to completely visible, the luminosity of the underlying object emerges at substantially different wavelengths. The energy distribution at these various stages are illustrated in figure 1.

This scenario for quasar formation provides a natural qualitative explanation for the evolution of quasars. Once galaxies form, the average rate of formation of quasars should be proportional to the space density of gas rich galaxies, which increases with redshift. Furthermore, the required merging of galaxies creates a natural cutoff in the observed redshifts of quasars, i.e. at a sufficient time after the galaxy formation epoch to allow such interactions to occur.

3.0 CONCLUSIONS

One of the most surprising results to emerge from the IRAS survey has been the existence of a substantial population of ultraluminous galaxies. We believe that these systems have shown the natural link between normal galaxies and quasars, with the merging of normal, gas rich galaxies creating the majority of quasars.

References

Becklin, E. E. and Wynn-Williams, G. C. 1987, "Star Formation in Galaxies," ed. C. Lonsdale-Persson, U. S. Government Printing Office, Washington, D. C..

Filippenko, A. and Sargent, W.L.W. 1985, *Ap.J. Suppl.*, **57**, 503.

Matthews, K., Neugebauer, G., McGill, J., and Soifer, B. T. 1987, *A.J.*, **94**, 297.

Norman, C. 1987, Proceedings of 3rd IRAS Conference.

Sanders, D.B., Soifer, B.T., Elias, J.H., Madore, B.F., Matthews, K., Neugebauer, G., and Scoville, N.Z. 1987, *Ap.J.*, in press.

Scoville, N. Z., Sanders, D. B., Sargent, A. I., Soifer, B. T., Scott, S. L., and Lo, K. Y. 1986, *Ap.J. (Letters)*, **311**, L47.

Soifer, B.T., Sanders, D.B., Neugebauer, G., Danielson, G.E., Lonsdale, C.J., Madore, B.F., and Persson, S.E. 1986, *Ap.J. (Letters)*, **303**, L41.

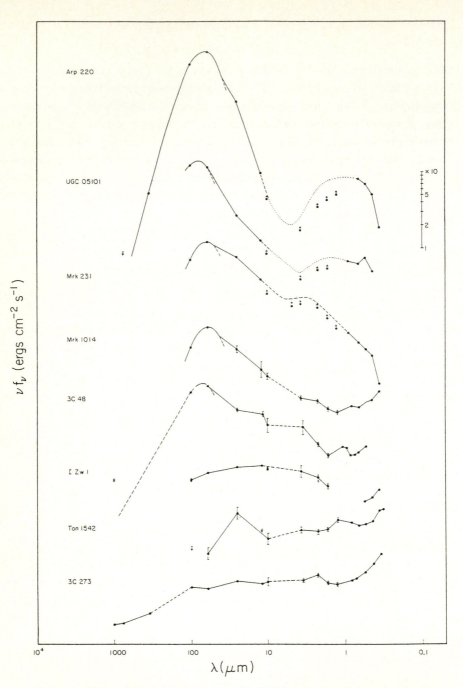

figure 1. The evolution of the energy distribution during the stages of quasar evolution from "infrared loud" (top) to "UV loud" (bottom) systems. The earliest, most dust enshrouded stage is represented by Arp 220, the middle, still "infrared loud," stage by Mrk 231, and the late "UV loud" stage by the radio quiet quasar Ton 1542 and the radio loud quasar 3C273.

STARBURSTS: NATURE AND ENVIRONMENT

Colin. A. Norman

Johns Hopkins University
Homewood Campus
Baltimore, MD 21218
and
Space Telescope Science Institute
3700 San Martin Drive
Baltimore, MD 21218

ABSTRACT

The physical mechanisms responsible for producing starbursts are outlined including interactions, mergers, bars, and sinking gas-rich satellites. The question of why a burst occurs is analysed as well as the problem of the disparate timescales of the interaction, burst and dynamical times.

It is shown that there are three principal phases of starbursts associated with dynamical processes. Most starbursts will remain in the first normal starburst phase but given sufficiently large central gas masses the second very luminous stage related to a central bar instability can be achieved. In some cases a final ultra luminous phase occurs associated with the self-gravity of a massive central gas cloud. Here an active nucleus can form associated with a dense stellar cluster formed in the burst.

This final fascinating phase has been treated in detail including the growth of a central black hole, the formation of broad emission line clouds from red-giant envelopes photoionized from the outside by the central continuum source, collisions between red giants, the evolution of the stellar population in the central star cluster, and the cosmological implications for the number density and luminosity evolution of quasars, active galactic nuclei and starbursts.

The context and relevance of starburst galaxies with respect to these important astrophysical phenomena is shown. We can learn much from concentrating on the starburst physics relevant to galaxy formation and galaxian evolution along the Hubble sequence. The intergalactic medium can be significantly enriched by the prodigious metal enhanced outflows from starburst systems driven by OB winds and supernovae. It can also be heated and ionised by the energy and momentum input from the combined active galactic nuclei and starburst galaxies. Quasar absorption lines may well arise in these huge circumgalactic flows.

1. INTRODUCTION

It has become clear that starburst galaxies are triggered into that mode by some external influence such as an interaction with a companion, or even a merger in the most extreme cases. Barred systems, and other nonaxisymmetrically distorted galaxies such as those with interacting satellites also show the starburst effect to various degrees. Burst timescales are usually estimated to be quite short $\lesssim 10^7 - 10^8$ years as inferred from total gas consumption rates even with a truncated initial mass function. Interaction and merging time scales are an order of magnitude longer $\sim 10^8 - 10^9$ yr. Standard estimates for significant gas flows in normal galaxies are yet again an additional order of magnitude longer $\sim 10^{10}$ yr. The resolution of these obvious timescale discrepancies and the physics of interacting, barred, merging and other related systems are discussed in the following section.

Three phases of the starburst phenomenon emerge naturally from the theoretical framework presented here. The observational basis centres on the measurements of extraordinary gas masses—typically $10^{10} M_\odot$ within 500 pc in the central region of starbursts using a number of independent techniques (Scoville 1987, Becklin 1987, Solomon 1987, Mezger 1987, Sofue 1987 and Carlstrom 1987)! The dynamical clue to all this is the increasing signficance of the self-gravity of the gas usually referred to in the current literature as the ratio of gas mass to dynamical mass. The first stage is the normal starburst mode where due to the physical driving mechanisms discussed in Section II gas flows inwards, builds up a large surface mass density and an enhanced star formation rate ensues. When sufficient mass concentration builds up in the centre—and the observation verify that this *does* happen, then the central gas mass tend to become unstable to the classic bar instability (Ostriker and Peebles 1973). Depending in detail on the spatial variation of the gas mass distribution, this will occur at a ratio of gas mass to dynamical mass of order 10 to 30 percent. Rapid angular momentum transport is expected to ensue as well as a much enhanced star formation rate. After a timescale of order $\lesssim 10^7$ years i.e. approximately ten dynamical times in the central regions a totally *self-gravitating* central gas mass will result. This is the ultra luminous phase. Rapid collapse takes place in this third ultra luminous phase. The star formation rate is expected to rise drastically and as discussed this may lead to the formation of an active galactic nucleus. Note that *very few* galaxies will attain this state since such remarkable central gas masses are rare—although, to emphasize the point once again—in the ultra luminous cases such masses are *observed*. These various phases are studied in Section 3.

This third phase of starburst evolution is most interesting. It seems that for the first time, realistic, observed, initial conditions have been found that are relevant to the formation of a massive central star cluster $\sim 10^{10} M_\odot$ and associated black hole. An analysis of this situation has been performed in collaboration with Nick Scoville. We take a massive coevally generated central star cluster and study the evolution of the star cluster, the production of red giants, the stellar mass loss, the growth of a central massive black hole, the formation of broad line clouds by photoionisation of giant envelopes from the outside by the central continuum source, the formation of logarithmic line profiles and many other aspects. The model relates the starburst and active galactic nucleus phenomenon in a quantitative way. A fundamental prediction is that stars should eventually be seen associated with active galactic nuclei. Identification of giants as broad line clouds is one way, the possible association of megamasers with this broad line cloud system is another. The post-active phase is an important one to study observationally since, for example, A-type spectra may be seen in the very central region of, say, weak Seyferts that are in a low state of activity. These concepts are discussed in Section 3.

The starburst galaxies are centrally related to our ideas on the formation of galaxies and their subsequent evolution. A quite respectable view (Toomre 1977) is that the formation of round systems such as elliptical galaxies and bulges is a secondary process that results from interactions between massive galaxies in the first instance and rather unequal mass galaxies in the second case. Usually this merging and interacting process is envisaged to occur early in the galaxy formation phase at a redshift of, say, a few and it probably involves the collision of gas-rich protodisks such as those observed recently by Wolfe *et al.* (1986). However, at the current epoch, luminous starburst galaxies are probably low level and late forming examples of this phenomenon of bulge building and elliptical galaxy production. Furthermore we can bring many techniques to bear to determine the physical properties of the galaxies at the present epoch using them as laboratories for studying processes which mainly occur at much earlier epochs. The outflows observed from starbursts are of order 1–100 $M_\odot$ yr^{-1} extending to tens of kiloparsecs. Significant metal enrichment is occurring as well as energy and momentum input into the circumgalactic and intergalactic media. Hot, enriched halos around freshly minted ellipticals can be formed and the intracluster and intergalactic media can be very substantially enriched. The universe at redshift one to five is becoming a most interesting and observationally accessible place. Quasar absorption lines are one particularly fascinating field and the origin of metal rich absorption line system have already been suggested to be related to starbursting Magellanic systems and dwarf galaxies (York *et al.* 1986, Ikeuchi and Norman 1987). These starburst outflows seem, however, to be even more plausible candidated with their large filling factors and their filamentary, metal-rich knots— particularly if they are much more numerous at earlier times. These matters are discussed in §5. A summary and conclusions are given in Section 6.

2. PHYSICAL MECHANISMS FOR GAS INFLOW

Here we shall study three specific physical mechanisms for the transport of substantial amounts of gas into the central regions of galaxies. Firstly we shall study the disk capture of gas-rich dwarf galaxies.

2.1 Disk capture of gas-rich dwarf galaxies

Cold disks are extremely responsive to perturbations by even relatively low mass satellites. This in turn means that satellite orbits can be strongly affected by cold disks. A detailed study of the sinking satellite process has been made by Quinn and Goodman (1987) who find that a reasonable estimate of the satellite sinking time is

$$\tau_s \sim 4 \times 10^9 \left(\frac{V_{circ}}{220 \text{ km s}^{-1}}\right) \left(\frac{10^9 \, M_\odot}{M_s}\right) \left(\frac{r_s}{10 \text{ kpc}}\right)^2 \left(\frac{3}{\ln \Lambda}\right) \text{ yr}$$

where V_{circ} is the flat rotation curve velocity of the disk and M_s and r_s are the mass and galactocentric distance of the satellite and $\ln \Lambda$ is the standard cut-off term. The effects are quite subtle, incorporating resonances and horseshoe orbits, as well as straight dynamical friction. Substantial disk damage can occur in the late stages which can result in significant bulge heating in the central parts. There is certainly a *lack* of gas rich dwarfs available to be accreted by luminous, L_*, galaxies at the current epoch. There were undoubtedly more of these objects at redshifts greater than or of order unity but to account for the current starburst population it is probably necessary to manufacture them in an interaction process whereby bridges and tails are formed that then break up into small, bound subunits that can then be accreted—the net result being a real mass transfer. Note that considerable amounts of gas must be involved here—at least of order $\gtrsim 10^8 \, M_\odot$!

For the more powerful systems, direct mergers seen to be the only way. Galaxies are made of collisionless stars and multi phase interstellar media and can, in fact, undergo quite inelastic collisons if the relative velocity is less than 110% of their internal velocity dispersion. In the subsequent violent relaxation process where potential fluctuations are of order unity, approximately half the relative motion is soaked up by internal energy, and each particle has roughly equal probability of being anywhere in the available phase space. Bound clouds will be thrown toward the center and suffer cloud-cloud collisons thereby increasing this binding energy. A density law $p \sim r^{-\alpha}$ where α varies between 2 and 3 is usually found (van Albada 1982, Carlberg, Lake and Norman 1986). A comparison of simulations and actual observation of mergers gives one great confidence that the merging process is occurring in very luminous and ultra luminous starburst galaxies (Joseph and Wright 1985).

The general form for interaction with companions that drive wave and bars has been studied by (Norman 1985, 1987a, b, c, Combes 1987). Given a general non-axisymmetric distortion the time scale for gas to flow inwards is given by

$$\tau_{inflow} \sim \frac{1}{2m^2\gamma} \left(\frac{\Omega(r)}{\Omega - \Omega_p} \right) \left(\frac{E_{total}}{E_{wave}} \right)$$

where $\Omega(r)$ is the angular velocity curve, m is the number of arms, E_{wave}/E_{total} is the ratio of wave to total energy and γ is a parametrisation of the dissipation rate due to shocks, and cloud-cloud collisions. This wave induced drag exceed the viscous inflow purely due to collisions if $E_{wave}/E_{tots} \gtrsim (1/\sqrt{3})(h/mr)$ when h and r are the scale height and radius of the disk. A simple formula that can be used as a rule of thumb is

$$\tau_{inflow}(10 \text{ kpc}) \sim 10^8 \left(\frac{\text{Total energy}}{\text{Wave energy}} \right) \text{ yr}$$

and thus for normal Sc's we have $\tau_{inflow} \sim 10^{10}$ yr but for systems with wave amplitudes of order $\sim 30\%$ resulting from a very large distortion due to interactions etc. we have a time scale of $\sim 3 \times 10^8$ yr!! Note that, as discussed by Lubow (1987), even for a 10–15% gas to star ratio all the response is in the *gas*—not the stars!!

3. THE THREE PHASES OF STARBURST GALAXIES

Let us first look at the relevant timescales. For massive central starbursts the burst time scale is $\sim 10^7 - 10^8$ yr. The time scale to accumulate gas in the central region from the work of Section 2 is $\sim 10^8 - 10^9$ years which can for, strong interaction and perturbations, make the dynamical timescale associated with such interactions sufficiently short. Why then is there a burst? Is there a real threshold effect? Recent observations lead us to this being a very strong possibility. For the very luminous galaxies (Scoville *et al.* 1987, Scoville 1987, Becklin 1987, Mezger 1987), the smaller system II Zw 31 (Sage and Solomon 1987) and the well studied nearby system M82 (Sofue 1987, Carlstrom 1987) values of gas mass inside, say 500 pc, are found to be of order 0.1–0.3 the total interior dynamical mass. This then leads us to the almost inevitable conclusion that a bar instability will form in the gas. Robust criteria to use here are the Ostriker-Peebles condition $T/W \sim 0.14$ (Ostriker and Peebles 1973; see also Efstathiou *et al.* 1982) where T is the rotational kinetic energy and W is the total potential energy. Upon bar formation, rapid angular momentum transfer will occur due to, for example, the interaction of the strongly non-axisymmetric bar with the background potential and the time scale for further contraction is roughly estimated to be (Weinberg and Tremaine 1983)

$$\tau_{decay} \sim 1 - 10(M_{halo}/M_{bar})t_{dyn} \sim 3 \times 10^6 - 10^7 \text{ yr!}$$

Thus, these *observed* central gas masses are inherently bar unstable when $M_{gas}/M_{dynamical} \gtrsim 10 - 30\%$.

After of order $\sim 10^7$ year the gas will then have concentrated by a factor of, say, three to ten, and then a significant factor of the original gas mass will become *totally* self gravitating (depending on the specific spatial distribution of the stars and gas). This central gas mass will then decouple from the galactic potential and, as discussed in the next section may well lead to the formation of an active galactic nucleus associated with the central starburst.

In each of these three stages a substantial change in the luminosity generated per unit gas mass is expected as shown in Fig. 1. Also shown is the ratio of gas mass to dynamical mass.

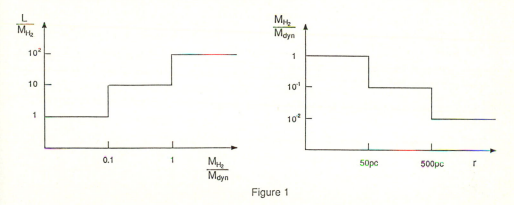

Figure 1

4. ULTRALUMINOUS STARBURSTS AND ACTIVE GALACTIC NUCLEI

The ultra luminous starburst phase is most fascinating particularly for its relation to active galactic nuclei. We have analyzed the fate of a massive central star cluster of $\sim 10^9 - 10^{10} \, M_\odot$ within a galactocentric radius of 10 pc (Scoville and Norman 1987). We have assumed efficient star formation and analyzed the evolution of the star cluster for a range of initial mass functions and upper and lower mass cut-offs. Specific calculations involve the production of the red giants, the mass loss from the star, the fate of that lost mass, most of which goes to feed a black hole, the growth of a central massive black hole and the role of stellar collisions. Interesting effects result from the influence of the strong central continuum radiation source on the envelope of red giants giving a strong *externally* induced mass loss. It is remarkable that the ionisation structure, density and size for these stellar envelope parameters is such that these externally ionised red giants can account quite naturally for the broad emission line cloud. At later times of order $\sim 10^7 - 10^8$ yrs the covering factor of red giants is $\sim 10\%$ with a large number, of order, $10^6 - 10^7$ red giants. A black hole of mass $10^8 - 10^9\ M_\odot$ has grown by this stage and the circular velocity of the clouds is typically of order 5,000 km/s^{-1}, at distances of order $\sim 10^{16}$ cm. The red giant envelopes *do* collide after $\sim 10^5$ yr but are replenished rapidly on a timescale less than $\lesssim 10^2$ yr. Much early work is related to this (Zeldovich and Novikov 1964, Spitzer 1971, Begelman and Rees 1973, Bailey 1980, Shull 1983, Mathews 1986) but here *at last* we have good initial conditions for the active galactic nuclei and black hole scenarios. This model can explain the centrality of the burst. It is a very real prediction that dense central stellar clusters should be seen associated with active nuclei. A good place to look may be in the immediate post activity phase where a dense stellar cluster with an A-type spectral may be found. Low luminosity example of this may be observable in the cores of nearby galaxies such as M31, M32 and M81.

5. STARBURSTS, GALAXY FORMATION AND EVOLUTION, AND QUASAR ABSORPTION LINES

There are a number of significant, mainstream topics in astrophysics to which the starburst phenomenon is relevant. The relationship with active galactic nuclei has been made clear in the previous section. Let us now turn to the study of galaxy formation and evolution in this context. Recalling that starbursts seem to arise in collision or interaction of gas rich objects there are three points that seem quite relevant here. The universe at redshift of order $z \sim 2$ is covered with high column density $N_H \approx 10^{20}$ cm^{-2} objects discovered by Wolfe *et al.* (1986). These authors deduce the covering factor to be $\Omega_{covering} \sim 20\%$ and the amount of mass involved to be of order the entire luminous mass of the Universe at the present epoch to within a factor of ~ 2. The favoured interpretation is that they are gas-rich proto disk of about five times the size of ordinary disks. Assuming this interpretation is correct, these objects can and definitely will interact and their interactions may account for the phenomena of fuzz around quasars and some of the distorted emission line images seen recently at high redshift (Djorjovskii *et al.* 1987, Chambers *et al.* 1987). The connection to starbursts is clear, namely, these objects when interacting *are* starbursts.

The second point is that gas rich disks may in fact be *the* building blocks of galaxies. This is one view (another is that the more fundamental units are gas rich dwarfs; Silk and Norman 1981) which we adopt here. In such a galaxy formation model, bulges of spirals and ellipticals are formed from the interaction of these disks over a range of relative masses of the interacting components. Collisions of massive objects can and probably will result in elliptical galaxies. Here recall Arp 220 itself. This galaxy will become an elliptical. This elliptical formation process will most likely be far more common at high redshift and potentially even more violent. The association of galaxy formation, starbursts and the epoch of quasar formation and rapid quasar evolution is explainable in the context of such a model. Bulges are built in this model by rather less violent interaction with lower mass objects falling into the center and giving rapid bulge heating and relaxation there. A typical example perhaps is M82. This galaxy is growing a bulge.

The third point is more general. Evolution tends to go not *smoothly* but in a series of *catastrophes*. It is these catastrophes that have far more effect than a much lower level, smoother and more continuous process. The star*burst* phenomenon is such a process.

Studies of local epoch starbursts may give significant observational clues to the bulge building, elliptical galaxy making process that is a major component of the formation of galaxies and their evolution along the Hubble sequence.

The quasar absorption lines associated with metal rich systems have long been thought to be associated with the extended halos of galaxies. A long-standing problem with this point of view has been the large extent of metal rich halo material required to surround galaxies at high redshift. A more recent problem (Danly, Blades and Norman 1987) is that most quasar absorption lines are *not* like halos of galaxies at the current epoch. Significant evolution must be invoked. Recent observations of huge extended flows around starburst systems may solve both these problems. The rate of injection of enriched mass is of order 10^2 $M_\odot$ yr^{-1}. The structures are huge, of order $\sim$10–100 kpc and consist of low ionization filaments with small velocity widths or b-values for the clumps (Heckman *et al.* 1984). These are undoubtedly more common at high redshift and are natural candidates for the QSO absorption line systems.

A most interesting test here would be to observe some quasars behind local starburst outflows and see if these absorption line systems would then satisfy the stringent tests described by Danly *et al.* (1987). Of course this is a major project for the Hubble Space Telescope.

6. SUMMARY AND CONCLUSION

There are three phases of a starburst galaxy.

I. Normal Starburst Phase

The timescale is $\sim 10^8 - 10^9$ yr. Interaction and merging gives mass transfer to the center, the central surface density rises and the star formation rate consequently increases.

II. Very-Luminous Starburst Phases

The timescale is $10^6 - 10^7$ yr. The crucial parameter here is the ratio of gas mass to dynamical mass $M_g/M_{dyn} \sim 0.1 - 0.3$. Detailed observations of rotation curves and ratios of gas mass to dynamical mass as a function of radius are crucial here.

III. Ultra-Luminous Starburst Phase

A massive black hole and central star cluster forms. The gas is self-gravitating at the initiation of this phase. The crucial observation is to find the stars associated with this active nucleus. The broad line clouds may be red giants irradiated from the outside inwards. A starburst cluster with say an A-type spectrum may be seen in the post active phase of such a nucleus.

Detailed high resolution observations of local epoch starbursts can help us understand the physics of the gas dynamic processes involved in galaxy formation, including bulge formation and Hubble sequence evolution, driven by interactions . Elliptical galaxy formation driven by mergers is probably observable in, say, Arp 220 right now. Metal-rich quasar absorption lines could be produced by the prodigious outflows of metal-rich material ejected to large distances 10–100 kpc in the starburst-related, galaxy formation epoch at redshifts of order a few.

It is a pleasure to thank Nick Scoville for much encouragement and a stimulating scientific collaboration. Others who have contributed to this work in a helpful and creative way include Eric Becklin and Bob Joseph, as well as many other participants at the meeting.

REFERENCES

Bailey, M. E. 1980, *M.N.R.A.S.*, **191**, 195.

Becklin, E. 1987 in Galactic and Extragalactic Star Formation, ed. R. Pudritz and M. Fich (Reidel: Dordrecht), in press.

Begelman, M. C. and Rees, M. J. 1978, *M.N.R.A.S.*, **185**, 847.

Carlstrom, R. 1987 in Galactic and Extragalactic Star Formation, ed. R. Pudritz and M. Fich (Reidel: Dordrecht), in press.

Chambers, K., Miley, G. K. and van Breugel, W. 1987, *Nature*, in press.

Combes, F. in Galactic and Extragalactic Star Formation, ed. R. Pudritz and M. Fich (Reidel: Dordrecht), in press.

Danly, L., Blades, C. and Norman, C. A. 1987, preprint.

Djorgovskii, S., Strauss, M. A., Perley, R. A., Spinrad, H. and McCarthy, P. 1987, *A. J.*, **93**, 1318.

Efstathiou, G., Lake, G. and Negroponte, J. 1982, *M.N.R.A.S.*, **199**, 1069.

Heckman, T. M., Armus, L. and Miley, G. K. 1987, *A. J.*, **92**, 277.

Ikeuchi, S. and Norman, C. A. 1987, *Ap. J.*, **312**, 485.

Lubow, S. 1987, in preparation.

Mathews, W. G., 1986, *Ap. J.*, **305**, 187.

Mezger, P. 1987 in Galactic and Extragalactic Star Formation, ed. R. Pudritz and M. Fich (Reidel: Dordrecht), in press.

Norman, C. A. 1985 in *Galaxy Formation and Evolution*, ed. J. Audonze and J. Tran Thank Van, NATO ASI Series C, Mathematical and Physical Science, **117**, 327.

Norman, C. A. 1987a, in *Star Formation in Galaxies* eds. T. X. Thuan and T. Montmerle, in press.

Norman, C. A. 1987b, in Galactic and Extragalactic Star Formation, ed. R. Pudritz and M. Fich (Reidel: Dordrecht), in press.

Norman, C. A. 1987c, *Nature*, in preparation.

Ostriker, J. P. and Peebles, P. J. E. 1973, *Ap. J.*, **186**, 467.

Quinn, P. J. and Goodman, J. 1986, *Ap. J.*, **309**, 472.

Sage, L. J. and Solomon, P. M. 1987, *Ap. J. (Letters)*, **321**, L103.

Scoville, N. Z., Sanders, D. B., Sargent, A. I., Soifer, B. T., Scott, S. L. and Lo, K. Y. 1986, *Ap. J. (Letters)*, **311**, L47.

Scoville, N. Z. 1987 in Galactic and Extragalactic Star Formation, ed. R. Pudritz and M. Fich (Reidel: Dordrecht), in press.

Scoville, N. Z. and Norman, C. A. 1987, *Ap. J.*, submitted.

Shull, J. M. 1983, *Ap. J.*, **264**, 446.

Silk, J. and Norman, C. A. 1981, *Ap. J.*, **247**, 59.

Sofue, I. 1987 in Galactic and Extragalactic Star Formation, ed. R. Pudritz and M. Fich (Reidel: Dordrecht), in press.

Spitzer, L. 1971 in *Galactic Nuclei* ed. D. O'Connell, North Holland, p. 443.

Toomre, A. 1977 in Evolution of Galaxies and Stellar Populations, ed. B. Tinsley and R. Larson (Yale University Press) p. 401.

van Albada, T. S. 1982, *M.N.R.A.S.*, **201**, 939.

Weinberg, M. D. and Tremaine, S. 1983, *Ap. J.*, **264**, 364.

Wolfe, A., Turnshek, D., Smith, H. E. and Cohen, R. D. 1986, *Ap. J. Suppl.*, **61**, 249.

York, D. G., Dopita, M., Green, R. and Bechtold, J. 1986, *Ap. J.*, **311**, 610.

Zel'dovich, Y. B. and Novikov, I. D. 1964, *Dokl. Acad. Nauk, SSSR*, **158**, 811.

STAR FORMATION IN NORMAL GALAXIES

Bruce G. Elmegreen
IBM Thomas J. Watson Research Center
P.O. Box 218, Yorktown Heights, N.Y. 10598 USA

ABSTRACT

The formation of giant cloud complexes in galaxies is discussed. Spiral galaxies without density waves have approximately the same star formation rates and CO abundances as the average for galaxies with waves. This implies that the primary effect of the wave is to organize the existing molecular clouds into a global spiral pattern. A secondary effect of the wave is to trigger the formation of more molecular clouds and stars, but this triggering may be limited to only the strongest waves, as in M51. Strong waves apparently trigger more cloud formation than weak waves because the gas in a strong wave has time to dissipate and collapse gravitationally before it leaves the spiral arms. Strong waves should also have molecular shock fronts (dust lanes), unlike weak waves.

1. STAR FORMATION IN GIANT CLOUD COMPLEXES

Most of the star formation in normal galaxies occurs in giant regions measuring several hundred parsecs on a side. In galaxies with density waves, these regions are usually in the spiral arms. In galaxies without density waves, they are scattered throughout the disk (Kennicutt and Hodge 1980; Seiden and Gerola 1982; Rumstay and Kaufman 1983; see review in Elmegreen 1987a).

Giant regions of star formation are usually found associated with equally large HI and CO cloud complexes (Elmegreen and Elmegreen 1983). A galaxy the size of ours typically contains 20 to 100 of these clouds, each of which contains between $10^6 M_\odot$ and several times $10^7 M_\odot$ of gas. In M33, M101, M81, M31, M106 and in the outer part of our Galaxy, the largest clouds are mostly atomic (Wright, Warner and Baldwin 1972; Newton 1980; Allen, Goss and van Woerden 1973; Allen and Goss 1979; Viallefond, Allen and Goss 1981; Viallefond, Goss and Allen 1982; Rots 1975; Emerson 1974; Unwin 1980a,b; Bajaja and Shane 1982; van Albada 1980; Henderson, Jackson and Kerr 1982). In at least the inner regions of M51 (Lo *et al.* 1987) and IC342 (Lo *et al.* 1984), and in some parts of M83 (Allen, Atherton and Tilanus 1986), the largest clouds are mostly molecular. In the inner 5 kpc region of our Galaxy, the largest clouds are approximately half molecular (Elmegreen and Elmegreen 1987). The cloud masses and sizes are all very similar from galaxy to galaxy and within a galaxy, but the molecular fraction per cloud varies by a large factor (> 10).

The masses, separations and virialized velocity dispersions of the largest clouds in galaxies are consistent with their formation by mild gravitational instabilities in the galactic gas disks (Elmegreen 1979; Cowie 1981; Viallefond, Goss and Allen 1982; Elmegreen and Elmegreen 1983; Jog and Solomon 1984; Balbus and Cowie 1985; Tomisaka 1987; Elmegreen 1987b, hereafter E87b; Balbus 1988). The instability in a magnetic, shearing galaxy grows at essentially the conventional Jeans growth rate, which is the same rate as for a non-magnetic, non-rotating gas disk. The rates are the same because the magnetic field resists the Coriolis force (E87b). Unlike the non-shearing instability, however, the available growth time is limited by shear, or by the flow time in the case of a spiral wave (Balbus and Cowie 1985).

The shear time is the inverse of the Oort *A* constant. This time is usually too short in the bright optical parts of spiral galaxies to give large density enhancements from gravitational instabilities. The result instead is a continuous growth and decay of shearing spiral wavelets, as discussed by Goldreich and Lynden-Bell (1965) and Toomre (1981). The shear time can be much longer in other regions of galaxies, such as the inner parts where the rotation curve may become solid body, and in small, irregular galaxies, where the rotation curve is solid body throughout. In these regions of low shear, the gravitational instability should grow without impediment to produce large and dense cloud complexes. This may explain the origin of giant star-forming regions in the central parts of spiral galaxies and in irregular galaxies. The only requirement for such intense star formation is that there is enough gas to make the Jeans length smaller than the size of the region.

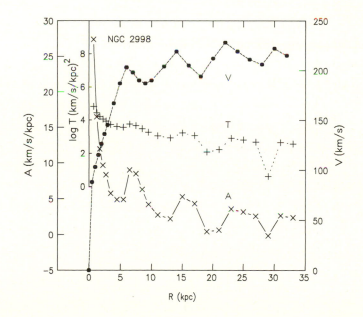

Figure 1 - The rotation velocity, V, Oort shear parameter, A, and tidal acceleration per unit length, T, are plotted as a function of radius for the galaxy NGC 2998 (from Elmegreen 1987c).

A third region where shear can be very low is inside density wave spiral arms, where the flow pattern makes the circular velocities somewhat solid body (E87b). This can be seen from the rotation curves of edge-on spiral galaxies. The circular velocity increases inside the arms, and then decrease between the arms, giving a flat rotation curve on average. This velocity pattern is illustrated by NGC 2997 (Rubin, Ford and Thonnard 1980), for which the radial dependence of the rotation velocity, V, Oort constant, A, and tidal acceleration per unit length, T, are plotted in Figure 1. The strong decreases in A and T in the spiral arms imply that the rotation curve is temporarily solid body there.

It is possible that what limits the growth of gravitational instabilities in spiral arms is the arm flow-through time and not the shear time, because shear is generally low throughout the arm. This suggests that strong density waves, which have long flow-through times, can trigger denser cloud complexes than weak density waves (§3.1).

High density complexes are also formed if the gas is very dissipative. Then the incident kinetic energy from shear will be removed by dissipation. Because the initial angular momentum from shear is also removed by magnetic fields (E87b), shear will have essentially no effect when the gas is highly dissipative. Giant cloud complexes may be able to form anywhere when dissipation is strong, including the disks of galaxies without density waves. If a density wave is present, then these clouds will form in the spiral arms because of the lower tidal disruptive force and the higher density and magnetic field strengths there. If a density wave is not present, they may form throughout the disk.

The amount of dissipation that occurs in a spiral arm flow-through time may determine the density enhancement of a perturbation that grows by a gravitational instability in the arm. If the gas dissipates significantly before it flows out of the arm, then dense molecular cores and intense star formation could result. If the gas cannot dissipate in the flow time, then the instability only collects pre-existing clouds together; there may be only a small density enhancement and little excess star formation. In both cases, giant beads of star formation would appear along the spiral arms. The primary difference between the two cases would be that the highly dissipative case gives a greater star formation rate per unit gas mass than the weakly dissipative case. Dissipative and non-dissipative instabilities may have some implications for triggered star formation in strong and weak density-wave arms, respectively, as discussed in §3.1.

2. STAR FORMATION RATE AS A FUNCTION OF SPIRAL ARM MORPHOLOGY

To within a factor of 2, galaxies have approximately the same star formation rates per unit area whether or not they contain a prominent spiral density wave (Elmegreen and Elmegreen 1986; McCall and Schmidt 1986; see review in Elmegreen 1987a). They also have approximately the same average CO emission per unit area, and the same CO integrated linewidths from the disks (Stark, Elmegreen and Chance 1987). Irregular galaxies with no global pattern from a density wave have prominent star formation too (Hunter and Gallagher 1986). These observations imply that density waves are not necessary for molecular cloud formation or star formation. The waves usually do not trigger a noticeable excess of star formation compared to what the galaxy would have produced without the wave.

Strong density waves apparently produce more of an effect on the star formation rate or CO abundance than weak density waves, although the average rate for galaxies with strong or weak waves is the same as for galaxies without waves. M51, for example, has an unusually strong near-infrared spiral arm strength, reaching a 10 to 1 arm/interarm contrast in the outer regions (Elmegreen and Elmegreen 1984). Its CO surface brightness is twice as large as the average CO surface brightness for other grand-design spiral galaxies of that Hubble type (Stark, Elmegreen and Chance 1987), and the local star formation rate is modulated at the 50% level by the spiral arms (Lord, Strom and Young 1987). M81 and M33 have weak near-infrared grand design spirals, with arm/interarm contrasts of at most ~3; their CO surface brightnesses are relatively low. Such weak-arm galaxies are expected to show less spiral arm modulation of the star formation rate.

The primary influence of a density wave on the CO abundance and star formation rate in a galaxy appears to be one of organization: the wave places most of the gas and star-forming regions in the spiral arms because of the flow pattern, and it enhances the formation of giant clouds and star-forming complexes by collecting together pre-existing clouds. A secondary effect of density waves, which has been measured only for the strongest waves so far, is the formation of new molecular clouds and a net excess of star formation. Perhaps more sensitive observations will reveal this triggering at a lower level in weak-arm galaxies.

It follows that the primary mechanism of star formation in normal spiral galaxies is small-scale molecular cloud formation driven by local processes, as in the swept-up shells around OB associations (e.g., McCray and Kafatos 1987). Density waves then organize the resulting clouds and star formation sites into a global pattern, and mild gravitational instabilities and spiral-arm cloud coagulation collect some of the gas into giant complexes (Tomisaka 1987; Roberts and Steward 1987). A possible 50% modulation on this mechanism comes from density-wave triggering of new molecular clouds.

3. LARGE-SCALE TRIGGERING MECHANISMS

3.1. The Importance of Dissipation

Star formation begins only after interstellar gas dissipates a large fraction of its kinetic and magnetic energy. The time available for this dissipation may be the shear time, or the spiral arm flow time, or any gas compression time scale. If the available time is less than the dissipation time, then high density clouds cannot form, and the compressed region will return to its original low density when the compression ends.

The dissipation rate for cloud-cloud collisions in the interstellar medium is approximately half the collision rate, $0.5n_c\sigma_c c$, for cloud density n_c, collision cross section σ_c, and velocity dispersion c. Because $n_c = \rho/M_c$ for average density ρ and cloud mass $M_c = 4\pi\rho_c R_c^3/3$, with cloud density ρ_c and radius R_c, and because $\sigma_c = \pi R_c^2$ and $\rho = f\rho_c$ for cloud filling factor f, it follows that

$$0.5n_c\sigma_c c \simeq 0.5fc/R_c. \tag{1}$$

The shear rate is

$$A = -0.5r\frac{d\Omega}{dr} \simeq 0.5\Omega, \tag{2}$$

where this latter result is for a flat rotation curve. Evidently the dissipation of kinetic energy in a density perturbation occurs before shear destroys the perturbation if

$$\frac{fc}{R_c} > \Omega. \tag{3}$$

In the solar neighborhood, $c\sim 5$ km s^{-1}, $R_c\sim 5$pc and $f\sim 0.04$, so regions with $\Omega < 0.04$ km s^{-1} kpc^{-1} allow rapid dissipation. Locally, $\Omega\sim 0.025$ km s^{-1} kpc^{-1}, so this condition is barely satisfied. This implies that the *average* interstellar medium can form only loosely bound cloud complexes, if it can form them at all in the presence of shear.

The rate of shear is smaller than average in the center of a density wave spiral arm. If the arm/interarm contrast ratio is $\mathscr{A}$, then the shear rate in the arm is (E87b)

$$A = \frac{\Omega}{\mathscr{A} + 1}. \tag{4}$$

The arm flow-through rate is

$$\omega_{flow} = \frac{2(1 + \mathscr{A})(\Omega - \Omega_P)}{\mathscr{A}} \tag{5}$$

for pattern speed Ω_P. This rate is taken equal to 2π divided by the arm-to-arm flow time, which is $\pi/(\Omega - \Omega_P)$ for a two arm spiral, and divided by the fraction of the time spent in the arm, which is $\mathscr{A}/(1 + \mathscr{A})$. The ratio of the shear rate to the flow-through rate is therefore

$$\frac{A}{\omega_{flow}} = \frac{\mathscr{A}\Omega}{2(\mathscr{A} + 1)^2(\Omega - \Omega_p)} \simeq \frac{\mathscr{A}}{(\mathscr{A} + 1)^2}, \tag{6}$$

where this latter equality is because $\Omega - \Omega_p \sim \Omega/2$ halfway out to the optical edge of a galaxy with a flat rotation curve. For large $\mathcal{A}$, the flow rate becomes much larger than the shear rate, and the flow time becomes the limiting factor in determining the amplitude of the perturbation growth.

Now, suppose f in a spiral arm scales with density, and so with $\mathcal{A}$ as $f \sim 2\mathcal{A}f_o/(1 + \mathcal{A})$ for azimuthally averaged filling factor f_o. Then the dissipation rate is larger than the arm flow-through rate if

$$\frac{f_o c}{R_c} > \frac{0.5\Omega(1 + \mathcal{A})^2}{\mathcal{A}^2}. \tag{7}$$

This inequality is more easily satisfied for large $\mathcal{A}$ than small $\mathcal{A}$, but only by a factor of 4 at most. Low amplitude waves ($\mathcal{A}$ small) allow relatively little dissipation, so only mild and transient instabilities should result. High amplitude waves provoke a more substantial collapse of the gas, and possibly trigger an excess of star formation in the new molecular cores that form, but the overall effect is not very large. This may be why strong density waves, as in M51, seem to be associated with a slight (factor-of-two) excess in the global CO abundance, and a similar excess in the arm/interarm contrast of the star formation rate.

3.2. Dust–Lane Trapping of Incident Molecular Clouds

Another difference between strong and weak-arm galaxies is that strong-arm galaxies can have such dense shocks (dust lanes) that incident molecular clouds cannot penetrate them. Weak density waves should have lower-density shocks, and molecular clouds should be able to go though them in a ballistic fashion. The quantitative difference between strong and weak shocks in this sense can be determined as follows (from Elmegreen 1988).

The critical front compression factor for cloud trapping is on the order of 10. This follows from the thickness of the front L (measured parallel to the plane), the perpendicular thickness of the interstellar medium, H, the average column density through the galactic disk, σ_{ISM}, and the average column density through a giant molecular cloud, σ_{GMC}. If $\mathcal{C}$ is the compression factor of the interstellar medium in a front, then the column density through the front measured parallel to the plane is $\mathcal{C}\sigma_{ISM}L/H$. This column density exceeds σ_{GMC} if $\mathcal{C} > (\sigma_{GMC}/\sigma_{ISM})(H/L) \sim \sigma_{GMC}/\sigma_{ISM}$, because $L \sim H$ for a dust lane. The column densities through giant molecular clouds in the Galaxy are all approximately $170 M_\odot$ pc^{-2} (Solomon *et al.* 1987), so if $\sigma_{ISM} \sim 20 M_\odot$ pc^{-2} for a typical galaxy, then $\mathcal{C} > 8.5$ gives a dust lane with a transverse column density greater than the column density of a molecular cloud. Dense fronts ($\mathcal{C}$ large) should therefore trap incident giant molecular clouds, and weak fronts should let them pass through.

It follows that the first transition region that occurs in a cloudy gas that enters a *weak* density wave should be a collision front for diffuse clouds. The dust lane that forms from these diffuse clouds should be on the inner edge of the spiral arm, where Roberts (1969) found the shock front in a continuous fluid. The molecular clouds flow through this dust lane and collect downstream, in a ridge near the center of the arm, as in the N-body simulations of ballistic cloud motions by Combes and Gerin (1985), Roberts and Steward (1987), and others. In the case of a weak wave, the molecular ridge should be somewhat loose and irregular, forming a broad spiral arm of molecular clouds and associated HII regions. This result gives the correlation between arm width and shock strength discussed by Roberts, Roberts and Shu (1975).

Individual molecular clouds should go right through a weak diffuse cloud front without noticeably disrupting it. For a typical spiral density wave shock velocity $v \sim 20$ km s^{-1}, the time interval between molecular cloud impacts at each location in a dust lane is $\lambda_m/20$ km s^{-1} = 10^8 years for molecular cloud mean free path $\lambda_m \sim 2000$ pc. The hole that is created in a dust lane by a typical molecular cloud ($R < 50$ pc) is much smaller than the height of the dust lane, so the disruption per collision should be small. Moreover, continued accretion by the dust lane should fill in the hole before the next molecular cloud hits it. The time for a dust lane to be built up by diffuse cloud collisions is the column number density of diffuse clouds in the dust lane, multiplied by λ_d/v for diffuse cloud mean free path λ_d. The column density of a dust lane corresponds to several magnitudes of extinction in a weak-arm galaxy, and the column density of a diffuse cloud corresponds to approximately $1/2$ magnitude of extinction, so the column number density of diffuse clouds in a weak dust lane is 3 to 10. The diffuse cloud mean free path is ~ 100 pc, so if the perpendicular component of the density wave speed is 20 km s^{-1}, then $\lambda_d/v \sim 5 \times 10^6$ years. It follows that the time for a diffuse cloud dust lane to be built up is 1.5 to 5 $\times 10^7$ years. This is shorter than the molecular cloud collision time, so the holes created in dust lanes by molecular cloud impacts should continuously close up by accumulation of diffuse clouds. The transverse motion of high pressure gas in the dust lane will also help close the holes.

Strong arm spiral galaxies should have only one cloud collision front and a very opaque dust lane, because the shock is so dense that both diffuse clouds and molecular clouds are forced to stop there.

This expected difference between molecular and atomic dust lanes is in agreement with the available observations. Grand-design spiral galaxies such as M81 and M33 have relatively weak near-infrared arms and mostly atomic dust lanes (Rots 1975; Newton 1980). Other galaxies with atomic dust lanes are M31 (Emerson 1974), and M106 (van Albada 1980), which also appear to have weak or faint arms. Strong-arm galaxies such as M51 and M83 (Talbot, Jensen and Dufour 1979) have molecular dust lanes (Lo *et al.* 1987; Allen, Atherton and Tilanus 1986).

Molecular dust lanes should also be distinguished on optical photographs of galaxies because they will show up in both the B and I bands as a result of their large opacity. Atomic dust lanes may be too low in opacity to show up on I band photographs, although they should look the same as molecular dust lanes on B band photographs. The molecular dust lane in M51, for example, is very prominent on an I band photograph (Elmegreen 1981).

REFERENCES

Allen, R.J., Goss, W.M., and van Woerden, H. 1973, *Astron.Astrophys.*, **29**, 447.

Allen, R.J., and Goss, W.M. 1979, *Astron.Astrophys.Suppl.*, **36** , 135.

Allen, R.J., Atherton, P.D., and Tilanus, R.P.J. 1986, *Nature*, **319** , 296.

Bajaja, E., and Shane, W.W. 1982, *Astron.Astrophys.Suppl.*, **49** , 745.

Balbus, S.A. 1988, *Astrophys.J.*, in press.

Balbus, S.A. and Cowie, L.L. 1985, *Astrophys.J.*, **297**, 61.

Combes, F. and Gerin, M. 1985, *Astron.Astrophys.*, **150**, 327.

Cowie, L.L. 1981, *Astrophys.J.*, **245**, 66.

Elmegreen, B.G. 1979, *Astrophys.J.*, **231**, 372.

Elmegreen, B.G. 1987a, in *Star Forming Regions*, ed. M. Peimbert and J. Jugaku, (Dordrecht: Reidel), p. 457.

Elmegreen, B.G. 1987b, *Astrophys.J.*, **312**, 626 (E87b).

Elmegreen, B.G. 1987c, in *Physical Processes in Interstellar Clouds,* ed. G.E. Morfill and M. Scholer (Dordrecht: Reidel), p. 1.

Elmegreen, B.G. 1988, *Astrophys.J.*, submitted.

Elmegreen, B.G., and Elmegreen, D.M. 1983, *Monthly Not.Roy.Astron.Soc.*, **203** , 31.

Elmegreen, B.G., and Elmegreen, D.M. 1986, *Astrophys.J.*, **311**, 554.

Elmegreen, B.G., and Elmegreen, D.M. 1987, *Astrophys.J.*, **320**, in press.

Elmegreen, D.M. 1981, *Astrophys.J.Suppl.*, **47**, 229.

Elmegreen, D.M., and Elmegreen, B.G. 1984, *Astrophys.J.Suppl.*, **54**, 127.

Emerson, D.T. 1974, *Monthly Not.Roy.Astron.Soc.*, **169**, 607.

Goldreich, P., and Lynden-Bell, D. 1965, *Monthly Not.Roy.Astron.Soc.*, **130** , 97.

Henderson, A.P., Jackson, P.D., and Kerr, F.J. 1982, *Astrophys.J.*, **263** , 116.

Hunter, D.A., and Gallagher, J.S.,III. 1986, *Pub.Astron.Soc.Pac.*, **98**, 5.

Jog, C., and Solomon, P.M. 1984, *Astrophys.J.*, **276**, 114.

Kennicutt, R.C., and Hodge, P.W. 1980, *Astrophys.J.*, **241**, 573.

Lo, K.Y., Berge, G.L., Claussen, M.J., Heiligman, G.M., Leighton, R.B., Masson, C.R., Moffet, A.T., Phillips, T.G., Sargent, A.I., Scott, S.L., Wannier, P.G., and Woody, D.P. 1984, *Astrophys.J.(Letters)*, **282** , L59.

Lo, K.Y., Ball, R., Masson, C.R., Phillips, T.G., Scott, S., and Woody, D.P. 1987, *Astrophys.J.(Letters),* **317** , L63.

Lord, S.D., Strom, S.E., and Young, J.S. 1987, in *Star Formation in Galaxies*, ed.C.L. Persson, (Pasadena, NASA), p. 303.

McCall, M.L., and Schmidt, F.H. 1986, *Astrophys.J.*, **311** , 548.

McCray, R., and Kafatos, M. 1987, *Astrophys.J.*, **317**, 190.

Newton, K. 1980, *Monthly Not.Roy.Astron.Soc.*, **190**, 689.

Roberts, W.W. 1969, *Astrophys.J.*, **158**, 123.

Roberts, W.W., Jr., Roberts, M.S., and Shu, F.H. 1975, *Astrophys.J.*, **196** , 381.

Roberts, W.W., and Steward, G.R. 1987, *Astrophys.J.*, **314** , 10.

Rots, A.H. 1975, *Astron.Astrophys.*, **45**, 43.

Rubin, V.C., Ford, W.K., Jr., and Thonnard, N. 1980, *Astrophys.J.*, **238** , 471.

Rumstay, K.S., and Kaufman, M. 1983, *Astrophys.J.*, **274**, 611.

Seiden, P.E., and Gerola, H. 1982, *Fundamentals of Cosmic Physics*, **7**, 241.

Solomon, P.M., Rivolo, A.R., Barrett, J. and Yahil, A. 1987, preprint.

Stark, A.A., Elmegreen, B.G., and Chance, D. 1987, *Astrophys.J.*, **322** , in press.

Talbot, R.J., Jensen, E.B., and Dufour, R.J. 1979, *Astrophys.J.*, **229** , 91.

Tomisaka, K. 1987, *Pub.Astron.Soc.Japan*, **39**, 109.

Toomre, A. 1981, in *The Structure and Evolution of Normal Galaxies,* ed. S.M. Fall and D. Lynden-Bell, (Cambridge: University of Cambridge), p. 111.

Unwin, S.C. 1980a, *Monthly Not.Roy.Astron.Soc.*, **190**, 551.

Unwin, S.C. 1980b, *Monthly Not.Roy.Astron.Soc.*, **192**, 243.

van Albada, G.D. 1980, *Astron.Astrophys.*, **90**, 123.

Viallefond, F., Allen, R.J., and Goss, W.M. 1981, *Astron.Astrophys.*, **104** , 127.

Viallefond, F., Goss, W.M., and Allen, R.J. 1982, *Astron.Astrophys.*, **115** , 373.

Wright, M.C.H., Warner, P.J., and Baldwin, J.E. 1972, *Monthly Not.Roy.Astron.Soc.*, **155** , 337.

THE ASSOCIATION BETWEEN STELLAR BARS AND ENHANCED
ACTIVITY IN THE CENTRAL KILOPARSEC OF SPIRAL GALAXIES

N. A. Devereux

Institute for Astronomy, University of Hawaii
Honolulu, Hawaii 96822 USA

ABSTRACT

A total of 156 objects have now been observed in the course of a survey of 10 μm emission from the central 6″ of spiral galaxies. The larger data base substantiates the differences between barred and unbarred spirals first discussed at the 1986 IRAS conference. New 2.2 μm observations have permitted a determination of the central 2.2 μm luminosity. Preliminary results are discussed in the context of the different bar-central 10 μm luminosity association for early- and late-type spirals.

1. INTRODUCTION

The survey of 10 μm emission from the central 6″ of spiral galaxies is essentially complete. The sample and results are summarised in Devereux (1987a). The larger data base substantiates the 1986 preliminary result that the histograms of central 10 μm luminosity for early-type (Sb and earlier) barred and unbarred spirals are significantly different. The sense of the difference is that the central 10 μm luminosity of 40% of the barred spirals exceeds the maximum $\sim 10^9 L_\odot$ observed for unbarred types. The combination of ground-based and IRAS data has further revealed that the excess 10 μm luminosity is confined to the central ~ 1 kpc diameter region. The larger data base has also strengthened the correlation between compactness at 10 μm and the IRAS $S_{25\mu m}/S_{12\mu m}$ ratio first discussed at the 1986 IRAS conference. The correlation is illustrated in Figure 1.

2. DISCUSSION

2.1 Origin of the High Central 10 μm Luminosity

A 25 μm color excess, defined here as $S_{25\mu m}/S_{12\mu m} \geq 2.5$, is associated with both Seyfert and starburst nuclei (Lawrence et al. 1985). Indeed, many of the early-type spirals exhibiting a 25 μm color excess have been classified as such on the basis of optical spectra (see Figure 1). Therefore, the correlation illustrated in Figure 1 suggests a distinction between the IRAS $S_{25\mu m}/S_{12\mu m}$ color of 'nuclei' and 'disks'. As discussed in detail in Devereux (1987a), the 100,60 μm and 60,25 μm spectral indices, far-infrared luminosity, and space density of early-type spirals, segregated on the basis of either high compactness at 10 μm or a 25 μm color excess, are different from all other early-type spirals and yet similar to known Seyfert and starburst galaxies. Therefore, on the basis of the observational data examined to date, the origin of the high central 10 μm luminosity in

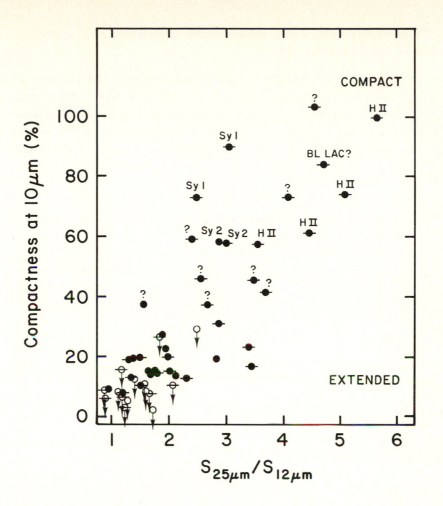

Figure 1. The correlation between the compactness at 10 μm and the IRAS $S_{25\mu m}/S_{12\mu m}$ ratio for early-type spirals. Galaxies regarded as active because they exhibit bright optical emission lines are indicated.

early-type barred spirals is consistent with either Seyfert activity or star formation or a mixture of both.

2.2 The Additional Parameter

Only 40% of the early-type barred spirals in this sample, constituting $\sim$10% of all known early-type barred spirals, exhibit excess central 10 μm emission. Therefore, the bar is necessary, but not sufficient, for high central 10 μm luminosity. Another, as yet undetermined, parameter, in addition to a stellar bar, is necessary for the bar–high central 10 μm luminosity phenomenon. Identifying this 'additional' parameter is clearly of great importance in establishing the specific role of the bar in enhancing activity in the central 1 kpc region of early-type spirals.

2.3 The Different Bar-Central 10 μm Luminosity Association
for Early- and Late-Type Spirals

The difference in the histograms of central 10 μm luminosity between barred and unbarred spirals is most evident for Sb and earlier types (Devereux 1987a). The apparently differing role of the bar in early- and late-type spirals may reflect an anticipated difference in the central mass density due to the contribution from a bulge component in early-type spirals. Recent 2.2 μm observations (Devereux 1987b) in fact support this suggestion. The 2.2 μm luminosity of the central (500 pc diameter) region of early-type spirals is typically larger than in late-type spirals as illustrated in Figure 2.

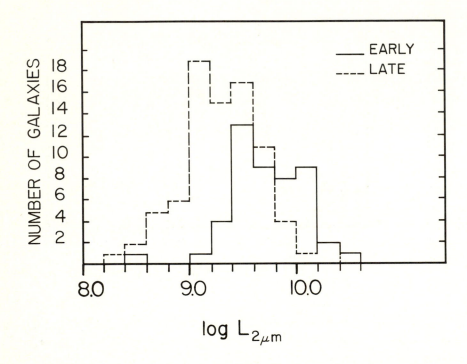

Figure 2. The histograms of central (500 pc diameter) 2.2 μm luminosity for early-type (solid line) and late-type (dashed line) spirals.

Adopting a mass-to-light ratio, $M/L \sim 0.7$ at 2.2 μm, which is appropriate for the central region of spiral galaxies (Devereux et al. 1986), one finds that the 2.2 μm luminosity corresponds to a stellar density averaged over the central 500 pc diameter region of typically $\geq 30 M_\odot/\text{pc}^3$ for early-type spirals. Such high-mass densities are inferred for only a small percentage, ~30%, of late-type spirals. The full consequence of the central mass density on the physical role of the bar, however, is not yet understood. The central mass density may not be the only parameter to be considered with regard to the different bar-central 10 μm luminosity association observed for early- and late-type spirals. Elmegreen and Elmegreen (1985) have noted differences in the structure of bars between early- and late-type spirals that may also be relevant.

3. CONCLUSIONS

The association between stellar bars and enhanced central 10 μm luminosity is now a firmly established result for early-type (Sb and earlier) spirals. Our understanding of the specific role of the bar, although unclear at present, may benefit greatly from both theoretical and observational studies. In particular, we need detailed theoretical models of the response of gas to bar forcing that incorporates realistic structural parameters inferred from the most recent observations of Elmegreen and Elmegreen (1985) and Devereux (1987b). Further, new observations are needed to identify the 'additional parameter' that distinguishes the 'active' from the 'inactive' early-type barred spirals.

4. ACKNOWLEDGMENTS

I am most grateful for the enthusiastic support of this work provided by Gareth Wynn-Williams, Eric Becklin, and Steve Eales. I would also like to thank the staff at IPAC, in particular Carol Lonsdale, Walter Rice, and George Helou, without whose assistance this work would not have been possible. This study was supported in part by NSF grant AST 84-181197 and in part under the IRAS extended mission program by JPL contract 957695.

References

Devereux, N. A., Becklin, E. E., and Scoville, N. Z. 1986, *Astrophys. J.* **312**, 529.

Devereux, N. A., 1987a, *Astrophys. J.*, in press.

Devereux, N. A., 1987b, in preparation.

Elmegreen, B. G., and Elmegreen, D. M., 1985, *Astrophys. J.* **288**, 438

Lawrence, A., Ward, M., Elvis, M., Fabbiano, G., Willner, S. P., Carleton, N. P., and Longmore, A., 1985, *Astrophys. J.* **291**, 117.

GLOBAL PROPERTIES OF STAR FORMATION IN SPIRAL GALAXIES

T.N. Rengarajan and K.V.K. Iyengar

Tata Institute of Fundamental Research
Homi Bhabha Road, Bombay 400005, India

ABSTRACT

Star formation in spiral galaxies has been studied by making use of
21 cm observations, observations of H band (1.6 μm),   blue band
magnitudes and far-infrared data from IRAS. It is found that the
luminosities in the various bands are well correlated with the
dynamic mass of the galaxies. From these, it is inferred that the
star formation rate (SFR) and the IMF are about the same averaged over
$\sim 10^7$ and $\sim 10^9$ years, whereas SFR averaged over the life time of the
galaxy is higher. Also more massive galaxies have had more star
formation in the past. There is inconclusive evidence for a
correlation between far-IR luminosity and M_{HI} the mass of neutral
hydrogen.

1. INTRODUCTION

Global properties of star formation in spiral galaxies can be inferred
by studying the correlations amongst the various properties of
galaxies like their masses, gas contents and luminosities in different
bands. In the present study, we make use of information from
catalogues of 21 cm observations, H magnitude observations and IRAS
observations. The dynamic mass M_G of the galaxy can be inferred from
W_{20} , the 20% width of the 21 cm line and is used as an important
parameter. The luminosities in the far-IR, blue and H bands are used
as indicators of SFR over $\sim 10^7$ yrs, $\sim 10^9$ yrs and over the life time
of the galaxy respectively.

2. DATA
For 21 cm observations we made use of the catalogues of Fisher and
Tully (1981; FT) and Lewis, Helou and Salpeter (1985; LHS) and for H
magnitudes the catalogue of Aaronson et al. (1982; AHM). For all
galaxies that have W_{20} measurements and have optical diameter < 20',
we searched for counterparts with definite flux densities in both 60
and 100 μm bands, in the IRAS Catalog (Lonsdale et al. 1985).

Distances were normalised to H_O = 100 km s^{-1} Mpc^{-1}. Relative distances given in AHM were converted to absolute distances by taking the distance to Virgo Cluster as 10.7 Mpc. Values of M_G, M_{HI}, FIR, the 40-120 μm flux in W m^{-2}, H and B magnitudes were taken from the respective Catalogues.

3. ANALYSIS

The results of the study of correlations amongst the different quantities are shown in Table 1 which also lists the parameters of least squares fits. In Figure 1 we show (a) B_o(abs), the absolute B magnitude, (b) H_o(abs), the absolute H magnitude, (c) log $(L_{IR}/L_\odot)$ and (d) log $(M_{HI}/M_\odot)$ plotted against log $(M_G/M_\odot)$. In general, there are good correlations between M_G and luminosities in various bands, the relationship being of the form $L \propto M_G^\alpha$. For L_B and L_H the correlations are the well known optical and infrared Fisher-Tully relations between luminosities and W_{20}, but with M_G as a variable. For the far-IR band also, there is a good correlation between L_{IR} and W_{20}. The correlations between L_{IR} and M_G are very similar for the data of different Catalogues as well as for sub-groups of small diameter (<5') galaxies. Further, it is found that if we use the mass of warm dust grains instead of L_{IR} given by $M_d \propto L_{IR}/\int B(\nu,T_c)\epsilon(\nu)\,d\nu$ where T_c is the dust temperature and $\epsilon(\nu) \propto \nu$, the correlation with M_G is tighter. It is known that there is a good correlation between L_{IR} and M_{H2}, the mass of molecular hydrogen (Rengarajan and Verma, 1986). However, as seen from Table 1, the evidence for correlation between L_{IR} and M_{HI}, is inconclusive.

4. DISCUSSION AND CONCLUSIONS

The luminosity in a given band results from contributions from stars of different masses and depends on the IMF of stars and fraction of bolometric luminosity radiated in the band. In the far-IR, the effective contribution is from stars of mass > 20 $M_\odot$ while for the blue band it is from 2-5 $M_\odot$ stars. For H band, the contribution is dominantly from old giants (Aaronson, Huchra and Mould, 1979). The similar dependence of L_{IR} and L_B on M_G implies that the SFR and IMF averaged over ~10^7 and ~10^9 years are the same. The higher power dependence of L_H on M_G implies increased SFR in the past as well as increased SFR for more massive galaxies.

We show in Table 2, the logarithmic mean values of M_G, L_B/M_G, L_H/M_G, L_{IR}/M_G and M_{HI}/M_G for different morphological types. Since

Table 1

Fits for Log Y = A + α Log X

X	Y	Sample	α	r^2	$(\Delta \text{ Log Y})_{rms}$
M_G	L_B	FT	0.95 ± 0.03	0.92	0.26
M_G	L_H	AHM	1.34 ± 0.04	0.95	0.20
W_{20}	L_{IR}	AHM	3.27 ± 0.19	0.65	0.36
M_G	L_{IR}	AHM	0.98 ± 0.05	0.78	0.35
M_G	L_{IR}	LHS	0.76 ± 0.07	0.68	0.45
M_G	L_{IR}	FT	0.92 ± 0.04	0.75	0.40
M_G	M_d	FT	1.12 ± 0.04	0.84	0.35
H[a] (Mag)	FIR (W m^{-2})	AHM	-0.29 ± 0.02	-0.79	0.28
M_G	M_{HI}	FT	0.78 ± 0.03	0.84	0.30
M_{HI}	L_H	AHM	1.13 ± 0.07	0.72	0.44
M_{HI}	L_{IR}	FT	0.89 ± 0.06	0.65	0.46
FIR (W m^{-2})	HI (Jy kms^{-1})	FT	0.54 ± 0.06	0.43	0.44

[a] Magnitude used in place of Log X

Table 2

Mean Values vs Morphology (Units Arbitrary)

	Types 1,2,3		Types 4,5,6		Types 7,8,9,10	
	MEAN	RMS	MEAN	RMS	MEAN	RMS
Log $(M_G/M_\odot)$	10.98	0.47	10.71	0.42	10.01	0.59
Log (L_B/M_G)	-3.22	0.23	-3.14	0.25	-3.04	0.44
Log (L_H/M_G)	-4.1	0.15	-4.28	0.21	-4.54	0.20
Log (L_{IR}/M_G)	-1.56	0.35	-1.47	0.32	-1.57	0.31
Log (M_{HI}/M_G)	-0.88	0.27	-0.63	0.25	-0.45	0.20

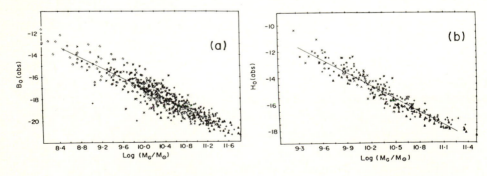

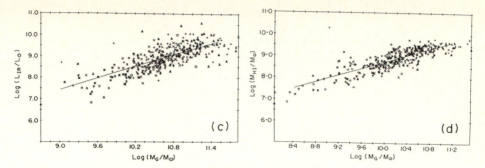

Figure 1. The plots of (a) absolute blue magnitude, (b) absolute H magnitude, (c) the far-IR luminosity, (d) neutral hydrogen mass against the dynamic mass of spiral galaxies. Symbol for morphological types: ⊙ 0,1; ▲ 2,3; + 4,5; ✕ 6,7; * 8,9; ◆ 10,11; ⚠ ambiguous subclasses.

$\alpha_B = \alpha_{IR} \cong 1$, the luminosity per unit mass is almost a constant for the B and far-IR bands, while with $\alpha_H > 1$, it decreases as morphological type increases (mean mass decreases). Similarly since $\alpha_{HI} < 1$, the fractional mass in gas increases. The increase of L_H/M_H and decrease of M_{HI}/M_G are reflections of larger star formation using up more of the available gas. Most of the correlations observed seem to be primarily dependent on M_G. Although mass is a very important parameter governing star formation, the factors responsible for morphological appearance are yet to be understood.

References

Aaronson, M., Huchra, J., Mould, J., 1979. Astrophys. J., 229, 1.

Aaronson, M., Huchra, J., Mould, J.R., Tully, R.B., Fisher, J.R., Van Woerden, H., Goss, W.M., Chamaraux, P., Mebold, U., Siegman, B., Berrimann, G., Persson, S.E., 1982. Astrophys. J. Suppl. Ser., 50, 241 (AHM).

Fisher, J.R., Tully, R.B. 1981. Astrophys. J. Suppl. Ser., 47, 139 (FT)

Lewis, B.M., Helou, G., Salpeter, E.E., 1985. Astrophys. J. Suppl. Seri., 59, 161.

Lonsdale, C.J., Helou, G., Good, J.C., Rice, W., 1985. Catalogued Galaxies and Quasars observed in the IRAS Survey, JPL, D-1932.

Rengarajan, T.N., Verma, R.P., 1986. Astron. Astrophys., 165, 300.

Rickard, L.J., Harvey, P.M., 1984. Astron. J., 89, 1520.

RADIO AND OPTICAL STUDIES OF A COMPLETE SAMPLE OF IRAS GALAXIES

R.D. Wolstencroft*, S.W. Unger[+], A. Pedlar[†], J.N. Heasley[+],

Q.A. Parker*, J.W. Menzies[X], A. Savage,*, H.T. MacGillivray*,

S.K. Leggett*, W. Gang* and R.G. Clowes*

* Royal Observatory, Edinburgh, Scotland

+ Royal Greenwich Observatory, Herstomonceux Castle, East Sussex, England

† Nuffield Radio Astronomy Laboratories, Jodrell Bank, England

+ Institute for Astronomy, University of Hawaii, USA

x South African Astronomical Observatory, Capetown, South Africa

ABSTRACT

Radio maps, spectra and CCD images have been obtained for almost all the 158 objects in a complete sample of IRAS galaxies. The linear relation between radio and far–infrared luminosity is valid over the complete luminosity range (up to $L_{IR} = 1.8 \times 10^{12} L_\odot$, $H_0 = 75$ km s^{-1} Mpc^{-1}). The majority of the 10 most luminous galaxies show evidence of tidal disruption and have companion galaxies at projected separations which range between 6 and 138 kpc. The second most luminous galaxy, IRAS 00275–2859, is a quasar with a spectrum reminiscent of a broad line absorption quasar and with a band of absorption running across the object: its infrared to radio luminosity ratio is identical to that of the typical IRAS galaxy.

1. INTRODUCTION

We have recently completed a program of optical identification of all sources in the IRAS Point Source Catalogue in the south galactic polar cap ($b \gtrsim 60°$) down to a limiting magnitude of B =21. The number of sources of various types in this large SGP sample, and in a smaller subset of this sample that was identified in an earlier study (Wolstencroft et al., 1986), are as follows:

SGP Sample	Area (deg^2)	All sources	Stars	Galaxies	Empty field objects
Small	304	312	148	154	10
Large	2600	2800	1480	1255	65

Follow–up studies are nearing completion for the small sample and are just beginning for the large sample. The results for the small sample will be described in this paper. Of the 10 empty field objects 4 have been shown to be galaxies just outside the 95% confidence error ellipse (Wolstencroft et al., 1987) and thus the total number of galaxies in the small sample is 158.

The following observations have been obtained: spectra with typical resolutions of 5Å to provide redshifts and emission line strengths; CCD images in broad bands (B,R) of the highest luminosity galaxies; and 20 cm radio continuum maps.

2. THE INFRARED LUMINOSITY FUNCTION (SMALL SAMPLE)

We have determined redshifts for 85% of the galaxies so far, with median and maximum values of z = 0.03 and 0.33. The far-infrared luminosity has been calculated from flux densities at 60 and $100 \mu m$ using

$$\frac{L_{IR}}{L_\odot} = \frac{4\pi\ r^2\ (3.25\ F_{60} + 1.26\ F_{100})\ 10^{-14}\,wm^{-2}}{3.827 \times 10^{26} w}$$

where r, the distance to the galaxy, is derived assuming H_0 = 75 km s^{-1} Mpc and q_0 = 0. The histogram of L_{IR} peaks at 3×10^{10} $L_\odot$, with the most luminous galaxy being IRAS 00441$-$2221 (L_{IR} = 1.83×10^{12} $L_\odot$). A provisional infrared luminosity function has been deduced for our 85% complete sample. At high luminosities ($>3 \times 10^{10} L_\odot$) $\Phi(L_{IR}) \propto L_{IR}^{-2.4}$ in agreement with the slope found by Lawrence et al. (1986) for a comparable sample in the NGP. There is a change to a shallower slope at $L_{IR} \approx 10^{10} L_\odot$ but the value of this slope is uncertain because there are relatively few galaxies in our sample with L_{IR} $<3 \times 10^9 L_\odot$. The number of ultraluminous galaxies (L_{IR} $>10^{12} L_\odot$) is also low, 4 out of 130 or 3.1% of the sample, so that $\Phi(L_{IR})$ is uncertain at the highest luminosities. For the large SGP sample, which contains 1255 galaxies, we predict that it will contain 33 ultraluminous galaxies which should enable us to derive a reliable luminosity function down to $10^{13} L_\odot$. It is worth noting that the fraction of ultraluminous galaxies depends sensitively on H_0: for H_0 = 75 (50) km s^{-1} it is 3.1 (8.5)%.

3. RADIO PROPERTIES

All 154 galaxies and 10 empty fields were mapped at 20cm with the B/C array of the VLA during 1986. 131 (85%) of the galaxies were detected above 1mJy with 120 being classified as compact and 11 as extended based on visual inspection of the maps. 38 of the galaxies with optical diameters ≲20 arc sec were also mapped (at 20 cm) with the A/B array (~4 arc sec FWHM beam): 78% show the same flux in the two arrays suggesting that the radio flux may be centrally concentrated in many of these galaxies.

The radio luminosites, L_{RAD}, for our sample cover a wide range from 5×10^{20} to 2×10^{24} w Hz^{-1}. The correlation between L_{RAD} and L_{IR} holds extremely well over the entire range of luminosity and is consistent with $L_{RAD} \propto L_{IR}^{1.0\pm0.1}$. This correlation agrees with the result found by Hummel et al. (1987) for a complete sample of 65 Sbc galaxies of relatively low luminosity (20.3 < log L_{RAD} (w Hz^{-1}) <22.7) who obtain $L_{RAD} \propto L_{IR}^{1.1\pm0.1}$. For half of the sample

$\log(L_{IR}/L_{RAD})$ lies within 0.15 of the mean value, and this narrow dispersion is the same at all luminosities. The correlation between radio and far-infrared flux densities was first established by de Jong et al. (1985) and Helou et al. (1985) for somewhat heterogeneous samples: note however that the correlation between radio and far-infrared luminosities and flux-densities is not necessarily identical for a sample of objects with a wide range of distance. The fact that we find the correlation is still valid for galaxies with $L_{IR} > 10^{12} L_{\odot}$ is an important new result. It implies that the same mechanism is operating over the entire luminosity range and that galaxies with a strong nuclear non-thermal (and often variable) radio source are either absent or very minor constituents of this sample. One object in our sample, IRAS 00275-2859, is a quasar (see below) and has L_{IR}/L_{RAD} exactly equal to the mean value for the sample: it is presumably radio quiet. As many authors have discussed, the correlation may be tied to massive stars with short ($\sim 10^7$yr) lifetimes, which at early epochs heat grains in their vicinity and later provide a reservoir of relativistic electrons via supernova remnants; if this is so then the narrowness of the observed dispersion suggests that many molecular clouds are contributing to the total emission and are triggered into star formation activity at random phases to produce a steady state global emission. To achieve the highest luminosities it may be necessary for the triggering of molecular clouds to be less random and more bunched in phase, perhaps by galaxy-galaxy interactions, increasing the chance that a galaxy may be observed with radio and infrared luminosities different from their time averaged steady state values, and thus increasing the dispersion in the L_{IR}/L_{RAD} ratio. It may be that the present sample is not large enough to show this effect, but VLA observations of the ~ 33 ultraluminous galaxies in the large sample might do so. A full description and interpretation of these results will be presented elsewhere (Unger et al., 1987).

5. THE HIGH LUMINOSITY IRAS GALAXIES

The properties of the 10 most luminous galaxies in our sample are listed below:

IRAS Name	$L_{IR}/L_{\odot}$	z	B	L_{IR}/L_B	F_{60}/F_{100}	Double?
00441-2221	1.83×10^{12}	0.314	17.3	11	0.41	O
00275-2859	1.45×10^{12}	0.279	18.3	23	0.85	–
23515-2917	$>1.18 \times 10^{12}$	0.334	19.3	>42	>0.47	O,R
00406-3127	1.07×10^{12}	0.246	20.0	119	0.72	R
01358-3300	8.20×10^{11}	0.199	20.3	190	0.74	O
01199-2307	8.12×10^{11}	0.156	18.0	38	1.12	O,R
00456-2904	7.41×10^{11}	0.110	18.1	78	0.76	O
23515-2421	4.93×10^{11}	0.154	17.7	17	0.86	O
00148-3153	4.87×10^{11}	0.105	18.7	93	0.45	O
00335-2732	4.56×10^{11}	0.073	17.2	46	1.44	O

Five of these galaxies are more luminous than Arp 220 which for $H_o = 75$ km s^{-1} Mpc^{-1}, $q_o = 0$ has $L_{IR} = 8.0 \times 10^{11}$ $L_{\odot}$. As indicated in the table, all except one of these objects has a close companion (separation $\leqslant 20$ arc sec) which is an optical (O) companion in 8 cases, both optical and radio (R) in 2 cases (which coincide in position) and radio only in 1 case. The object which is not obviously a double is IRAS 00275–2859, a quasar. The majority of the optical doubles show evidence of tidal disruption in CCD images indicating they are physically close with projected separations in the range 6 to 138 Kpc (median value = 42 Kpc). Source counts above 1 mJy suggest that the radio source 17 arc sec from 00406–3127 is unlikely to be a chance positional coincidence. If a recent interaction is the source of the high luminosity of these galaxies, the earliest possible time of closest approach of these companion galaxies for a median projected separation of 42 Kpc occurred 2.8 x 10^8 yr ago. Although this is consistent with model calculations of the time between closest approach and maximum infrared luminosity of a few x 10^8yr (Byrd et al., 1986; 1987; Noguchi and Ishibashi, 1986), in the case of larger separations this argument is more difficult to defend. One possibility is that one of the companion galaxies is undergoing a merger which cannot be resolved at the distance of these galaxies.

IRAS 00275–2859 is the only one of the 10 most luminous objects which is not obviously double. As reported earlier this is a quasar: it was discovered independently by Vader and Simon (1987) and by Wolstencroft et al.(1987). CCD imaging in the R band during November 1986 with the University of Hawaii 2.2m telescope in conditions of sub arc sec seeing show that the source is 8 arc sec in diameter and elongated at a position angle of about 117˚, with a dark absorbing band running N–S, ie approximately along the minor axis; it is somewhat similar in appearance to Arp 220. Our spectrum with a resolution of about 3Å between 4750 and 6250Å shows that the Hβ emission line is very broad (3500 km s^{-1} FWHM) with an absorption trough 180Å wide on its blue side reminiscent of broad line absorption quasars. Narrow lines are also present but are not double at our spectral resolution. A full description of these results will be presented elsewhere (Wolstencroft et al., 1987).

REFERENCES

Byrd, G.G., Valtonen, M.J., Sundelius, B., and Valtaoja, L. 1986. *Astron. Astrophys.* **166**, 75.

Byrd, G.G., Sundelius, B., and Valtonen, M. 1987 *Astron. Astrophys.* **171**, 16.

Helou, G., Soifer, B.T. and Rowan–Robinson, M. 1985. *Astrophys. J.* **298**, L7.

Hummel, E., Davies, R.D., Wolstencroft, R.D., van der Hulst, J.M., and Pedlar, A. 1987. *Astron. Astrophys.* (submitted).

Jong, T. de, Klein, U., Wielebinski, R., Wunderlich, E. 1985. *Astron. Astrophys.* **147**, L6.

Lawrence, A., Walker, D., Rowan–Robinson, M., Leech, K.J., and Penston, M.V. 1986. *Mon. Not. R. astr. Soc.* **219**, 687.

Noguchi, M., and Ishibashi, S. 1986. *Mon. Not. R. astr. Soc.* **219**, 305.

Unger, S.W. et al. 1987. In preparation.

Vader, J.P., and Simon, M. 1987. *Nature* **327**, 304.

Wolstencroft, R.D., Savage, A., Clowes, R.G., MacGillivray, H.T., Leggett, S.K. and Kalafi, M., 1986 *Mon. Not. R. astr. Soc.* **223**, 279.

Wolstencroft, R.D. Unger, S.W., Pedlar, A., Heasley, J.N., Parker, Q.A., Menzies, J.W. Savage, A., MacGillivray, H.T., Leggett, S.K., and Clowes, R.G. 1987. Proceedings of the 22nd Rencontre de Moriond Workshop on Starbursts and Galaxy Evolution, Les Arcs, France, March 1987 (in press).

Wolstencroft, R.D. 1987. In preparation.

OPTICAL AND FAR INFRARED PROPERTIES OF A 60μm FLUX LIMITED SAMPLE OF IRAS GALAXIES

J. Patricia Vader
Department of Astronomy, Yale University
Box 6666, New Haven, CT 06511, U.S.A.

M. Simon
Astronomy Program, Department of Earth and Space Sciences
S.U.N.Y. at Stony Brook, Stony Brook, NY 11794-2100, U.S.A.

ABSTRACT

The optical luminosity function (OPLF) of a 60μm flux-limited sample of IRAS galaxies is determined and compared to that of an optically selected sample. Of the 92 IRAS sources considered, 7 have no or more than one possible optical within the position error ellipse. The most luminous object is a quasar with an interacting host galaxy system.

1. OPTICAL IDENTIFICATIONS AND LUMINOSITY FUNCTIONS

The IRAS source content of 7 fields with a total area of 152.8 sq. deg. and covered by IIIa-J plates from a set used by Kirshner et al. (1979 KOS) to determine the OPLF of optically selected galaxies has been examined. Selection of IRAS galaxies with high and medium quality flux densities ≥ 0.5 Jy at 60μm yields 92 objects, of which 50 are catalogued galaxies. For 7 of the unidentified objects the optical identification is dubious. The 4 IRAS sources whose position error ellipse contains no optical galaxy candidate on the KOS plates are shown in Fig. 1. The 3 IRAS sources with more than one optical candidate are shown in Fig. 2.

J magnitudes were measured from the KOS plates. Redshifts are missing for 12 of the optically faintest objects ($m_J \geq 17.5$), which have the largest infrared to optical luminosity ratios, so that our data are biased against the infrared most luminous objects ($L_{60} > 10^{12} L_\odot$). The infrared luminosity function (Fig. 3a) derived for the 80 galaxies with redshifts (dots) shows indeed a deficiency of IR luminous objects relative to other determinations (Lawrence et al. (1986), solid line; Soifer et al. (1986), open circles). The shape of the OPLF (Fig. 3b) should not be affected by the redshift incompleteness. The OPLF of IRAS galaxies (dots) as compared to that of field galaxies (KOS, line) indicates that in the range $-18 > M_J \geq -22$ IRAS galaxies represent about 15% of field galaxies. The turnover at $M_J > -19$, which becomes more pronounced when only objects with $L_{60}/L_J \geq 1$ (53 objects, open circles) are considered, suggests a deficiency of optically low-luminosity objects among IRAS galaxies. A full account of this work is given by Vader and Simon (1987a).

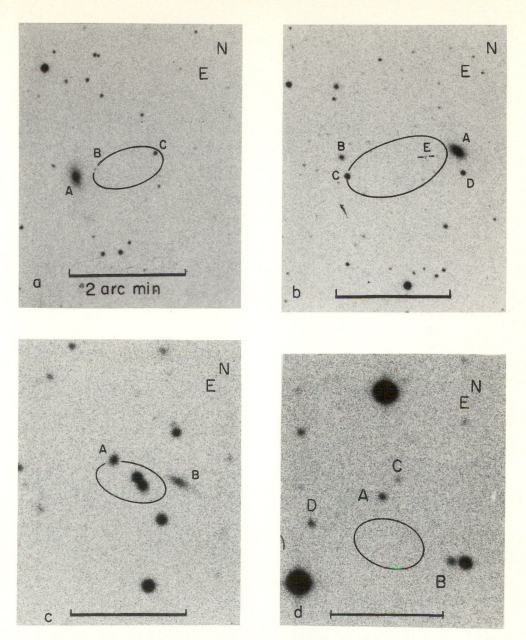

Fig. 1: IRAS sources with no fuzzy optical candidates inside the total (IRAS + optical) position error ellipse. Object A is the adopted optical counterpart of the IRAS source. Objects inside the ellipse are either of stellar appearance on the KOS plate, or spectroscopically confirmed stars, or at the plate limit. Quoted redshifts are galactocentric. (a) 08309+6433: A is a galaxy (11401 km/s), B is very faint, C looks stellar; (b) 08379+6753: A is a galaxy (11243 km/s), B looks fuzzy and C and D stellar, E is near the plate limit; (c) 03075- 1139: A and B are galaxies (11677 and 11797 km/s), the two objects near the center of the ellipse are stars; (d) 03075-0953: A, B, C, and D all look fuzzy (no spectra available).

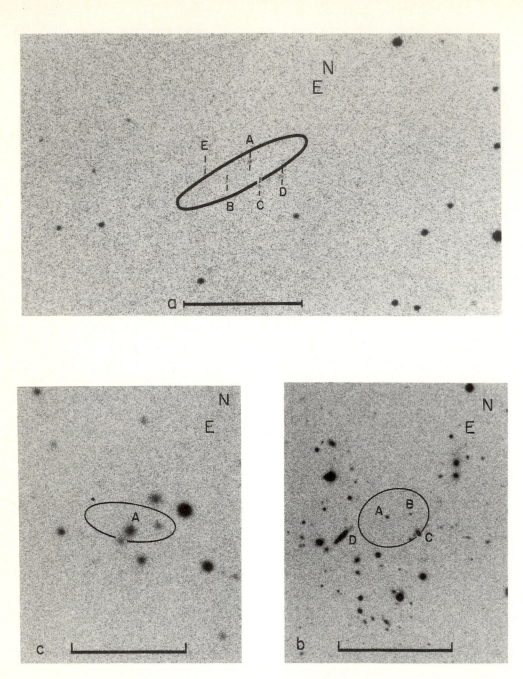

Fig. 2: IRAS sources with more than one optical candidate. A is the adopted optical counterpart. (a) 13416+2614: all identified objects are near the plate limit (no spectra available); (b) 08175+6822: possibly a point-source piece of a complex field (SES1 = 2 flag in the IRAS Point Source Catalog). No spectra available for A, B (both faint), and C(fuzzy). D is a galaxy (17356 km/s); (c) 02565+1114: A is a galaxy (25493 km/s). Except for the two brightest, all nearby objects look fuzzy.

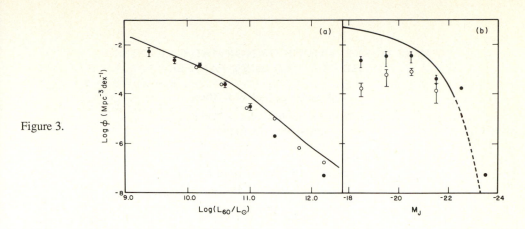

Figure 3.

The highest luminosity point in the IRAS OPLF is contributed by the infrared-luminous (L_{60} = 1.3×10^{12} $L_\odot$) quasar IRAS 00275-2859 (Vader and Simon 1987b). It ranks among the strongest Fe II emitting active nuclei and belongs to an interacting system of two galaxies (Vader et al. 1987). VLA observations of July 1987 (A configuration) show that 00275-2869 is an unresolved radio source at both 6 and 20 cm, with spectral index of -0.25. Wolstencroft et al. (1988) show that it obeys the tight correlation between the 20 cm and 60 μm luminosities of IRAS galaxies. The known properties of quasar 00275-2859 are consistent with those of one of the two basic classes of active nuclei. Given the distinct spectroscopic characteristics of the host galaxies associated with these 2 classes (Boroson et al. 1985), a red spectrum without emission lines and dominated by a stellar continuum would be predicted for the host system of 00275-2859. On the other hand, if its far-infrared radiation is emitted by the host galaxy, an emission-line spectrum typical of that of most IRAS galaxies is expected. Spectroscopy of the host system of 00275-2859 is needed to resolve this issue and to help establish whether the source of the far-infrared emission is the active nucleus or star formation.

References

Boroson, T.A., Persson, S.E. & Oke, J.B., 1985. Astrophys. J. 293, 120.

Kirshner, R.P., Oemler A. & Schechter, P.L., 1979. Astron. J. 84, 951 (KOS).

Lawrence, A., et al. 1986. Mon. Not. R. Astr. Soc. 219, 687.

Soifer, B.T. et al., 1986. Astrophys. J. (Letters) 278, L71.

Vader, J.P. & Simon, M., 1987a. Astron. J., in press.

Vader, J.P. & Simon, M., 1987b. Nature 327, 304.

Vader, J.P., Da Costa, G.S., Frogel, J.A., Heisler, C.A. & Simon, M., 1987. Astron. J., in press.

Wolstencroft, R.D. et al., 1988, this volume.

IRAS OBSERVATIONS OF NORMAL GALAXIES:
THE UGC REDSHIFT SAMPLE

C.J. Lonsdale and W.L. Rice

IPAC, California Institute of Technology
Pasadena, CA 91125

and G.D. Bothun

University of Michigan, Dennison Building
Ann Arbor, MI 48109

ABSTRACT

We have used a large sample of nearly 2000 UGC galaxies with known redshifts and IRAS 60 and 100 μm detections to identify some fundamental correlations between the far infrared (FIR) and optical properties of normal galaxies and to assess the principal dust heating mechanisms that allow a galaxy to be detected by IRAS. We conclude (1) that the 60/100 μm flux density ratio is a useful discriminant of the heating source, (2) that for most normal galaxies the FIR emission is powered by the general diffuse interstellar radiation field (ISRF), and (3) that the overall intensity of the ISRF and the size of the galaxy are the most important parameters driving the FIR luminosity. Thus, the FIR luminosity of a typical spiral is not a good measure of the recent massive star formation rate, but is rather a natural consequence of a constant star formation rate over several Gyr in an optically thin disk of gas and dust.

1. INTRODUCTION

The distance-independent FIR/B flux ratio of galaxies exhibits an extremely large dynamic range, which introduces a level of ambiguity, depending on sample selection, in properly identifying the power sources. Although the luminous, high IR/B luminosity galaxies are the most spectacular, by far the majority of galaxies have much more modest FIR properties. We have studied a large optically selected sample - IRAS-detected UGC galaxies - in order to investigate the IR power sources of the more populous 'normal' galaxies. The UGC is well matched to the IRAS Point Source Catalog, the two having similar median redshifts (Bothun, Lonsdale and Rice 1987). Full details of this work, which is a follow up to the two temperature model of Lonsdale and Helou (1987; LH), are presented by Bothun, Lonsdale and Rice (1987).

2. RESULTS AND DISCUSSION

We have separated out all galaxies in the sample with a priori evidence for a starburst or active nucleus and compared their distributions of IR/B and 60/100 ratio to those of the rest of the sample. We find a much clearer distinction between the two sub- samples in the 60/100 ratio than in IR/B, which indicates that 60/100 is the more useful parameter for determining the FIR power source. This agrees well with studies within the Galaxy (eg. Boulanger and Perault 1987), which show that proximity to hot stars is required to produce high 60/100 ratios. We therefore adopt the two-temperature model of LH as a working hypothesis. The overall distribution of 60/100 for normal galaxies is heavily skewed to low temperatures, characteristic of the cirrus-like

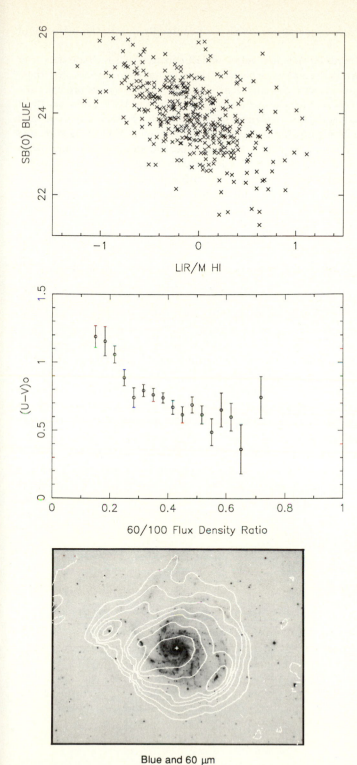

Figure 1. Blue surface brightness ($mag/arcmin^2$) plotted against the ratio of the far infrared luminosity to the HI mass (solar units).

Figure 2. (U-V) color plotted against 60/100 flux density ratio.

Figure 3. 60 μm map of M101 from Rice et al. (1987), superimposed on a blue photograph.

Blue and 60 µm

cool component, with a tail to high temperatures which is undoubtedly due to the influence of active star forming regions.

What is the principle power source for this relatively cool FIR emission that is characteristic of the majority of spiral galaxies? An important piece of information comes from the fact that there is no evidence at all that the IR-emitting regions of IRAS-detected normal spirals are optically thick: we find no dependence of either the blue surface brightness or the UBV colors on IR/B ratio. Moreover, as illustrated in Figure 1, the relationship between blue surface brightness and LIR/MHI (the ratio of FIR luminosity to HI mass in solar units) goes in the wrong direction for an optically thick interpretation. In fact, Figure 1 is in excellent qualitative agreement with the models of Draine and Anderson (1984) for the Galactic cirrus emission in which the dust/gas ratio and ISRF temperature are held constant, and the FIR emissivity per H-atom is driven by the intensity of the ISRF. That is not to say that the dust/gas ratio and ISRF temperature are unimportant; they may well be responsible for much of the scatter about the mean relationship in Figure 1.

A second telling relationship is illustrated in Figure 2, where the optical (U-V) color is plotted against the 60/100 flux density ratio. The FIR color responds to the temperature of the ISRF even at (U-V) colors redder than 0.8 where there is essentially no contribution to the ISRF from OB stars (Kennicutt 1983). Since only very shallow relationships between either IR/B, or LIR, and 60/100 are present in this sample, this implies that a red ISRF (negligible OB star contribution) is quite capable of producing a healthy FIR emission. This emission scales directly with disk mass, leading to luminous FIR emission for big disks. Also, 60/100 responds to the (U-V) color at the very blue end where OB stars do contribute to the ISRF, which supports the conclusion that OB star formation gives rise to FIR- emitting regions that are on average optically thin, because the UV-blue photons are seen to be escaping the galaxy.

Support for our conclusions that diffuse, ISRF-heated emission dominates the FIR luminosity of most normal spirals comes form the Large Galaxy Atlas of Rice et al. (1987), an example from which is shown in Figure 3. The FIR distributions tend to be smooth, with IR sizes very comparable to the optically measured size.

References

Bothun, G.D., Lonsdale, C.J. and Rice, W.L. 1987, A.J. submitted.

Boulanger, F., and Perault, M. 1987, these proceedings.

Draine, B.T., and Anderson, N. 1984, *Ap.J.*, **292**, 494.

Kennicutt, R.C. 1983, *Ap.J.*, **272**, 54.

Lonsdale, C.J., and Helou, G. 1987, *Ap.J.*, **314**, 513.

Rice, W.L., Lonsdale, C.J., Soifer, B.T., Neugebauer, G., Kopan, E.L., Lloyd, L.A., de Jong, T., and Habing, H.J. 1987, Ap.J.Supp., in press.

SEPARATION OF NUCLEAR AND DISK COMPONENTS IN IRAS OBSERVATIONS OF SPIRAL GALAXIES

Roger Ball

Space Sciences Laboratory

University of California

Berkeley, CA 94720, USA

and

K.Y. Lo

Department of Astronomy

University of Illinois

Urbana, IL 61801, USA

ABSTRACT

We have analyzed IRAS survey addscans of relatively nearby, large spiral galaxies by fitting to a model in which an axisymmetric, inclined disk is superposed on a central point source. A focal plane model of the IRAS detector response is integrated into the analysis. Preliminary results indicate that the technique is useful for spatial decomposition of IRAS fluxes for normal spirals.

1. INTRODUCTION

Within the past year, the study of IRAS data on normal galaxies has been strengthened by the appearance of increasingly sophisticated analyses which have begun to lead toward a more mature understanding of the dust emission from spirals than can be expressed by a single number such as a naively derived, global star formation rate. For example, recent work by Helou and Persson (Helou 1986; Persson and Helou 1987) has pioneered a more thoughtful examination of the spectral characteristics of IRAS galaxy observations, an approach that has been elaborated at some length by several authors at this conference (Boulanger 1987; Cox and Mezger 1987; Persson, Bothun, and Rice 1987). We report on a complementary approach to the IRAS database, which deals with the spatial decomposition of the IRAS flux.

Our goal is to distinguish between extended emission associated with galactic disks and point-like sources in the nuclei of nearby spirals of large angular extent. Because of the limited angular resolution of the IRAS detectors, our operational definition of a "point" source is one smaller than a few tens of arc seconds, depending on the IRAS band and the galaxy's signal to noise ratio. Thus when the term "nucleus" is used in this contribution, we are implicitly referring to the "near-nuclear zone" which is likely to include the flux from active star formation regions in the galaxy's inner kpc or so, in addition to the optically unresolved nucleus.

This work is motivated by the question of whether the physical processes that drive star formation in galactic nuclei may differ from those that dominate in the disks. Specifically, it has often been suggested that radial inflows of dense gas to the nucleus of a spiral galaxy might trigger a surge of star formation there. Such inflows could be caused by the action of a bar or by interaction with a companion (*e.g.* Lo *et al.* 1984; Keel *et al.* 1985; Ball *et al.* 1985; Hawarden *et al.* 1986; Sanders *et al.* 1987). It remains to be shown that the enhancement of star forming activity in candidate galaxies actually is concentrated to the nucleus, in our broad definition of the word. One approach to this problem is to try to evaluate the level of star forming activity in the centers of galaxies from ground-based data and then compare the results to IRAS data (*e.g.* Devereux 1987; Hawarden 1987). We have tried instead to see what can be done by exploiting the spatial information of the IRAS data themselves.

2. METHOD

Our analysis relies on a spatial decomposition of the observed infrared emission by fitting the IRAS data to a very simple model galaxy. Thus, our approach is not a true deconvolution technique and does not aim to improve drastically on the inherent spatial resolution of IRAS, but rather to remove the effects of the unusual (from the traditional astronomer's point of view) spatial sampling of the sky that IRAS carried out, with its oblong detector shapes and arbitrary scanning orientation. It is crucial to the success of this analysis that the actual responses of the physical detectors aboard the spacecraft be included as accurately as possible; otherwise, the solution for the combined disk plus point source system is likely to be dominated by the errors in the point source fit. Our method therefore incorporates the technique, sometimes called "focal plane modelling," of building the available data on the detector array itself into the analysis. We plan to make the program available to the general community of IRAS data users once it has been more extensively tested.

We have implemented this idea with a computer program that models the fluxes from a series of one-dimensional survey scans across each galaxy, the IPAC ADDSCAN data product. The program uses the simplest possible model for a spiral galaxy. The infrared surface brightness is assumed constant on (inclined, elliptical) annuli, so that the model galaxy has inherent azimuthal symmetry. There is an additional contribution to the flux from an unresolved point source, whose position relative to the nominal IRAS position is found in a preliminary step. At each point where IRAS has sampled the galaxy's flux, the output signal is given by a convolution of the detector response function over the part of the galaxy inside the limits of the detector mask at that time.

We use a set of detector maps computed and made available to us by M. Moshir of IPAC. The full map (64 grid points in-scan, ~ 40 cross-scan) was used for calculating the response to the point source, but for convolution with the galaxy disks a smoothed version was computed with fine sampling at the cross-scan edges of the detector, where the slope of the response is steepest, and less resolution elsewhere for computational efficiency.

The application of the program has so far been restricted to the 60 μm and 100 μm bands. Although the shorter bands offer possibilities of higher resolution, and the 25 μm band in particular seems an excellent star formation indicator (Hawarden 1987), the problem is complicated at shorter wavelengths by the presence of many point sources in a typical field. Since these do not have accurate *a priori* positions, the program to subtract them must be made more sophisticated than our current version.

The convolution is performed over the model galaxy and an idealized detector which is divided into a number of constant sensitivity boxes. The solid angle of overlap with each annulus is computed for every box in turn. By choosing an in-scan box dimension commensurate with the total detector size, one only has to do the geometrical part of the calculation once and store results. The result is a matrix of least squares coefficients that specify the observed flux at each point. Because the solution for the point source position has been handled separately, a linear least squares technique can be used to solve for the nuclear and disk fluxes simultaneously.

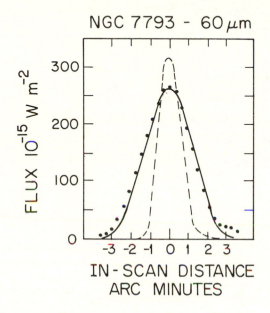

Figure 1. A single scan over NGC 7793 at 60 microns. *Dots*, observed data. *Solid line*, best fit including axisymmetric, inclined disk. *Dashed line*, best fit for point source only. The galaxy is clearly extended.

3. RESULTS

Figure 1 shows an example of running the program on a single scan of the galaxy NGC 7793 at 60 μm. Included are two fits to the data, one which uses only a point source and the adopted model with disk included. The disk component clearly contributes substantially to the flux density, about half the total in this case. As one can see from this example, the simple model used here can represent scan data quite well in many cases, indicating that the axisymmetric disk approximation is reasonable for many galaxies and that a useful separation of the extended disk emission from the nuclear flux is possible in cases of adequate signal to noise ratio.

Our plan is to carry out this analysis for a fairly large sample of 100 − 200 galaxies of sufficient angular diameter. Many of these galaxies have also been observed at 1 μm with a CCD, to search for stellar bars. To date, our work with the IRAS data has been directed primarily to developing the software tools to do this analysis, and only a small sample of about 15 objects has yet been studied. The following trends have emerged in this small data set, but it remains to be seen whether they are confirmed by the complete sample. Most of the galaxies studied are measurably extended, with the fractional luminosity of the unresolved nucleus varying from 20 to 80 percent at 60 μm. The barred spirals are more centrally concentrated than unbarred galaxies of the same angular size, but possible selection effects must still be evaluated. If this result is borne out, it would support the idea that bars promote nuclear star formation. No significant trends with Hubble type or (H I) gas content are apparent in this small sample.

One very important test to be carried out in the near future is the analysis of synthesized data with noise added, for the purpose of determining quantitative limits for the separation of disk and nuclear components by our program.

4. SUMMARY

A technique for spatially decomposing the IRAS signals from nearby spiral galaxies has been developed, using a detailed model of the instrumental response of the IRAS focal plane detectors. Preliminary results suggest that the dominance of the near-nuclear zone in the long-wavelength IRAS data varies significantly from galaxy to galaxy and that an increase in central concentration may be correlated with the presence of a stellar bar.

Acknowledgements

We are grateful to Mehrdad Moshir and John Good of IPAC for help with this project, and to Lee Mundy of Caltech for providing a very helpful data conversion program at short notice. This work was supported by a grant from the NASA IRAS Data Analysis Program.

References

Ball, R., Sargent, A.I., Scoville, N.Z., Lo, K.Y., and Scott, S.L. 1985, *Astrophys. J.,* **298**, L21.

Boulanger, F. 1987, these proceedings.

Cox, P., and Mezger, P.G. 1987, *ibid.*

Devereux, N. 1987, *ibid.*

Hawarden, T. 1987, *ibid.*

Hawarden, T.G., Fairclough, J.H., Joseph, R.D., Leggett, S.K., and Mountain, C.M. 1985, in *Light on Dark Matter: Proceedings of the First IRAS Symposium,* ed. F.P. Israel (Dordrecht: Reidel), p. 455.

Helou, G. 1986, *Astrophys. J.,* **311**, L33.

Keel, W.C., Kennicutt, R.C., Jr., Hummel, E., and van der Hulst, J.M. 1985, *Astr. J.,* **90**, 708.

Lo, K.Y., Berge, G.L., Claussen, M.J., Heiligman, G.M., Leighton, R.B., Masson, C.R., Moffet, A.T., Phillips, T.G., Sargent, A.I., Scott, S.L., Wannier, P.G., and Woody, D.P. 1984, *Astrophys. J.,* **282**, L59.

Persson, C.J., Bothun, G.D., and Rice, W.L. 1987, these proceedings.

Persson, C.J., and Helou, G. 1987, preprint.

Sanders, D.B., Young, J.S., Scoville, N.Z., Soifer, B.T., and Danielson, G.E. 1987, *Astrophys. J,* **312**, L5.

A POST IRAS VIEW OF ACTIVE GALAXIES

A. Lawrence
School of Mathematical Sciences,
Queen Mary College,
Mile End Road,
London E1 4NS

ABSTRACT

Only a few percent of IRAS detected galaxies are clearly AGN. Colour selected lists produce AGN more efficiently, but miss most of the AGN we previously knew about. IRAS selected AGN samples are biased towards Type 2 Seyferts, indicating the importance of dust absorption and re-emission. Most high luminosity IRAS galaxies appear to be starbursts rather than AGN, but a population of completely dust enshrouded AGN may be dominating at the very highest luminosities. A large minority ($\sim 40\%$) of high luminosity IRAS galaxies appear to be in interacting or merging systems. IRAS measurements of AGN have greatly improved the frequency coverage of the wide ranging continuum emission, allowing to us to distinguish several emission components.

1. INTRODUCTION

I would like to concentrate on four areas where IRAS has had a major impact on active galaxy studies. Firstly, IRAS has been a good tool for simply finding Active Galactic Nuclei (AGN), especially Type 2 Seyferts, but the best technique for digging them out is not so obvious. Secondly, it seems that IRAS may have discovered a new class of AGN, quasars buried in dust. Again, it is not trivial to unearth these, as it seems that most high luminosity objects are "starbursts". Thirdly, study of of IRAS galaxies has re-excited our interest in interacting galaxies and their role in promoting energetic phenomena, either of the Seyfert or of the Starburst kind. Finally, IRAS measurements have filled a big gap in our knowledge of the overall continuum flux distributions for AGN, and thus have strengthened our knowledge of the components making up the AGN continuum.

Many people have been working in these areas, as we shall see, but of course I will refer most often to relevant work done here at QMC with various collaborators. Several samples of IRAS galaxies are involved. **S1** : A complete redshift survey for 303 IRAS galaxies in the NGP (Lawrence *et al* 1986). **S2** : An all sky redshift survey of IRAS galaxies, with one-in-six random sampling. This is in progress, involving Rowan-Robinson, Lawrence, Saunders, Efstathiou, Kaiser, Ellis, Frenk, and Parry. These first two samples are primarily for cosmological purposes, but along the way contain data of astrophysical significance. **S3** : For more detailed studies of IR-luminous galaxies, we defined a subset of S1, with $z > .01$, S(60 μm) > 2 Jy, or S(60 μm) > 1 Jy and L60 > 11.5. (Here and hereafter, L60 is $\log(\nu L_\nu)$ at 60 μm, expressed in solar units, and calculated using $H_0 = 50, \Omega = 1$). Most of these (61) galaxies have L60 > 11, in contrast to the general content of S1, which is dominated by normal spirals, with typical L60 $\sim$ 10. Collaborators in this work are Lawrence, Rowan-Robinson, Leech, Penston, Terlevich, Wall, and Jones. **S4** : Rowan-Robinson and Crawford (1987) have studied a sample of 227 catalogued galaxies detected in all four IRAS bands. This contains a fairly even mixture of Seyfert galaxies, starburst galaxies, and normal spirals. Finally I shall also make reference to **X1** : a sample of 37 X-ray selected AGN, the subject of an ongoing multi-wavelength program, with collaborators as in Ward *et al* (1987).

2. AGN CONTENT OF THE POINT SOURCE CATALOGUE

2.1 AGN in flux limited surveys

Flux limited samples are of prime importance for two reasons : (i) they result from an objective search method, and (ii) they have a relatively simple "selection function", so that recovery of population properties from sample properties is not difficult. Selection of IRAS sources at 60 μm at high galactic latitudes finds almost exclusively galaxies. Only about 5-6% of these however are clearly Seyfert galaxies, as judged from the presence of high excitation emission lines (S1 : Leech *et al* 1987). This is not very different from the general incidence of Seyfert nuclei in spiral galaxies (Keel 1983). Most have HII region like lines. The IRAS PSC thus contains $\sim 10^3$ Seyferts, but a lot of hard work is required to sift them out. There are very few new quasars. (Pointed observations of well known quasars are discussed in Neugebauer *et al* 1986). Three new quasars have been found in examination of flux limited subsets of the PSC (Beichmann *et al* 1986; Vader and Simon 1987; and a third, as yet unpublished, in our sample S3). The sparsity of new quasars is simply a reflection of the characteristic depth of the survey ; furthermore there are good reasons for avoiding any distinction, at least in sample selection, between AGN where we can see the galaxy and those where we can't (see discussion in Lawrence 1987). Probably the best way to find all of the brightest AGN is flux selection at 12 μm, which is more less equivalent to the four-colour sample, S4. An important aim should the gathering of good quality spectroscopy for all of such a sample.

2.2 Colour selection of AGN.

De Grijp *et al* (1985) proposed selection on the basis of 60-25 μm colour, with $0.5 < \alpha < 1.25$, this range excluding Galactic objects at the warm end, and normal galaxies and starbursts at the cool end. (Throughout I shall use spectral index α in the sense implied by $S_\nu = K\nu^{-\alpha}$). They find 80% of such objects to be Seyfert galaxies. Similar results are found by Carter (1985) and De Robertis and Osterbrock (1985). If however we ask how many previously known Seyferts pass the colour test, the results don't seem so efficient. Consider the X-ray sample, X1. This sample is dominated by Type 1 Seyferts, 28/37 of which are detected by IRAS (Ward *et al* 1987). Only 10 pass the colour test ; 8 are too cold, with IRAS colours dominated by the host galaxy and/or accompanying starburst ; 10 are too warm, being dominated by a thermal component, which sometimes peaks at 25 μm. Next we can consider the sample of Seyferts found in the CfA redshift survey. IRAS detected 37/48 of these (Edelson, Malkan and Rieke 1987). Only 8 pass ; 24 are too cold, and 5 are too warm. In this sample, there are many Type 2 Seyferts, and the IRAS colours of these are often strongly thermal and relatively cool (Miley *et al* 1985 ; Kailey and Lebofsky, this volume). Colour selection then is probably the best method for simply producing a long list of Seyferts for further study, but such a list may be unrepresentative, and difficult to treat statistically. In recent ongoing work, De Grijp *et al* (1987) have extended their colour range, to $0.0 < \alpha < 1.5$. This should pick up a larger fraction of Seyferts, but an even larger number of "polluting" objects.

2.3 Relative numbers of Type 1 and Type 2 objects

In traditional Seyfert lists (e.g Weedman 1977), which were largely UV excess selected, Type 1 objects outnumber Type 2 objects by 4 to 1. In our IRAS flux limited list however (S1) Type 2 objects outnumber Type 1s by ~ 2 to 1. In the "warm" lists discussed above, the Type 2s are winning by 3 to 1. In the CfA sample, selected by parent galaxy magnitude, Type 1 and 2 objects occur in similar numbers. This strongly suggests that the "true" ratio is 1:1, and that most Type 2 Seyferts are obscured Type 1 Seyferts, re-radiating absorbed energy thermally in the IR (c.f. Rowan-Robinson 1977; Lawrence and Elvis 1982; Lawrence 1987). In this picture, Type 2 Seyferts must be obscured in the line of sight, to lose the strong optical-UV continuum, but they must also have, as a class, higher globally averaged opacity, to explain the bias towards Type 2 Seyferts with IR selection. Study of the classic nearby Type 2 Seyfert, NGC 1068, suggests that we may still see, weakly in scattered light, both the BLR (Antonucci and Miller 1985) and hard X-rays (Elvis and Lawrence 1987).

3. HIGH LUMINOSITY IRAS GALAXIES - BURIED QUASARS ?

3.1 The occurrence of large IR luminosities

Early excitement was caused by the discovery of two galaxies, ARP 220 and NGC 6240, with IR luminosities of the order $10^{12} L_\odot$ (Soifer *et al* 1984; Wright *et al* 1984). Further such objects were soon found (Allen *et al* 1985), and the first complete surveys showed that 3% of IRAS galaxies have L60 > 12 (Lawrence *et al* 1986 ; Soifer *et al* 1986). However until recently, the debate on the nature of such objects has remained centred on ARP 220 and NGC 6240. To put such luminosities into perspective, note that normal Shapley Ames spirals have L60 = 9-11, emitting typically about 20% of their blue luminosity reprocessed into the far-infrared (De Jong *et al* 1984; Lawrence *et al* 1986; Rowan-Robinson *et al* 1987). About a third of IRAS galaxies have L60 > 11 and almost certainly need some extra energy source, rather than being simply large or dusty galaxies. The IRAS emission from many of these galaxies will be dominated by the extra energy source. At L60 > 12, this will almost certainly be the case. To get a feel for this luminosity, note that 3C 273 has L60 = 12.58 . The obvious candidates for the energy source are (i) an active nucleus, and (ii) large numbers of recently formed hot massive stars.

Spectroscopically however, neither ARP 220 nor NGC 6240 look either Seyfert like, or HII region like (Heckman *et al* 1983; Fosbury and Wall 1979). This leads to the idea that the energy source itself is not seen, being buried in dust, and thus produces most of its energy output in the IR. The idea of a starburst buried in dust in the nuclear regions of galaxies has of course been around for some time (see review by Rieke and Lebofsky 1979). In the case of buried Seyferts, we may ask what the difference is between such postulated objects and Type 2 Seyferts.... we will indeed ask this later. How common is this phenomenon ? By comparison with the IRAS properties of sample X1, Lawrence *et al* (1986) find that, for a given IR luminosity, high luminosity IRAS galaxies are ∼ 50 times as common as Type 1 Seyferts. However if all the energy is absorbed and reprocessed, we should compare space densities at the same bolometric luminosity. This is very difficult, as we don't know the shape of the AGN continuum in the XUV, but Soifer *et al* (1986) make a reasonable guess, and find that high luminosity IRAS galaxies and Seyfert galaxies are roughly equally common. On the buried quasar hypothesis then, about 50% of AGN are buried in dust.

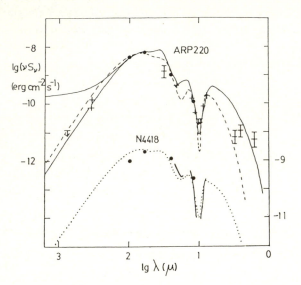

Fig. 1 Model fits to IRAS colours and other data for ARP 220 and NGC 4418, from Rowan-Robinson and Crawford (1987). Dashed line : starburst inside cloud with $A_v=20$, and additional foreground extinction $A_v=78$. Solid line : quasar inside cloud with $A_v=40$. For the lower curve, the power law has been cut off at 100 μm.

3.2 Quasars or Starbursts ?

So are there hidden quasars, or just hot stars, in ARP 220 and NGC 6240 ? We should look sceptically at the arguments that have been raised. The first three arguments are to do with optical depth. (i) The IRAS colours are different from most galaxies, active or otherwise, probably because of large optical depth. Fig. 1 shows two models of ARP 220, calculated by Rowan-Robinson and Crawford (1987) : one is an obscured starburst, the other an obscured quasar. Shortward of 100 μm, one cannot tell the difference. Longward of 100 μm, it depends on whether the underlying quasar power law has turned over or not. (ii) The extinction is certainly large in ARP 220 - near IR continuum shape implies $A_v \sim 10$ (Rieke *et al* 1985) and the 10 μm silicate feature implies $A_v \sim 50$ (Becklin and Wynn Williams 1987). Optical and near IR spectroscopy of NGC 6240 implies $A_v \sim 3$ (De Poy *et al* 1986). All this tells us nothing about the energy source however, except that it is well hidden. (iii) The "IR excess", that is, the ratio of IR luminosity to the luminosity in Lyα, as deduced from other recombination lines, is far larger than seen in Galactic HII regions (De Poy *et al* 1986). Where foreground extinction is clumpy, and/or dust is mixed with the ionised gas, extinction values derived even from near-IR emission lines may be a severe underestimate, and dust may compete for ionising photons. Furthermore, in large star-forming complexes, as opposed to individual HII regions, many OB stars may be completely hidden in their parent molecular clouds, and new stars of lower mass will contribute to dust heating, but not to ionisation. From first principles therefore, the IR excess doesn't tell us much about the underlying energy source, but more about its environment. De Poy *et al* (1986) argue nonetheless that the IR excess is even larger than seen in some well known nearby starburst galaxies, and closer to that of NGC 1068, thus empirically supporting the quasar hypothesis. However, a safe statement on the IR excess of starbursts should await more detailed study of IRAS selected objects.

The next two arguments are to do with the nature of emission lines. (iii) Joseph *et al* (1984a) have argued that strong H_2 emission (probably due to shocks) in both ARP 220 and NGC 6240, empirically supports the starburst hypothesis. However on first principles it's not clear why shocks should be more important in starburst than in AGN, and in fact Kawara *et al* (1987) have claimed that AGN are actually stronger H_2 emitters. (iv) De Poy *et al* (1987) have, after many hours of integration, found a $Br\alpha$ line in ARP 220 with FWHM = 1300 kms^{-1}, certainly much broader than the lines normally seen from HII region nuclei (2-500). This is then strong evidence for a quasar in ARP 220, but still not conclusive - in an optically thick starburst, it is possible that (a) turbulent velocities become large, and (b) before we see such a line, it has been broadened by multiple scattering (Joseph, private communication).

The final two arguments are morphological. (v) Rieke *et al* (1985) claim that both ARP 220 and NGC 6240 are extended at 10 μm, requiring a distributed heating source, i.e. a starburst. Becklin and Wynn-Williams (1987) dispute this finding for ARP 220, claiming that the 10 μm size is < 1.5″($\sim$ 500pc), consistent with either hypothesis. (vi) Norris *et al* (1985) find a triple radio source in ARP 220, reminiscent of those found in Seyfert galaxies. Such a signature is unaffected by reddening. However more recent work (Norris 1987) indicates that the weak third component may be spurious. A double source is still consistent with Seyfert nature. However, given that ARP 220 appears to be a merger, a double source may also represent star formation in each nucleus. Perhaps a safer bet is to look for a compact (pc scale) flat spectrum core. These are of course normal in radio loud AGN, but may also be present at low flux levels in radio quiet AGN (Preuss and Fosbury 1983).

Overall then, the evidence is circumstantial and far from conclusive. The best bet so far is that ARP 220 is a quasar and NGC 6240 is a starburst. What however is generally the case for luminous IRAS galaxies ?

3.3 Spectroscopic properties of complete samples of high luminosity galaxies

New observations of IR-luminous galaxies are presented in this volume by Soifer, by Persson, and by Becklin. Here I will stress results from our own samples, S1 and S3, especially recent IPCS observations of the latter (Leech *et al* 1987; further work in preparation). The first finding is that the IR excess is usually very large. The observed $L_{IR}/L_{H\alpha}$ is typically around 4000, give

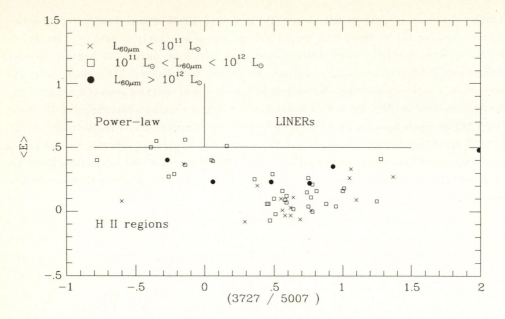

Fig. 2 Emission line diagnostic diagram for IRAS galaxies in sample S3. Data are reddening corrected using observed $H\alpha/H\beta$ and include an estimated absorption correction for $H\beta$

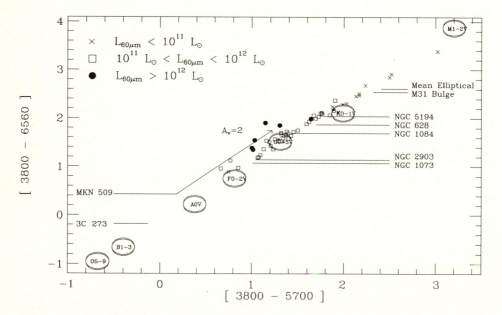

Fig. 3 Colour-colour diagram for IRAS galaxies in sample S3. The colours are flux ratios at the indicated wavelengths, expressed in magnitudes. The location of stellar colours are indicated in ellipses. Colours of galaxies are indicated at the ends of lines with galaxy names at the opposite end. A reddening vector is drawn from the MKN 509 point.

or take a factor of ten. The extinction correction is very uncertain, but the IR excess is usually several times larger than seen in Galactic HII regions, but only occasionally as high as in ARP 220 and NGC 6240. For the extra energy source in the high luminosity galaxies, some combination of the problems discussed in section 3.2 (clumpy extinction, dust mixed with ionised gas, embedded stars) must be very important. From the measured $H\alpha/H\beta$, the typical A_v is ~ 3. In simple spherical models, the IR excess, the $H\alpha/H\beta$, and the IRAS colours are very hard to explain simultaneously (Crawford 1987). The simplest fix is to postulate that in most directions opacity is very large, but that the dust cover is broken by holes for about 1-10% of the surface area. Alternatively, in the starburst picture, most of the OB stars are hidden deep in a thick cloud, but 1-10% are in relatively low opacity regions - possibly those on the edge of the cloud have blown away the dust cover, as in the "blister" model (e.g. Habing and Israel 1979).

An important point here is that any holes will allow the energy source, hidden in the line of sight, to reveal itself by the nature of the emission lines produced in gas ionised through the holes, as these will dominate the weak reddened lines from optically thick regions. Fig. 2 shows a diagnostic diagram, in the manner of Baldwin, Phillips and Terlevich (1981), for sample S3. Almost all look like HII regions. The typical excitation level is higher than normally seen in Galactic HII regions, which centre round $\langle E \rangle \sim 0$, but well below the power law line. Only a few objects are clearly Seyferts, but noticeably the highest luminosity objects have higher $\langle E \rangle$. However, a private communication from R.Terlevich suggests that the BPT (1981) dividing lines may not be correctly placed, in which case there may be as many as many as 7 Seyferts in sample S3. Fig. 3 shows the optical continuum colour-colour diagram. Objects with L60 < 11 have the colours of bulge stars. Those with 11 < L60 < 12 have a range of colours, similar to those in nearby Sc galaxies, consistent with enhanced star formation. The galaxies with L60 > 12 tend to lie above this locus. The lower two groups can be explained by a mixture of bulge starlight and reddened O stars. Some of the most luminous objects may also be explained this way, but many are better explained by reddened AGN. (In either case, $A_v \sim 2-5$). Note that for starbursts, the reddening applies only to that fraction of the O stars which are relatively clear, not the deeply hidden majority.

It would seem likely then, that for objects with 11 < L60 < 12 the majority are starbursts that are mostly heavily obscured, but with optically thin parts. For objects with L60 > 12, a larger fraction are likely to be reddened quasars.

4. HOW MANY IRAS GALAXIES ARE INTERACTING ?

Pre-IRAS studies have consistently shown that interactions have a clear statistical link with both star formation activity, and Seyfert activity (e.g. Larson and Tinsley 1978; Joseph *et al* 1984b; Joseph and Wright 1985; Dahari 1984; Keel *et al* 1985), in the sense that we see a larger than expected minority of disturbed galaxies. Renewed interest in these questions came with the discovery of ARP 220 and NGC 6240, and the realisation that both seemed to be mergers (Soifer *et al* 1984; Wright *et al* 1984). IRAS selected galaxies are not generally interacting (Lawrence *et al* 1986) but we must examine the unusually luminous objects. Reports concerning incomplete and small samples of the most luminous IRAS galaxies have suggested that a very large fraction of these are interacting or merging (Allen *et al* 1985; Sanders *et al* 1986).

To get a clear answer to this question, we obtained CCD images of all the objects in sample S3 (Lawrence *et al* 1988, in preparation). These have been subjectively classified on a scale from 0 to 6, depending on the closeness and brightness of neighbours, and the presence or absence of disturbed structure, such as tidal tails. The sample is self controlled in the sense that we can compare the properties of low and high luminosity galaxies. The basic result can be simplified as follows. For L60 < 11 only 10% of IRAS galaxies are interacting or merging, but at higher luminosities the figure is 40%. There is a problem however, in that for most of the pairs, both galaxies will have been in the IRAS aperture, leading to a bias towards pairs in a flux limited survey. Correction for this effect is not completely obvious, but the most pessimistic correction, dividing the luminosity of all pairs by a factor two, leads to new figures of 20% interacting for L60 < 11, and 30% for L60 > 11. What is the case for "normal" galaxies anyway ? We are in the process of studying a control sample with a similar range of apparent magnitudes, but meanwhile, a rough idea can be gained by asking, what fraction of Shapley Ames galaxies, in the North, is also present in the ARP or VV catalogues ? The answer is 14%. This is similar to the figure for low luminosity IRAS galaxies, but significantly smaller than the figure for high luminosity IRAS galaxies.

Our finding then is that dynamical interaction is of great importance for the IRAS galaxy phenomenon, but is unlikely to be the most common trigger. There are two possible weaknesses with our study. (i) We could have missed many dwarf companions. However, these are probably very common anyway - testing their causal significance is going to need extremely careful work. (ii) Perhaps the fraction of interactions and mergers is even higher at L60 > 12 - such objects are the most distant and have the poorest images, so that relatively faint disturbed structure might have been missed. Overall however, we sound a note of caution amidst the din of enthusiasm for interactions.

5. FLUX DISTRIBUTIONS OF AGN

5.1 IRAS colours

The 100, 60, and 25 μm colours of well known AGN were examined by Miley *et al* (1985). Type 1 objects have, on average, $\alpha_{100-60} \sim \alpha_{60-25} \sim 1$. Type 2 objects are steeper as a class, and show curvature in that $\alpha_{100-60} < \alpha_{60-25}$, suggesting relatively cool thermal emission. Rowan-Robinson (1986) and Helou (1986) looked at the colours of normal spirals, and found a locus of points, suggesting pure disc (cirrus) emission at one end, and pure HII region emission at the other end. Rowan-Robinson and Crawford (1987; RRC), in studying the complete four colour sample, S4, found the most powerful diagram to be S_{60}/S_{25} vs S_{25}/S_{12}. Here a two branched locus is seen, suggesting that the colours of galaxies can be explained by a linear mixture of of three components : disc, starburst, and Seyfert. RRC actually decompose the members of S4 on this basis, and furthermore produce models to explain each component. (i) The disc component is modelled empirically as the sum of two blackbodies, (ii) the starburst point can be fitted very well by a hot star embedded in a spherical dust shell with $A_v \sim 20$, (iii) the Seyfert point can be fitted well by a power law component with $\alpha = 0.7$ inside a dust shell with $A_v = 0.2$. (This adds a thermal bump to the power law, peaking at 25 μm in νF_ν).

However, RRC also show that some Seyferts require an optically thick component, and as we see in Fig. 1, very optically thick quasars and starbursts can be hard to distinguish. Type 2 Seyferts tend to lie further along the mixing line towards the "starburst" component. Some of this line may then be due to variation in optical depth rather than mixture with a starburst. Finally, as we shall see in the next section, the underlying power law probably has a slope, not of 0.7, but of 1.1, and becomes self-absorbed near 100 μm.

5.2 From the far IR to the ultraviolet

IRAS measurements, when joined to ground based near IR and optical data, and IUE measurements, cover a factor 1000 in frequency. The expanded vision certainly helps. Edelson and Malkan (1986) performed this task for a heterogeneous collection of Seyferts, and subsequently Edelson, Malkan and Rieke (1987) studied a well defined sample, the CfA Seyferts, though without the optical-UV data. (Hereafter this team is EMR87). Meanwhile, Ward *et al* (1987) and Carleton *et al* (1987), hereafter WC87, have studied the X-ray selected sample, X1. Between these two groups, reasonably clear conclusions have emerged.

Type 2 AGN seem to be dominated by thermal emission and starlight (EMR87). Very few have UV or X-ray measurements. Type 1 AGN fall into three kinds, shown in Fig. 4 (WC87). Type A probably represents a universal underlying continuum that is flattish in νF_ν, with added "wiggles" : (i) thermal dust emission peaking at 25-60 μm, as in the RRC model, (ii) a "5 μm bump" of unexplained origin (EMR87) , and (iii) the famous blue bump, possibly the low energy tail of an accretion disc (Malkan and Sargent 1982). Type B is a reddened version of Type A; these objects have larger $H\alpha/H\beta$ and X-ray $/H\alpha$ (WC87). The extinction is typically $A_v \sim 3$, so we lose much of the optical-UV continuum, but can still see weak broad lines and hard X-rays, i.e these are the X-ray NELGs. Comparing the X-ray and IR levels, it seems that what we lose in the blue is partially returned in the IR, indicating globally higher opacity rather than just in the line of sight. Finally, Type C are objects where the nucleus is lost in the radiation from the parent galaxy and/or accompanying starburst.

With all these wiggles and features, it would seem hard to spot the underlying power law claimed by both WC87 and EMR87, but WC87 claim that simply finding a "baseline"by picking the lowest νF_ν point in the region 1-100 μm, and joining this to the X-ray level, defines a power law with a very small dispersion of slopes, centred on 1.1, very similar to the average slope through the IR region in Type A objects. EMR87 further point out that the power law has almost always started to turn over by 100 μm. If this is due to self-absorption, the implied size of about a light day is similar to the expected size of the hypothesised accretion discs (EMR87).

5.3 The effects of dust on apparent form

The sceptical reader will have noticed that I have invoked high dust opacity to explain three kinds of object that all look rather different - X-ray NELGs, with hard X-rays and broad lines, Type 2 AGN, with high excitation narrow lines, and ARP 220 - like objects, with nothing much of anything except IR emission. However there are three likely degrees of freedom in the effects of dust : (i) the globally averaged opacity, (ii) the line of sight opacity, and (iii) the (un)covering

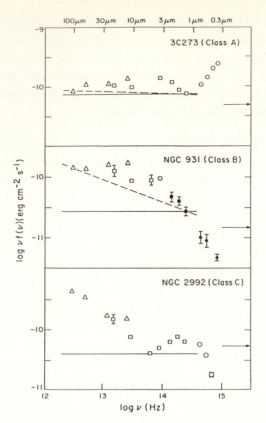

Fig. 4 Three characteristic types of IR to UV spectra for X-ray selected Seyfert galaxies. From Carleton *etal* (1987). The arrow indicates the level of 2 keV X-ray emission, and the lines indicate attempts to find a baseline power law, as explained in the text.

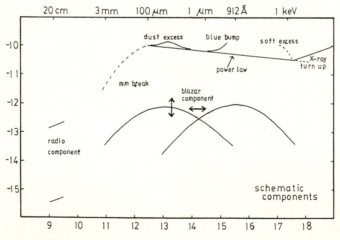

Fig. 5 A sketch of the components involved in the overall continuum spectra of AGN. From Lawrence (1987). The lower horizontal axis is frequency in Hz, and the vertical axis is $\log(\nu F_\nu)$ in arbitrary units.

factor, e.g. the number of holes, or the opening angle of a thick disc. We might need all three of these. Meanwhile, a tentative scheme might go something like this :

With A_v negligible in the line of sight, we see a classic Type 1 AGN, as found by UVX surveys. Small amounts of dust in other directions can produce the thermal "wiggles". With $A_v = 2 - 3$ in the line of sight, we see a "NELG", as found in hard X-ray surveys - steep optical-UV continuum, weak broad lines, and hard X-rays essentially undiminished. The increased thermal IR component often seen in such objects argues that increased line of sight opacity usually goes with increased global opacity, but line of sight geometrical effects can also play a part (c.f. Lawrence and Elvis 1982; Antonucci and Miller 1985). With $A_v = $ tens, the optical-UV and BLR are completely gone, except possibly weakly in scattered light, depending on the geometry. With large enough accompanying gas opacity, hard X-rays can also be lost through a combination of Compton scattering and photo-electric absorption in the increased path length. However, to make a Type 2 AGN, there must be holes (or free directions more generally, e.g. perpendicular to a thick disc) through which to ionise a high excitation nebula. Finally, with $A_v = $ tens and no holes, we see something like ARP 220. The covering factor also determines the degree to which objects are over-represented in the IRAS catalogue, compared to Type 1 AGN.

5.4 Radio to X-ray continuum ; emission components

Efforts are in progress by the WC87 and EMR87 groups to map the continuum of X-ray selected and CfA selected AGN all the way from 10^9 to 10^{19} Hz. Enough is known already however to postulate a decomposition into components. Ideas of this kind have been discussed recently by Lawrence (1987). Fig 5. is a sketch of the various components likely to be present in relatively low opacity, genuine AGN (i.e. after discounting Type 2 Seyferts, ARP 220s, and starburst nuclei). Blazars, i.e. rapid variables with high polarisation, show continua that differ a lot from object to object, but tend to look like log Gaussians (c.f. Landau *et al* 1986; Perry *et al* 1987). The best bet is that this represents emission from a relativistic jet beamed towards us, and that such a component is present to some extent in all AGN. Type 1 AGN differ little from object to object except in relative radio power. In terms of power per decade, the radio emission is any case insignificant, and some sort of spectral break in the mm region is required. Shortward of this, overall power per decade is relatively flat, probably indicating an underlying power law, with added bumps : (i) thermal IR emission, probably from dust in the NLR, (ii) a rise in the optical-UV, mirrored by an X-ray "soft excess", possibly representing thermal emission from an accretion disc (Malkan and Sargent 1982 ; Elvis, Wilkes and Tananbaum 1985), and (iii) an X-ray bump, presumably descending at higher energies, the fashionable explanation for which is a pair plasma (e.g. Fabian *et al* 1986; Svensson 1987 and references therein). We still don't know which of the power law, the blue bump, and the X-ray bump, contain most of the energy, lacking as we do the vital XUV and γ-ray data.

Because of these different components, and the effects of dust on them, samples of AGN selected at various wavelengths are sensitive to differing degrees of freedom available to an active nucleus. An idea of how this works out is sketched in Table 1. The IRAS survey has produced the last of a suite of samples which together give us a much clearer picture of nuclear activity.

TABLE 1 : SENSITIVITY OF PROPERTIES TO SELECTION TECHNIQUE

Hard X-ray	power now	$A_v < 100$	inclination independent
UV excess	power now	$A_v < 1$	inclination dependent
Infrared	power over 1 kYr (NLR dust)	$A_v > $ few	inclination independent
Low frequency radio	power over 1 MYr (Mpc lobes)	$A_v = $ anything	inclination independent
High frequency radio	jet power $*$ velocity	$A_v = $ anything	highly inclination sensitive

REFERENCES

Antonucci, R.R.J., and Miller, J.S., 1985. *Astrophys.J.*, **297**, 621.

Allen, D.A., Roche, P.F., and Norris, R.P., 1985. *Mon.Not.R.astr.Soc.*, **213**, 67p.

Baldwin, J.A., Phillips, M.M., and Terlevich, R., 1981. *Pub.astr.Soc.Pacif.*,**93**, 5.

Beichmann, C.A., *et al* , 1986. *Astrophys.J.*, **308**, L1.

Becklin, E.E., and Wynn-Williams, C.G., 1987. In *"Star formation in galaxies"*, ed. C.J. Lonsdale Persson, NASA conference publ. 2466.

Carleton, N.P., Elvis, M., Fabbiano, G., Willner, S.P., Lawrence, A., and Ward, M.J., 1987. *Astrophys.J.*, in press.

Carter, D., 1985. *Astr.Express,* **1**, 61.

Crawford, J.C., 1987. PhD thesis, Univ. of London.

Dahari, D., 1984. Astron.J., 89, 966.

de Grijp, M.H.K., Miley, G.K., Lub, J., and de Jong, T., 1985. *Nature,* **314**, 240.

de Grijp, M.H.K., Miley, G.K., and Lub, J., 1987. *Astron.Astrophys.Supp.*, **70**, 95.

De Jong, T., *et al* , 1984. *Astrophys.J.*, **278**, L67.

De Poy, D.L., Becklin, E.E., and Geballe, T.R., 1987. *Astrophys.J.*, **316**, L63.

De Poy, D.L., Becklin, E.E., and Wynn-Williams, C.G., 1986. *Astrophys.J.*, **307**, 116.

De Robertis, M.M., and Osterbrock, D.E., 1985. *Pub.astr.Soc.Pacif.*, **97**, 1129

Edelson, R.A., and Malkan, M.A., 1986. *Astrophys.J.*, **308**, 59.

Edelson, R.A., Malkan, M.A., and Rieke, G.H., 1987. *Astrophys.J.*, in press.

Elvis, M.S., and Lawrence, A., 1987, in preparation.

Elvis, M., Wilkes, B.J., and Tananbaum, H., 1985. *Astrophys.J.*, **292**, 357.

Fabian, A.C., *et al* , 1986. *Mon.Not.R.astr.Soc.*, **221**, 931.

Fosbury, R.A.E., and Wall, J.V., 1979. *Mon.Not.R.astr.Soc.*, **189**, 79.

Habing, H.J., and Israel, F.P., 1979. *Ann.Rev.Astron.Astrophys.*, **17**, 345.

Heckman, T., *et al* , 1983. *Astron.J.*, **88**, 1077.

Helou, G., 1986. *Astrophys.J.*, **311**, L33.

Joseph, R.D., Wright, G.S., and Wade, R., 1984a. *Nature*, **311**, 132.

Joseph, R.D., Meikle, W.P.S., Robertson, N.A., and Wright C.S., 1984b. *Mon.Not.astr.Soc.*, **209**, 111.

Joseph, R.D., and Wright, G.S., 1985. *Mon.Not.Soc.astr.Soc.*, **214**, 87.

Kawara, K., Nishida, M., and Gregory, B., 1987. *Astrophys.J.*, submitted.

Keel, W.C., 1983. *Astrophys.J.*, **269**, 466.

Keel, W.C., Kennicut, R.C., Hummel, E., and van der Hulst, J.M., 1985. *Astron.J.*, **90**, 708.

Landau, R., *et al* , 1986. *Astrophys.J.*, **308**, 78.

Larson, R.B., and Tinsley, B.M., 1978. *Astrophys.J.*, **219**, 46.

Lawrence A., and Elvis., M.S., 1982. *Astrophys.J.*, **256**, 410.

Lawrence, A., Walker, D., Rowan-Robinson, M., Leech, K.J., and Penston, M.V., 1986. *Mon.Not.R.astr.Soc.*, **219**, 687.

Lawrence, A., 1987. *Pub.astr.Soc.Pacif.*, **99**, 309.

Leech, K.J., Penston, M.V., Lawrence, A., Walker, D., and Rowan-Robinson, M., 1987. *Mon.Not.R.astr.Soc.*, submitted.

Malkan, M.A., and Sargent, W.L.W., 1982. *Astrophys.J.*, **254**, 22.

Miley, G.K., Neugebauer, G., and Soifer, B.T., 1985. *Astrophys.J.*, **296**, L11.

Neugebauer, G., Miley, G.K., Soifer, B.T., and Clegg, P.E., 1986. *Astrophys.J.*, **308**, 815.

Norris, R.P., *et al* , 1985. *Mon.Not.R.astr.Soc.*, **213**, 821.

Norris, R.P., 1987. *Mon.Not.R.astr.Soc.*, submitted.

Perry, J.J., Ward, M.J., and Jones, M., 1987. *Mon.Not.R.astr.Soc.*, in press.

Preuss, E., and Fosbury, R.A.E., 1983. *Mon.Not.R.astr.Soc.*, **204**, 783.

Rieke, G.H., and Lebofsky, M.J., 1979. *Ann.Rev.Astron.Astrophys.*, **17**, 477.

Rieke, G.H. *et al* , 1985. *Astrophys.J.*, **290**, 116.

Rowan-Robinson, M., 1977. *Astrophys.J.*, **213**, 635.

Rowan-Robinson, M., 1986. In *"Star formation in galaxies"*, ed. C.J. Lonsdale Persson, NASA conf. publ. 2466.

Rowan-Robinson, M., and Crawford, J., 1987. *Mon.Not.R.astr.Soc.*, in press.

Rowan-Robinson, M., Helou, G., and Walker, D., 1987. *Mon.Not.R.astr.Soc.*, **227**, 589.

Sanders, D.B., *et al* , 1986. *Astrophys.J.*, **305**, L45.

Soifer, B.T., *et al* , 1984. *Astrophys.J.*, **283**, L1.

Soifer, B.T., *et al* , 1986. *Astrophys.J.*, **303**, L41.

Svensson, R., 1987. *Mon.Not.R.astr.Soc.*, **227**, 403.

Vader, J.P., Simon, M., 1987. *Nature*, **327**, 304.

Ward, M.J., Elvis, M.S., Fabbiano, G., Willner, S.P., and Lawrence, A., 1987. *Astrophys.J.*, **315**, 74.

Weedman, D.W., 1977. *Ann.Rev.Astron.Astrophys.*, **15**, 69.

Wright, G.S., Joseph, R.D., and Meikle, W.P.S., 1984. *Nature*, **309**, 430.

FAR INFRARED EMISSION OF TYPE 2 SEYFERTS

Walter F. Kailey and Marcia J. Lebofsky

Steward Observatory
University of Arizona
Tucson, Arizona 85721, USA

ABSTRACT

The 60 to 25 µm flux ratio of type 2 Seyferts is correlated with their emission line reddening in a way which is not consistent with a distributed heating source but agrees with predictions of a centrally heated dust shell model with a luminous UV power source.

1. INTRODUCTION

In the middle seventies it became known that Seyfert galaxies emit a significant portion of their bolometric luminosity in the mid- to far-infrared and that type 2 Seyferts are distinguished from type 1's by more steeply rising, curved infrared continua and higher IR to optical ratios. (e.g. Neugebauer et al. 1976).   In 1978 Rieke found a correlation between U-B color and steepness of the nonstellar continuum between 1 and 10 µm in Seyfert galaxies, which suggested that thermal emission from dust was important--at least in the redder objects. Since that time, the IR emission mechanism of most type 1 Seyferts has remained controversial, while that of type 2 Seyferts--generally believed to be thermal--has received little attention.

The IRAS database provides a unique opportunity to study the IR emission of type 2 Seyferts as a class, and there is good reason to do so.   The distinctive IR continua of type 2 Seyferts has already been mentioned.   Another important consideration is that the emission lines from type 2 Seyferts are all reddened to some degree, and the ratio of Hα to Hβ emission line flux provides a reliable estimate of the amount of reddening.   Thus we have a readily obtainable measure of at least one component of the dust in these objects.

Using the IRAS database and optical spectroscopy from our observations and from the literature, we have undertaken a systematic study of type 2 Seyferts detected by IRAS at 25 and 60 µm.   Our goal is to determine whether their IR emission is indeed due to dust and, if so, to investigate the location, heating source, and geometrical configuration of this dust.

2. PROCEDURE

Version two of the <u>IRAS Point Source Catalog</u> contains 74 known type 2 Seyferts detected at both 25 and 60 μm. Six of these are IRAS small extended sources, and they were excluded from further study, since they are nearby, low luminosity Seyferts whose 60 μm emission is likely to be contaminated by the host galaxy (but note that we do not exclude <u>a priori</u> the possibility of host galaxy emission in the remaining objects).

We have obtained Hα/Hβ ratios on 29 of the objects in this sample so far. Most are from Koski (1978) and our own observations. For a complete list of references, see Kailey and Lebofsky (1987; hereafter KL87). In addition, we have searched the literature for Hβ fluxes, host galaxy axial ratios, and X-ray fluxes for these objects.

3. RESULTS

The dependance of the Hα/Hβ ratio on host galaxy inclination in type 2 Seyferts is extremely weak; for 25 objects, the correlation coefficient is .38, and the rms scatter is ±.2 in log Hα/Hβ. Application of the inclination dependent extinction correction for Sa to Sb galaxies given by Holmberg (1975) is sufficient to eliminate the correlation. Keel (1980) found evidence for a source of extinction in type 1 Seyferts which is aligned with the host galaxy disk but is thicker than a normal spiral disk. This result does not appear to hold for type 2 Seyferts.

There is a good correlation between the Hα/Hβ ratio and the 25 to 60 μm flux ratio for type 2 Seyferts. When Hα/Hβ ratios corrected for the host galaxy inclination using the Holmberg (1975) relationship are used, the correlation coefficient is improved and the scatter is reduced. The figure on the next page shows inclination-corrected Hα/Hβ ratios plotted against the 25 to 60 μm flux ratio for type 2 Seyferts. The existence of this correlation has several immediate implications for the dust and far infrared emission in type 2 Seyferts.

The column density of dust in the line of sight to the narrow line region strongly affects the far IR emission. Any reasonable absorber which is no more than a few optical depths in the visible must be very thin at these wavelengths, and the effect must therefore be optically thin emission. The existence of significant preferred directions through the absorbing cloud is ruled out by the tightness of the observed correlation. Thus the absorber is distributed with roughly spherical symmetry about the narrow line region. The power in the dereddened emission line spectra is an order of magnitude below what is necessary to provide the 25 or 60 μm emission. Therefore the dust does not lie preferentially behind the narrow line clouds, unless their covering factor is near one. Edelson and Malkan (1986) found good agreement between emission line and continuum reddening estimators for a large sample of Seyfert galaxies, which supports this conclusion.

The far infrared luminosities of type 2 Seyferts can be used to constrain models of the UV-optical source which heats the dust. Yee (1980) has estimated the strength of the nonthermal continuum in a sample of galaxies

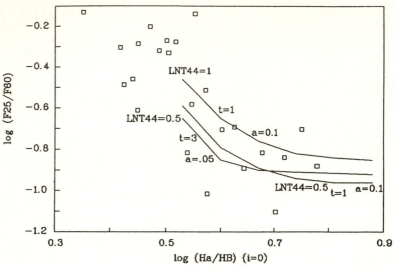

including 17 type 2 Seyferts using a least squares fit to a stellar continuum plus power law between 3000 and 9000 Å. Yee's continuum fluxes are well correlated with Hβ fluxes, and we have therefore used Hβ fluxes, where available, to derive similar estimates. We find that there is not sufficient energy in the dereddened UV continuum to power the 25 or 60 μm emission unless it is extrapolated upto .1 keV with a spectral index no steeper than -1/2. One is then forced to postulate a turnover before ˜.4 keV in order to be consistent with available hard X-ray observations. Since hydrogen and helium will compete very effectively with the dust for photons above the Lyman limit, a very luminous UV bump (possibly the well known 3000 Å bump) would seem the most likely heating source for the dust, if it is centrally heated.

Star formation provides a possible alternative power source. The IR emission could then be explained as a mixture of warm nuclear emission and cooler emission from the starburst. This model predicts a linear relationship between L_{60}, L_{25}, and τ, where L_{60} and L_{25} are 60 and 25 um luminosity densities and τ is the dust optical depth. A multivariable regression reveals significant dependence of L_{60} on both of these variables, but the reduced Chi squared is 5.2 with 25 points, which indicates that this model is a very poor description of the data.

This lead us to examine more closely a centrally heated dust shell model. We have assumed a power-law luminosity source with spectral index -1/2 between 2 um and .1 keV embedded in an H II region ($n_e=10^3$ cm^{-3}, $T=10^5$ K), which is in pressure equilibrium with the narrow line clouds. To reduce the number of model parameters, we set the inner radius of the dust shell equal to the Strömgren radius of this H II region. The luminosity of the nonthermal source between 3000 Å and 9000 Å is a free parameter (LNT44 x 10^{44} erg/s). The outer dust shell radius is set by the parameter t, the number of kpc per magnitude of extinction. We have assumed spherical graphite dust grains with radius a μm. We thus have a highly oversimplified three parameter model, which is nonetheless useful, as the emergent spectrum depends much more strongly on the dust optical depth than on the parameters over reasonable ranges of parameter space. Obviously, generalizations of this model are both feasible and desirable.

The model is governed by two integral equations, which determine the temperature structure of the dust and the emergent spectrum. The temperature solution was simplified using the Planck-averaged emissivities for graphite computed by Draine and Lee (1984). We used approximate analytic grain cross sections and an on-the-spot approximation (scattered photons are immediately absorbed at the point of scattering) in treating the UV-optical radiative transfer.

Further details of the computation, for which space is unavailable here, are given in KL87. Three of the resultant curves are plotted on the figure, where it can be seen that the predicted flux ratios are in impressive agreement with the observations, as is the general shape of the optical depth-IR color relationship.

References

Draine, B.T. & Lee, H.M. 1984, Ap. J., 285, 89.

Edelson, R.A. & Malkan, M.A. 1986, Ap. J., 308, 59.

Holmberg, E. 1975 in Galaxies and the Universe (University of Chicago Press: Chicago), IX, 129.

Kailey, W.F. & Lebofsky, M.J. 1987 (KL87), in preparation.

Keel, W.C. 1980, A. J., 85, 198.

Koski, A.T. 1978, Ap. J., 223, 56.

Neugebauer, G., et al. 1976 Ap. J., 205, 29.

Rieke, G.H. 1978, Ap. J., 226, 550.

Yee, H.K.C. 1980, Ap. J., 241, 894.

STARBURSTS IN INTERACTING GALAXIES

R D Joseph

Astrophysics Group, Blackett Laboratory,

Imperial College,

London SW7, England

ABSTRACT

We first review the evidence for IR activity in interacting and merging galaxies, based on 10 and 20 μm observations. We then investigate the 'starbursts vs. monsters' explanations for the energy source powering the IR activity in these galaxies. Using available IR photometry, and IR, optical, and ultraviolet spectroscopy, we show that, in general, a starburst induced by the interaction is by far the most plausible mechanism. We then apply simple starburst models to the IR data to derive some of the quantitative features of these starbursts. Finally, we argue that interaction-induced starbursts are likely to have occurred in the evolution of most galaxies, and we discuss some of the larger implications of this fact for extragalactic astrophysics.

1. INTRODUCTION

One of the major discoveries arising from the advent of IR astronomy is the realisation that the nuclei (i.e. the central few arcsec) of many gas-rich galaxies have IR luminosities orders of magnitude larger than can be produced by thermonuclear processes in a 'normal' stellar population. I take this as a working definition of 'IR activity' in galaxies. The IRAS survey has provided at least 10,000 new candidates for consideration as IR active galaxies.

It is now abundantly clear that many of the more luminous 'IRAS galaxies' are interacting or the remnants of a recent merger between two gas-rich galaxies (Joseph et al. 1984, Lonsdale, Persson & Matthews 1984, Cutri & McAlary 1985, Joseph & Wright 1985, Joseph 1986). It is also becoming evident that a large fraction of quasars show evidence of interactions (cf. Smith et al. 1986). CCD images of powerful radio galaxies also suggest that interactions are commonly associated with these galaxies (cf. Heckman et al. 1986). If interactions are frequently the common denominator in galaxies with high luminosities, one might expect that the underlying energy generation processes might also be similar in these galaxies. This argument would suggest that accretion onto a compact object, the mechanism generally thought to be responsible for the energy generation in quasars and powerful radio galaxies, would then also be operating in the powerful IRAS galaxies which appear to be interacting. On the other hand, there has generally been a tendency to interpret IR emission from galaxies in terms of a recent burst of star formation. Luminous interacting IRAS galaxies thus provide a particular example of the 'starbursts vs. monsters' controversy (Heckman 1983).

I should like to address this issue by investigating the central few arcsec of a sample of interacting galaxies using a variety of observational approaches. In this way we should be able to gain insight into the physics in the 'nuclei' (i.e. $\leq$ 1kpc) which is associated with interactions between galaxies, and produces powerful IR emission in them.

2. MID-INFRARED PHOTOMETRY OF INTERACTING GALAXIES

We (Wright et al. 1987) have carried out mid-IR (10 and 20 μm) observations for a number of the interacting galaxies which we had identified in an earlier study using JHKL photometry as likely to have large mid-IR luminosities (Joseph et al. 1984). In addition, we have collected from the literature all the other mid-IR photometry available on interacting galaxies. This provides a sample of 39 interacting or merging galaxies (listed in Wright et al. 1987) with 10 μm detections in 5-8 arcsec apertures. IRAS observations, with their large apertures, are not well suited to studies of such small regions in galaxies.

The continuum spectra for several of the interacting galaxies in this sample are shown in Fig. 1. It is evident that they are all very similar to each other and to that for NGC253, which is also shown. These spectra, which must peak in the far-IR, are characteristic of thermal radiation from dust grains which are at a variety of temperatures. They are similar to the spectra observed for galactic star formation regions. By comparison, the mid-IR spectra of quasars and Seyfert 1 galaxies are generally much flatter (Lawrence et al. 1985).

The IR luminosities of these galaxies are large. In Table 1 the 10 μm luminosities are compared with other classes of galaxies--Seyferts, classic starbursts, and bright spirals. It is evident that the interacting galaxies are substantially more luminous than the latter two types. Mergers are apparently more luminous than non-merging interacting galaxies by an order of magnitude, and about as luminous as Seyferts. As we suggested some time ago (Joseph & Wright 1985), mergers seem to be the high luminosity tail of the interacting galaxy luminosity distribution.

A mass-to-light ratio provides a good quantitative indication of the extent to which the luminosity of a galaxy may be accounted for in terms of the energy

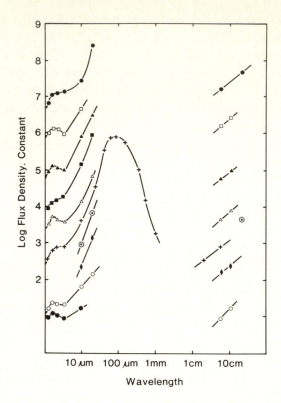

Figure 1. Continuum spectra of interacting and merging galaxies. The spectrum of N253 (+) is shown for comparison.

Table 1. Luminosities at 10 μm for various classes of galaxies.

Class of galaxy	N	Range $(L_\odot)$	Mean $(L_\odot)$
Interacting	24	$4 \times 10^7 - 7 \times 10^9$	2.5×10^9
Merging	9	$4 \times 10^9 - 5 \times 10^{10}$	2×10^{10}
Seyferts	50	$4 \times 10^8 - 10^{11}$	4×10^{10}
Starbursts: M82		10^9	
NGC253		6×10^8	
NGC2903		1×10^8	
Bright spirals	17	$10^5 - 7 \times 10^8$	2×10^8

generated by a normal population of stars (cf. Joseph 1986). The minimum M/L which can be maintained by thermonuclear energy generation for a Hubble time is M/L ~ 1 (in solar units, assumed hereafter). There are rotation curves available for ten of the galaxies in the sample, and it is possible to place upper limits on the masses of three more. In Table 2 the M/L_{IR} ratios, within either a 5 or 8 arcsec aperture, are listed for these galaxies. Here the total IR luminosity L_{IR} has been estimated as 15 that at 10 μm (cf. Scoville et al. 1983, Telesco & Gatley 1984). The M/L ratios for all the galaxies are substantially < 1.

Table 2. Mass-to-IR luminosities for interacting galaxies.

| Galaxy | | M/L_{IR} |
NGC	Arp	(solar units)
520	157	<0.01
1614	186	0.003
2798	283	<0.4
2992	245	0.001
3227	94	0.02
3256		0.01
3396	270	0.002
3395	270	0.007
4088	18	0.02
4194	160	0.03
5194		0.3
6240		<0.08
7714	284	0.01

We conclude, then, from the mid-IR data for the central few arcsec in this sample of interacting galaxies:

i) The IR luminosities are large, with mergers rivalling Seyfert galaxies in luminosity.

ii) The emission mechanism appears to be thermal emission from dust heated by absorption of harder radiation, as is seen in galactic HII regions.

iii) The M/L_{IR} ratios are too small to be maintained by energy generation by a normal population of evolved stars. Either there is recent star formation, with energy provides from a population of massive early-type stars, or there is a quasar-like source heating the dust, which re-radiates in the IR.

3. THE ENERGY SOURCE

There have been three proposals for the energy source which powers the IR activity in the central regions of interacting galaxies: the starbursts or monsters mentioned above, and, for the merging galaxies, the relative kinetic energy of the collision (Harwit et al. 1987). There is probably some contribution by all these processes, but the question is whether one might not frequently dominate. There are a variety of observational avenues by which we might be able to discriminate among these processes.

3.1 The Spatial Extent of the Mid-IR Emission

A dust grain at a distance R from a luminosity source will reach an equilibrium temperature T at which its thermal emission equals the rate of absorption of energy from the luminosity source. For typical grain parameters this gives

$$R \approx 50 T^{-5/2} L_s^{1/2} (L_\odot) \text{ pc}.$$

If there is a single central source of luminosity ~ 10^{10} $L_\odot$, i.e. a monster, it can heat dust grains to T ~ 300 K, i.e. hot enough to that they radiate at 10 μm, only out to a distance of ~ 3 pc. Thus, if there is measurable spatial extent at 10 or 20 μm, it cannot be due to heating by a single, central source i.e. a monster, and must be due instead to distributed sources of luminosity, i.e. a starburst.

This argument is complicated by the recent realisation that very small grains may be present which are not in thermal equilibrium with the energy sources which heat them. Such grains could be heated sufficiently by absorption of a single UV photon to reradiate at wavelengths of 10 μm, or even shorter (cf. Sellgren 1984). This possibility does not, however, uncut the use of extended 10 μm emission to distinguish between starbursts and monsters. Aitken & Roche (1985) have shown that the unidentified features at 7.7, 8.6 & 11.3 μm, which appear to be carried by the very small grains, and strikingly absent in the mid-IR spectra of Seyfert galaxies and quasars, presumably because they cannot survive the hard UV radiation field in the centres of these galaxies. On the other hand, Roche & Aitken (1985) also show that the spectral features associated with very small grains are common in starburst galaxies. Thus, if very small grains are responsible for any spatially extended mid-IR emission (which we interpret as evidence for a starburst), this would appear to be likely only if the galaxy were not dominated by a 'monster'. Thus the presence or absence of very small grains would not alter the conclusions.

For this sample of 39 galaxies, four interacting galaxies, N2798, N3227, N3690, and N5194, and five merging galaxies, N6240, N1614, N3256, N3310, and IC883, have measured spatial extent at 10 μm. Thus 25% of the sample is resolved at 10 μm. This suggests rather strongly that the 10 μm emission in these galaxies is not

powered by a single quasar-like nucleus; on the other hand, starbursts rather
naturally produce such spatial extent.

3.2 Optical Spectra

Another approach to determining the underlying energy source is to use diagnostic
diagrams for various optical line intensity ratios, such as those developed by
Baldwin, Phillips & Terlevich (1981). The required optical spectra are available for
~ 31 of the interacting galaxies in the sample, and we have plotted [OIII]/Hβ vs.
[NII]/Hα line intensity ratios for these in Fig. 2. The areas outlined in the
diagram are those which Baldwin et al. identified as regions of excitation
characteristic of HII regions, Liners, and Seyferts. For 21 of the interacting
galaxies, the excitation is characteristic of HII regions, 4 are like Liners, and 6
are like Seyferts. Thus the excitation spectrum is characteristic of HII regions in
two-thirds of the cases, and strongly supports the idea of an associated starburst.
In Section 5.3 I shall consider the possibility that 'Warmers', extreme Wolf-Rayet
stars whose existence is inferred by Terlevich & Melnick (1985), account for some of
the associations found between interactions and Seyfert-type activity. If there were
such a stellar evolutionary phase, it would be natural to find some excitation
characteristic of a harder spectrum present as the consequence of a starburst, and
therefore it would not be unlikely that the 10 non-HII galaxies have also experienced
a starburst which has now evolved to the 'Warmer' phase a la Terlevich & Melnick.

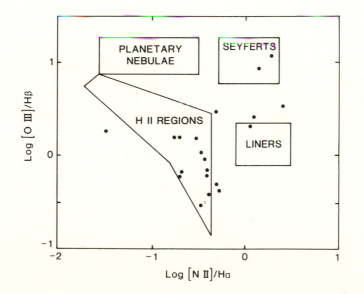

Figure 2. Optical line ratios for interacting galaxies.

3.3 Radio data

The characteristics of the radio emission from these galaxies provide another approach to distinguishing starbursts from monsters as the underlying energy sources in interacting galaxies. The hallmarks of radio emission from Seyfert-like nuclei are flat spectra and small, usually unresolved sizes. By contrast, steep spectra and extended nuclear radio emission would be expected for the supernovae and supernova remnants associated with a starburst.

Radio data is available for three-fourths of the sample of interacting galaxies (Wright et al. 1987). Virtually all of these have a steep spectral index, ~ -0.7. For ~ 80% of the sample the nuclear radio source is resolved, and the sizes are of the order of several arcsec. These radio features are not characteristic of 'monsters', but are qualitatively consistent with the supernova activity subsequent to a starburst.

3.4 Ultraviolet spectral features

The presence of strong UV emission lines due to CIV, HeII, and CIII is one of the hallmarks of Seyfert-type activity in galaxies. By contrast, one might expect starbursts to show absorption features due to CIV and SiIV, since these features are characteristic of giant and supergiant stars. In fact, SiIV absorption is luminosity dependent--strongest in supergiants and about zero in dwarfs (Wallborn & Panek 1984). We have devoted an observing shift on IUE to the most nearby example of an ultraluminous merging galaxy, N3256 (Joseph et al. 1986). The low resolution spectrum shows that the most prominent UV features are those due to SiIV and CIV in absorption, and there are no emission features typical of Seyfert-type activity. This is strong supporting evidence, from a rather different observational perspective, that there is strong starburst activity in this copybook example of a merging galaxy.

3.5 Infrared Spectral Features

A number of groups have been investigating the infrared spectra of interacting galaxies in the H & K windows (Fischer et al. 1983, Joseph et al. 1984, Rieke et al. 1985, DePoy et al. 1987, Joseph et al. 1987). Infrared spectroscopy is an almost completely undeveloped tool for investigating physical processes in galaxies and these kinds of studies have enormous potential.

Two results from analysis of our IR spectra of galaxies are particularly relevant to the question under consideration here. Firstly, we find the ratio of Brackett γ flux to IRAS 25 μm flux to be remarkably constant, and the same for classic starburst galaxies as for interacting and merging galaxies (Arp220 is the most discrepant galaxy in this group). Secondly, we find shocked H_2 emission which implies excitation rates of 30-30,000 $M_\odot$ of H_2 per year. For any reasonable lifetime for this excitation, this implies masses of shocked H_2 of 3×10^7 to 3×10^{10} $M_\odot$ of H_2 in these galaxies. Thirdly, the Brackett γ and H_2 emission are spatially extended over several kpc. Finally, we find deep stellar CO absorption features indicative of a large supergiant population. These features are all qualitatively suggestive of starbursts.

In summary, we have adduced observational evidence with the broad perspective provided by insights from variety of UV, optical, IR, and radio measurements. All of these observational approaches point toward the ubiquitous presence of starbursts in these luminous IR galaxies. It is difficult to rule out the proposal by Harwit et al. (1987) suggesting that the kinetic energy of the collision provides the IR luminosity we see the ultraluminous mergers; indeed, the correspondence between the extended Bracket γ and H_2 emission would provide some support for it. The direct evidence for the presence of recent star formation from the UV data and the CO absorption is probably the strongest argument against this idea. These data do not exclude the concurrent presence of a monster in some or all of these galaxies. However, it is difficult to avoid the conclusion that, in the large majority of cases, starbursts are the dominant energy sources driving the large IR luminosities we have shown to be a common feature of these galaxies. Apparently, interactions produce starbursts.

4 CHARACTERISTICS OF STARBURSTS IN INTERACTING AND MERGING GALAXIES

Using the simple analytical starburst models of Telesco and Gatley (1984), discussed in detail in Telesco (1985) we can investigate the quantitative astrophysical characteristics of starbursts in the central few arcsec of these galaxies. In these models the initial mass function (IMF), the luminosity per star, and the main sequence lifetime of the stars are all approximated by power-law functions of the stellar mass. A more detailed discussion is presented by Wright et al. (1987).

4.1 Star Formation Rates

We consider the formation of OBA stars in a starburst with a constant star-formation rate that has proceeded for a sufficiently long time so that equilibrium has been established and the death and birth rates of massive stars are equal. For the values of L_{IR} typical of interacting and merging galaxies (cf. Table 2), the ISM is being converted into early type stars in the central few arcsec at rates of $\sim 1\text{-}100 M_\odot$ yr^{-1}. By comparison, the star formation rate in the Galaxy, estimated from observations in the solar neighbourhood, corresponds to ~ 0.003 $M_\odot$ yr^{-1} in a region ~ 1 kpc in diameter (cf. Miller and Scalo 1979).

4.2 Duration of the Starbursts

The mass-to-light ratios provide a powerful constraint on the length of time during which the star formation rates derived above can be maintained, since the starburst will eventually consume all of the available gas. All of the M/L ratios in Table 2 are very small and imply short-lived starbursts. On the assumptions that 10% of the galaxy mass is gas, that the efficiency in turning gas into stars is 10%, and that massive stars ultimately return ~ 75% of their mass to the ISM, Wright et al. (1987) show that the M/L ratios of the interacting and merging galaxies in Table 2 imply that their interstellar gas will be consumed in very short timescales, ~ 5×10^5-10^7 yr. For most of the galaxies then, if the starburst is to be maintained for periods > 10^7 yr, fresh material must be supplied to the nuclear region on this timescale. The duration of a starburst nucleus must depend on the balance between interaction-driven infall of material and the outflow which will result from the large number of supernovae produced in the burst. Given the large fraction of interacting galaxies with starburst nuclei, it is evident that the interaction is continuing to supply fresh material for star formation on these timescales. In addition, the star formation efficiency must be greater than the ~ 5% observed in molecular clouds in the galaxy (Cohen & Kuhi 1979).

4.3 Starburst Initial Mass Function

The M/L ratio can also be used to constrain the lower mass cut-off to the IMF, since most of the mass of the starburst is in stars at the low-mass end of the IMF. Wright et al. (1987) have used the Telesco & Gatley model for a 10^7 year old starburst to estimate the total mass of stars formed in the burst for various lower mass cut-offs to the IMF. The results for all the galaxies in Table 2 for which M/L_{IR} ratios are available show that the total burst masses are about the same as the total mass estimated for the starburst regions, if the IMF extends to 0.1 $M_\odot$. However, the starburst mass cannot account for all of the observed mass in the nuclear regions since the galaxies must also have the normal evolved stellar population. If we assume that the fraction of a disc galaxy's mass in gas is ~ 10%, we can inquire what minimum lower mass cutoff to the IMF will produce a starburst mass equal to 10% of the observed mass. Wright et al. find in this case that the lower mass cutoff to the IMF must be ~ 3-6 $M_\odot$.

This simple analysis suggests that, like the canonical starbursts in NGC253 and M82 (Rieke et al. 1980, Kronberg et al. 1985), the extremely luminous starburst process induced in the nuclei of interacting galaxies is both extremely efficient and biassed towards massive stars.

5. WIDER ISSUES AND IMPLICATIONS

5.1 How prominent are interacting galaxies among luminous IR galaxies?

There have been a number of studies suggesting that interacting galaxies are
unusually luminous in the IR (Lonsdale et al. 1984, Cutri & McAlary 1985, Joseph
1986). In particular, Joseph & Wright (1985) argued that mergers of disc galaxies
produce ultraluminous IR galaxies. Supporting evidence for these ideas continues
to accumulate. Allen et al. (1985), in their studies of optically faint IRAS
sources, find many to be strongly disturbed or interacting galaxies. Soifer et al.
(1986, and in this volume) have identified a sample of the 15 most luminous IRAS
galaxies, and their CCD images indicate that most, if not all, are strongly
interacting. Lawrence et al. (in this volume) have undertaken CCD imaging for a
statistically complete sample of 61 high luminosity IRAS galaxies. They find that
the fraction of strongly interacting galaxies increases with 60 μm luminosity.

Joseph & Wright (1985) found evidence for an 'age effect' in the sample of
mergers studied which is related to this question. They classified their sample of
mergers into 'young', 'middle-aged', and 'old' on morphological criteria. They
found the 'middle-aged' group to have an average IR luminosity $\sim 10^{12}$ $L_\odot$, whereas
the other two categories were about 5 times less luminous. Some of the spread in
IR luminosities may be due to this age effect. It may be, therefore, that
luminosities $\sim 10^{12}$ $L_\odot$ may be more typical of the peak starburst activity in
mergers of large, gas-rich galaxies.

5.2 Spectral evolution of galaxies

On the most simple assumptions, the frequency of interactions between galaxies will
be proportional to the galaxies' peculiar velocities divided by the interaction
mean free path. This frequency will scale with redshift, z, roughly as $(1+z)^4$.
Thus, if we see about 5% of galaxies interacting now, most galaxies will have
experienced one interaction at $z \sim 1$. While this is probably too naive an
analysis, it indicates that we might expect to see the effects of interaction-
induced starbursts in the spectral and chemical evolution of most galaxies.

5.3 Relation to other forms of activity in galaxies

It is becoming increasingly evident that galaxy interactions are associated with
other classes of galactic nuclear activity, as Toomre & Toomre (1972) suggested 15
years ago. Seyfert galaxies frequently exhibit tidal distortions and have
companions (Balick & Heckman 1983, Keel et al. 1985). Lilly & Longair (1984)
describe evidence of the effects of starbursts at high redshifts based on the
optical-IR colours they find for 3CR radio galaxies. Smith et al. (1986) find that

in their deep CCD images of low redshift quasars, about half are hosted by morphologically peculiar galaxies. And recently Heckman et al. (1986) have shown that powerful radio galaxies are associated with morphologically disturbed galaxies.

One possible link between interaction-induced starbursts and some of these other forms of activity is provided by the proposed existence of 'Warmers' by Terlevich & Melnick (1985). These authors argue that such stars can account for most of the high excitation spectral features which characterise Seyfert galaxies and quasars. Since interaction-induced starbursts are apparently efficient in producing high mass stars, one can see that interactions might play a major role in producing these high excitation, high luminosity galactic nuclei.

Another suggestion which has been current is that the interactions provide fresh fuel for accretion onto a compact object lurking in the centre of a galaxy (e.g. Heckman et al. 1986). However, if interaction- induced starbursts do have the large spatial extent which seems to be typical, strong galactic winds driven by the ensuing supernovae should very quickly sweep a large central region free of gas, thereby depriving the 'monster' of any more 'food'.

Although it is clear that there is an association between interactions and non-thermal nuclear activity in galaxies, the physical and causal connections are not at all clear. However, it is likely that the powerful starbursts triggered by interactions are an important clue.

References

Aitken, D. K. & Roche, P. F., 1985. Mon. Not. R. astr. Soc., **213**, 777.

Allen, D. A., Roche, P. F. & Norris, R. P., 1985. Mon. Not. R. astr. Soc., **213**, 67P.

Baldwin, J. A., Phillips, M. & Terlevich, R., 1981. Pub. Astron. Soc. Pacific, **93**, 5.

Balick, B. & Heckman, T. M., 1982. Ann. Rev. Astr. Astrophys., **20**, 431.

Cohen, M. & Kuhi, L., 1979. Astrophys. J. Suppl., **41**, 743.

Cutri, R. M. & McAlary, C. W., 1985. Astrophys. J., **296**, 90.

DePoy, D. 1987. Star Formation in Galaxies, p. 701, ed. C. J. L. Persson, NASA, Conference Publication 2466.

Fischer, J., Simon, M., Benson, J., and Solomon, P. M. 1983, Astrophys. J. (Letters), **273**, L27.

Harwit, M., Houck, J. R., Soifer, B. T. & Palumbo, G. G. C., 1987. Astrophys. J., **315**, 28.

Heckman, T. M., van Breugel, W. J. M., Miley, G. K., & Butcher, H. R., 1983. Astr. J., **88**, 1077.

Heckman, T., Smith, E. P., Baum, S. A., van Breugel, W. J. M., Miley, G. K., Illingworth, G. D., Bothun, G. D. & Balick, B., 1986. Astrophys. J., **311**, 526.

Joseph, R. D., Meikle, W. P. S., Robertson, N. A. & Wright, G. S., 1984.

Mon. Not. R. astr. Soc., **209**, 111.

Joseph, R. D., Wright, G. S., & Wade, R. 1984, Nature, **311**, 132.

Joseph, R. D., 1986. _Light on Dark Matter_, p. 447, ed. F. P. Israel, Reidel, Dordrecht, Holland.

Joseph, R. D. & Wright, G. S., 1985. Mon. Not. R. astr. Soc., **214**, 87.

Joseph, R. D., Wright, G. S. & Prestwich, A. H., 1986. _New Insights in Astrophysics_, ESA SP-263, p. 597.

Joseph, R. D., Wright, G. S., Wade, R., Graham, J. R., Gatley, I. & Prestwich, A., 1987. _Star Formation in Galaxies_, p. 421, ed. C. J. L. Persson, NASA, Conference Publication 2466.

Keel, W. C., Kennicutt, R. C., Hummel, E. & van der Hulst, J. M., 1985. Astr. J., **90**, 708.

Kronberg, P. P., Biermann, P. & Schwab, F., 1985. Astrophys. J., **291**, 693.

Lawrence, A., Ward, M., Elvis, M., Fabbiano, G., Willner, S. P., Carleton, N. P. & Longmore, A., 1985. Astrophys. J., **291**, 117.

Lilly, S. J. & Longair, M. S., 1984. Mon. Not. R. astr. Soc., **211**, 833.

Lonsdale, C. J., Persson, S. E. & Matthews, K., 1984. Astrophys. J., **287**, 95.

Miller, G. E. & Scalo, J. M., 1979. Astrophys. J. Suppl., **41**, 513.

Rieke, G. H., Lebofsky, M. J., Thompson, R. I., Low, F. J. & Tokunaga, A. T., 1980. Astrophys. J., **238**, 24.

Rieke, G. H., Cutri, R., Black, J. H., Kailey, W. F., McAlary, C. W., Lebofsky, M. J., and Elston, R. 1985, Astrophys. J., **290**, 116.

Roche, P. F. & Aitken, D. K., 1985. Mon. Not. R. astr. Soc., **213**, 789.

Scoville, N. Z., Becklin, E. E., Young, J. S. & Capps, R. W., 1983. Astrophys. J., **271**, 512.

Sellgren, K., 1984. Astrophys. J., **277**, 623.

Smith, E. P., Heckman, T. M., Bothun, G. D., Romanishin, W. & Balick, B., 1986. Astrophys. J., **306**, 64.

Soifer, B. T., Sanders, D. B., Neugebauer, G., Danielson, G. E., Lonsdale, C. J., Madore, B. F. & Persson, S. E., 1986. Astrophys. J. (Letters), **303**, L41.

Telesco, C. M. & Gatley, I., 1984. Astrophys. J., **284**, 557.

Telesco, C. M., 1985. _Extragalactic Infrared Astronomy_, p. 87, ed. P. M. Gondhalekar, Rutherford Appleton Laboratory, Chilton, UK.

Terlevich, R. & Melnick, J., 1985. Mon. Not. R. astr. Soc., **213**, 841.

Toomre, A. & Toomre, J., 1972. Astrophys. J., **178**, 623.

Walborn, N. R. & Panek, R. J., 1984. Astrophys. J. (Letters), **280**, L27.

Wright, G. S., Joseph, R. D., Robertson, N. A., James, P. A. & Meikle, W. P. S., 1987. Mon. Not. R. astr. Soc., in press.

THE ROLE OF BARS IN STARBURST GALAXIES

T G Hawarden and C M Mountain
Royal Observatory, Blackford Hill
Edinburgh EH9 3HJ, UK

P J Puxley and S K Leggett
Astronomy Department, University of Edinburgh
Blackford Hill, Edinburgh EH9 3HJ, UK

ABSTRACT

Amongst spiral galaxies earlier than Scd, barred morphology is a necessary condition for the IRAS colours of the whole system to resemble those of an HII region. New radio and infrared spectroscopic observations demonstrate that the sources of the excess IRAS fluxes are concentrated near the nuclei of these galaxies but, being resolved on scales of arcseconds are probably regions of vigorous star formation and not active nuclei.

1.   INTRODUCTION

We have previously shown (Hawarden *et al.*, 1986) that for a complete sample of RSA systems with RC2 types between S..O/a and S..cd, only the barred systems have IRAS colours resembling those of HII regions. This continues to be true when the revised IRAS fluxes from the second edition of the Point Source Catalogue are used. We term the barred systems with steeply rising 12 to 25 μm spectra "h" galaxies, defined as those with $F(25)/F(12) > 2.22$. Those with flatter spectra we have labelled "l".

The association of barred morphology with an excess flux at 25 μm we attributed to warm dust in the complexes of HII regions which are common in the centres of barred spirals. Independant analysis of the IRAS data by Rowan–Robinson and Crawford (1986) and ground–based studies by Devereaux (1986) have also suggested that greatly enhanced rates of star formation are occurring in these systems.

These starbursts appear to be <u>caused</u> by the presence of the bar, probably by inflow and consequent accumulation of material at an inner resonance. It is interacting to note that the work of Combes & Gerin (1985) and Casoli *et al.* (this Conference) demonstrates that this is the same physical process as that which gives rise to the starbursts observed in interacting and merging galaxies, i.e., <u>all</u> starbursts may be caused by the presence of a rotating axisymmetric potential.

To confirm and quantify this scenario we have embarked on a programme of ground–based radio and infrared observations. The initial results of this programme completely support our proposed picture.

2.   THE RADIO OBSERVATIONS

2.1   The central sources at 20 cm in "h" and "l" galaxies.

A continuum study at 20 cm by Hummel (1980) showed a difference between barred (SB and SAB) and unbarred (SA) spiral galaxies on a 20 arc second scale: the barred systems tended to contain more powerful central sources.   Puxley *et al*. (1988, in preparation) have obtained additional observations at Westerbork which extend the overlap between Hummel's sample and our IRAS sample to 124 galaxies.

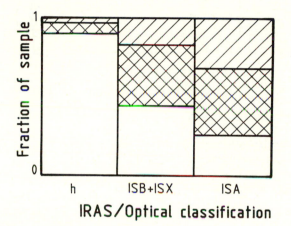

Figure 1.   The incidence of unresolved central radio sources (blank area), extended disc emission (hatched area) and non–detections (stippled) amongst galaxies from our IRAS samples which are accessible from Westerbork.

All but three of the 30 "h" galaxies have central sources which are unresolved in the 20 arcsec beam at 20 cm. Two of the three exceptions are nearby barred objects with known star–forming complexes larger than the Westerbork beam, NGC 3310 and NGC 4088.

A much smaller proportion of the "l" barred galaxies also exhibit central sources, although in total these are approximately as numerous as those in the "h" galaxies. This presents us with a puzzle: what are the starburst–like radio features in the non–starburst barred galaxies?

One possibility is that the radio emission arises from the enhanced magnetic field produced by the concentration of ISM near the centres of these galaxies, even when the density enhancement of the medium is insufficient to trigger a star formation episode.

Amongst the "h" SB galaxies the central sources contribute 30 to 40% of their total radio luminosity (as opposed to an average of only 10% for all the spirals in Hummel's sample). The central radio luminosities of these galaxies correlate extremely well with their point source 25 μm fluxes (Fig 2).

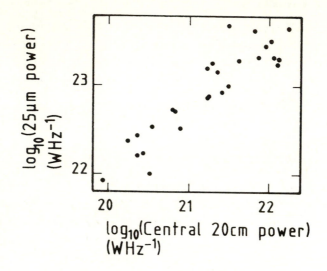

Figure 2.   The correlation between the 20 cm radio luminosities of the central sources and the point source catalog 25 μm fluxes for the "h" galaxies in Fig. 1.

2.2 VLA imaging of the central sources

The 20 cm results provide convincing evidence that the "starburst" IRAS colours arise in the vicinity of the nucleus, but the nature of the radio source is still to be unambiguously determined.   The ground–based 10 μm work of Devereaux *et al*., (1987) also indicates that the IR emission in these galaxies is centrally concentrated, but these authors consider that similar contributions could be coming from dust heated in Starbursts and from AGNs.   Once again we have had recourse to radio observations to investigate this.

VLA images, mostly at 6 cm, with resolutions of a few arcseconds, both from our own observations and from the literature, are now available for about half of the "h" and a few "l" galaxies in our sample. Fig. 3a shows an example, a VLA map of NGC 4536.   In almost all cases the radio sources in the galaxies for which we have data are resolved on scales between about 4 and 20 arcseconds (roughly 0.2 to 2 kpc, at the distances involved here).   As the emission is extended, it appears highly likely that its source in these objects is star formation. A few galaxies exhibit both an extended region of emission and an unresolved source, but even in these systems the central compact objects appear to make only a small contribution to the overall radio luminosity.

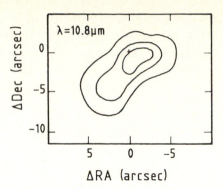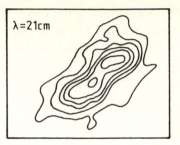

Figure 3 (a) VLA image at 20 cm of the central HII complex of NGC 4536 (from Condon *et al.*, 1982. (b) The same complex imaged at 10.8 μm by Telesco *et al.*, 1986.

3. IMAGING IN THE THERMAL INFRARED

We are warned by the presence of central radio sources in many galaxies without strong starbursts (a number of "l" SB systems, for example) that we cannot take for granted that the radio structure of our sources is a completely reliable guide to their infrared morphology. Mid-infrared maps of a sufficiently large and general sample are necessary to confirm this by direct comparison. Such maps are also useful in determining the true total luminosities of the complexes, free of contamination from any disc star formation complexes which may have been included in the large IRAS beams. Previous work (Telesco *et al.*, 1986) has already produced a number of thermal IR images for galaxies in the sample; that of NGC 4536 is shown in Fig. 3b. As in this case, all the 10 micron images have so far proved to be co-extensive with those in the radio, although the detailed structures in each are not identical. A programme to image a large fraction of our sample at 10 and, if possible, 20 μm using "Big Mac", (the Marshall Spaceflight Centre 20-channel bolometer array developed and operated by C M Telesco) is underway.

4. INFRARED RECOMBINATION-LINE SPECTROSCOPY

To investigate the age and evolutionary status of a starburst (e.g. Telesco, 1985) we need to know the total infrared luminosity also the total flux of ionising radiation in the complex which is an indicator of the number of unevolved upper main sequence OB stars present. We have therefore undertaken a programme of spectrometry at UKIRT, using the single-channel InSb spectrophotometer UKT9 to measure Brγ 2.166 μm hydrogen recombination line in a 20 arcsec beam and the 7-channel spectrometer CGS2 to observe Brα 4.052 μm line in a 5.4 arcsec beam.

Our programme comprises all those "h" and "l" galaxies in our sample with F(25)>= 1.5 Jy, at which level we expect to detect Brγ at 3 to 4 sigma in about an hour.

Again, this work will be described elsewhere (Puxley, *et al.*, in preparation) so we simply summarise the main results so far. Out of 21 "h" galaxies, Brγ has been detected in 17 while

Brα has been detected in 9 of the 10 systems with radio structure on 5 arcsec scales which have so far been examined in this line, including 2 of the 4 "h" galaxies in which Brγ was not seen. Amongst the 7 "l" galaxies observed to date Brγ has been detected in only one, NGC 3079 (Hawarden *et al.*, 1988 in preparation) which has a complex radio structure suggestive of a bipolar outflow from an active nucleus.

The detection of IR line emission in the "h" systems but not in the "l" galaxies with similar 25 μm fluxes clearly demonstrates a major difference in the sources of the IRAS emission in the two groups.

Very high internal extinctions in the undetected systems could account for this result. However the "l" galaxies, as a class, have lower IRAS luminosities – especially at 25 μm – than do the "h" systems and are much less numerous in our IR–flux limited spectroscopic sample. It is hard to believe that low IR luminosity can be strongly associated with a very high dust content.

The "l" galaxies could have a low level of ionising flux for their IR luminosity as is the case in Arp 220 (DePoy, Becklin & Geballe, 1987). However, none of the undetected "l" galaxies for which we have high resolution radio images are dominated by compact central sources as Arp 220 appears to be.

The spectroscopic results are most simply explained if the source of the IR luminosity in both "l" and "h" galaxies is star formation, which in the "l" objects is a disc phenomenon and therefore distributed so widely across the galaxy that too little of the associated ionised material is included in the small spectroscopic beam to contribute a detectable signal in the Br γ line. In contrast, of course, the bulk of this activity in the "h" galaxies falls close to the nucleus and within our observing aperture.

For a few galaxies in our sample the data set is nearly complete. One of these is M83, a portion of the K band spectrum of which is illustrated in Fig.4. From Telesco *et al.* (1986) we derive a total luminosity of $4 \times 10^9 L_o$. From Br α and Br γ observations, Turner, Ho and Beck (1987) (THB) infer extinctions A_v between 15 and 30 magnitudes. In combination with our Br γ flux from Fig. 5 ($19\pm4 \times 10^{-17}$ W m^{-2}) and the usual conventional assumptions we derive a burst age between 9×10^5 and 7×10^6 years for $A_v = 30$ and $A_v = 15$ mag, respectively.

These ages are more than an order of magnitude larger than those which THB derived when they took the estimated thermal radio emission into account. We hope to map this whole complex at Brα in the near future, which should greatly reduce the uncertainties in our result. A major end product of our programme will be ages on a consistent scale for all the starbursts in the galaxies of the spectroscopic sample.

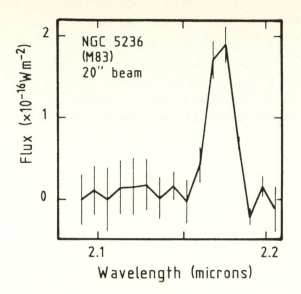

Figure 4. Continuum–subtracted CVF spectrum of M83 (NGC 5236) measured through a 19.6 arcsec aperture at a resolution of 0.019 μm, showing the Brγ line.

5. CONCLUSIONS

All the data available thus far are consistent with our general hypothesis that the large majority of galaxies with steep spectra between 12 and 25 μm owe this property to vigorous star formation in their central regions. The possession of a bar (i.e. membership of the morphological families SB or SAB in the RC2 system) is a necessary condition for the occurrence of this phenomenon.

Our low resolution radio results show that almost all the steep–spectrum "h" galaxies, but only a minority of the flat–spectrum "l" galaxies, have central sources less than 20 arcsec in diameter. Concentration of activity towards the centre is confirmed by near–IR spectroscopy of a 25 μm flux limited sample, amongst which a large majority of "h" galaxies, but few or none of the "l" systems, are detected at Brγ and/or Brα.

VLA Radio images at higher resolution show that in essentially all the galaxies thus far observed, the central sources are resolved on scales of a few arcsec. This is consistent with star formation, as opposed to an active nucleus, being the dominant contributor to the IR emission from these systems.

ACKNOWLEDGEMENTS

We thank the UK Panel for the Allocation of Telescope Time for observing time on UKIRT and the Time Allocation Committees for the VLA and the WSRT for access to these facilities. Gillian Wright, Dolores Walther, Thor Wold and Joel Aycock assisted us at various times at UKIRT. P J Puxley acknowledges support from an SERC studentship.

REFERENCES

Combes, F. & Gerin, M., 1985. Astr. Astrophys., 150, 327.

Condon, J.J., Condon, M.A., Gisler, G. & Puschell, J.J., 1982. Astrophys. J., 252, 102.

DePoy, D.L., Becklin, E.E. & Geballe, T.R., 1987. Astrophys. J., 316, L63.

Devereaux, N.A., 1986. 2nd International IRAS Conference, Pasadena.

Hawarden, T.G., Mountain, C.M., Leggett, S.K. and Puxley, P.J., 1986. Mon. Not. R. astr. Soc, 221,41P.

Hummel, E., 1981. Astr. Astrophys., 93, 93.

Rowan-Robinson, M. & Crawford, J., 1986. Light on Dark Matter, ed. Israel, F.P., Reidel, Dordrecht, Holland, p. 421.

Telesco, C.M., 1985. RAL Conference on Extragalactic Infrared Astronomy.

Telesco, C.M., Decher, R., Ramsey, B., Wolstencroft, R.D. & Leggett, S.K., 1986. 2nd International IRAS Conference, Pasadena.

Turner, J.L., Ho, P.T.P. & Beck, S.C., 1987. Astrophys. J., 313, 614.

Veron-Cetty, M.-P. & Veron, P., 1985. ESO Scientific Report No. 4.

ARE STARBURSTS THE RESULT OF THE FINE TUNING OF DYNAMICAL TIMESCALES ?

P.N.Appleton

School of Physics and Astronomy, Lancashire Polytechnic, Preston, U. K.

and

Curtis Struck-Marcell

Astronomy Program, Iowa State University, Ames, IA 50011, U.S.A.

ABSTRACT

When disk galaxies interact or collide gravitational fluctuations often generate density waves. We present cloud-fluid models of a particularly simple form of colliding galaxy, ring galaxies. Our models suggest that a certain degree of fine tuning between relevant timescales is required to obtain strong starbursts in the wave. Although the models are specific to ring galaxies, the results are applicable to other kinds of tidal interaction and to the early stages of certain types of galaxy merger. The results may explain why not all disturbed galaxies exhibit strong starburst activity.

1. INTRODUCTION

Interacting or merging galaxies may be responsible for triggering star burst activity in galaxies, giving rise to larger than normal far IR fluxes and higher IR colour temperatures as measured by IRAS. To gain an understanding of the processes which might be responsible for the starburst phenominon we have chosen to study a particularly simple form of colliding galaxy, the Ring galaxy (See for example Lynds and Toomre 1976). It is believed that ring galaxies are produced when a small companion passes close to the centre of a disk galaxy, generating near circular density waves in the disk. Although rare, these galaxies appear to show high levels of starformation activity similar to that of starburst systems but on an extended scale (See Jeske 1976; Appleton and Struck- Marcell 1987a).

It might be expected that in the early stages of a merger between two massive disk galaxies short lived rings may be formed as the potential fluctuates. It is interesting therefore that examples of possible ring galaxies are found in the high IR luminosity samples of Soifer et al and Wolstencroft et al (this conference proceedings).

2.1 THE MODELS

The models described here are based on the Oort picture of the interstellar medium (ISM) discussed by Scalo and Struck-Marcell (1984) and the full details of the model can be found in this and other papers referred to below. The cloud-fluid models treat the ISM a dilute gas composed of interstellar clouds the properties of which are governed by three major processes. These are;

1)cloud coalescence (or shredding) by the action of cloud collisions operating on a cloud-cloud collision time T_c; 2) the shredding of massive clouds in regions of star formation (this process is delayed by a time T_d to allow for stars to form and feedback energy and momentum into the ISM) and 3) the spatial flow of material from one computational cell to another, driven for example by the perturbation of the companion galaxy operating on a timescale T_{ex}.

In the absence of 3) the fluid behaviour is controlled by one principal parameter, namely the ratio T_d/T_c (see Scalo and Struck-Marcell 1986). If $T_d/T_c \ll 1$ the fluid was found to be stable. However when T_d/T_c approached unity, the solution to the fluid equations bifurcated leading to periodic bursts of star formation. This bifurcation condition is equivalent to a critical threshold density ρ_c above which bursts can develop.

2.2 INTERACTING SYSTEMS

In interacting systems spatial flows are important (Appleton and Struck-Marcell 1987b; Struck-Marcell and Appleton 1987, hereafter SMA). Even if the initial disks are well below the threshold density ρ_c required for the bifurcation to take place, density waves produced by the interaction can push regions of the galaxy over the bifurcation threshold. However to get a strong burst of star formation which can outshine the luminosity of the host galaxy, it is neccessary for the clouds in the wave to remain above the threshold long enough for the fluid oscillations to gain strength. If the clouds pass through the wave too quickly no burst will occur.

To ensure that a burst occurs, the dynamical timescale T_{ex} of the ring progagation (a timescale which depends mainly on the mass of the primary galaxy) must be tuned to the timescale of the cloud collisions T_c in the host galaxy. To demostrate this, Fig 1 and 2 show a time sequence of both density and starformation rates taken from some ring galaxy models descibed more fully in SMA. The figures show the radial behaviour of an initially flat 1-d cylindrically symmetric disk galaxy The centre of the target galaxy is at r = 0 and was originally set up in centrifugal equilibrium with its central softened mass. The disk is perturbed by a 1/5 mass companion passing through the centre just prior to the time shown in Fig. 1a and 2a. Both sequence show the growth of a density wave in the disk which moves away from the centre with time producing a ring. The 2 models shown here differ only in the initial cloud-cloud collision timescale of the target disk. In Figure 1 (Model B of SMA) the initial density of the disk is well below the critical density ρ_c. As the ring develops the threshold density is achieved by Fig 1b.

However the starformation rate does not give rise to a starburst because the fluid does not have enough time above the density threshold to develop the high amplitude bursts. Indeed, even at the later time shown in Fig 1c the starformation rate is only 4 times higher in the ring than the original target disk. The situation is markedly different in Fig 2 (Model D of SMA). Although the amplitude of the wave is the same as in the earlier model, the initial model was set up with a density much closer to the critical density (This is equivalent to saying the cloud collision time was very short in this galaxy before the collision compared with the model of Fig 1).

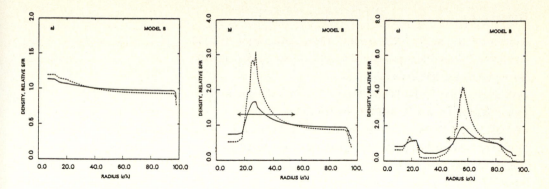

Fig.1 The radial distribution of density (solid line) and starformation rate (dotted line) at three times just after the passage of a companion through the centre. The bifurcation threshold $\rho_c = 1.3$ x initial density.

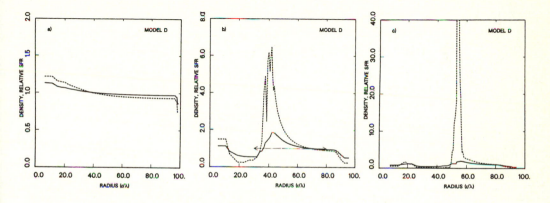

Fig.2 Same as in Fig. 1 except the value of $\rho_c = 1.01$ x initial density.

As the time sequence shows fluid elements spend much more time above threshold in this case and a true local burst occurs in the ring by Fig 2c. The net starformation rate rises to over 40 times the initial value and leads to a increase by a factor of 2 in the total star formation rate of the galaxy. The further evolution of the rings models showns the development of a second ring which nearly always shows starburst activity. This is because the second ring is rather shock-like and this appears to produce a strong burst in the cloud-fluid.

5. CONCLUSIONS

To achieve a strong burst of starformation in the ring galaxy models requires that the int erstellar medium of the interacting galaxy be "tuned" so that cloud-cloud collisions occur on timescales similar to dynamical interaction timescales, otherwise no significant burst will occur. The exception to this is in the second ring where a strong burst is always observed because of its shock-like nature. This necessity for fine tuning the dynamical timescales may go some way to explaining why not all extreme example of interaction give rise to strong starburst activity (e.g.Bushouse 1986).

References

Appleton,P.N. and Struck-Marcell,C.,1987a.*Astrophys.J.*,312,566.

Appleton,P.N. and Struck-Marcell,C.,1987b.*Astrophys.J.*,318,103.

Bushouse,H.A.,1986.*Astron.J.*,91,255.

Jeske,N.A.,1986. Ph.D. Thesis, University of California, Berkeley.

Lynds,R. and Toomre,A.,1976.*Astrophys.J.*,209,382.

Struck-Marcell,C. and Appleton,P.N.,1987.*Astrophys.J.*(in Press). (SMA)

Scalo,J.M. and Struck-Marcell,C.,1984.*Astrophys.J.*,276,60.

Scalo,J.M. and Struck-Marcell,C.,1986. *Astrophys.J.*,301,77.

IRAS OBSERVATIONS OF AN OPTICAL SAMPLE
OF INTERACTING GALAXIES

Susan A. Lamb

Department of Physics and Astronomy
University of Illinois at Urbana-Champaign
Urbana, Illinois 61801

Howard A. Bushouse, Michael W. Werner, Bruce F. Smith

NASA-Ames Research Center, MS 245-6
Moffett Field, California 94035

ABSTRACT

IRAS observations of an optically selected sample of closely interacting disk galaxies and comparisons with a control sample of isolated disk galaxies indicates that the interaction leads to enhancements in infrared luminosity, in L_{IR}/L_B, and in $L_{IR}/L_{H\alpha}$. The IR flux is well correlated with the $H\alpha$ and blue fluxes for both the interacting and isolated galaxies. Unique to the interacting sample is a class of very luminous systems which have unusually high L_{IR}/L_B values. The average 60–100μm color temperatures of the interacting galaxies are higher than those of the isolated disk galaxies.

1. INTRODUCTION

Galaxy collisions and mergers can have a dramatic impact upon the morphology and subsequent dynamical evolution of galaxies, as has been shown both from dynamical simulations (e.g. Toomre and Toomre, 1972; Miller and Smith, 1980) and from comparison of these results with the morphology of actual galaxy pairs and possible merger products. In recent years it has also become apparent that galaxy-galaxy collisions can often lead to enhanced levels of star formation activity, particularly in the nuclear regions of galaxies (Bushouse, 1986 and references therein). The *IRAS* survey has served to point out the prevalence of this phenomenon at IR wavelengths (see, e.g., *Star Formation in Galaxies*, 1987).

Here we concentrate on the phenomenon of enhanced star formation in strongly interacting galaxy pairs. We have chosen to investigate the infrared fluxes and colors of a sample of close pairs of disk galaxies for which good optical observations are available. This allows a comparison between the interacting sample and a control sample of isolated disk galaxies, and between optical and infrared indicators of star formation for the interacting sample.

2. SAMPLE SELECTION AND IRAS OBSERVATIONS

Our sample of interacting galaxies is taken from Bushouse (1986) and consists of 108 colliding pairs and several on-going mergers. This is a morphologically selected sample, containing only pairs of galaxies that exhibit features unmistakably associated with strong tidal interactions (e.g. tidal tails and bridges). Membership in the sample has also been limited to systems that show some evidence of a stellar disk. These galaxies were chosen without regard to any previous knowledge of such parameters as optical colors, spectral characteristics, group or cluster membership, or level of radio emission.

IRAS fluxes were obtained for 75% of the interacting pairs by use of the ADDSCAN utility at IPAC. Point Source Catalogue fluxes were used for the remaining systems. The small angular separations of the galaxies within each pair prevents the individual galaxies from being resolved in any of the four *IRAS* bands. Therefore when quoting luminosities for the interacting galaxies we will actually refer to one-half the total observed for a pair, in order to facilitate comparisons with individual isolated galaxies. The median distance to the interacting galaxies is 89 Mpc (assuming $H_0 = 75$ km s^{-1} Mpc^{-1}) and the median blue luminosity is $\sim 1.3 \times 10^{10}$ $L_\odot$. Only one definite Seyfert galaxy (UGC 6527d) was discovered among the sample of interacting pairs.

A control sample of 83 isolated disk galaxies, types Sa through Im, was chosen from the list of Kennicutt and Kent (1983). Point Source Catalogue fluxes were used for the entire control sample, and therefore the selected galaxies were limited to systems that have major axis diameters less than 4' so that their PSC fluxes are reasonably accurate. Known Seyfert galaxies have been omitted from the control sample primarily because we are interested in observables related to "normal" star formation activity and also because of the rarity of Seyferts among the interacting pairs. The median distance to the isolated galaxies is 19 Mpc and the median blue luminosity is $\sim 8 \times 10^9$ $L_\odot$. Thus we note that $L_B(\text{interactors})/L_B(\text{isolated}) \sim 1.7$.

3. INFRARED FLUXES AND COLORS

Far-infrared (FIR) luminosities have been computed for the galaxies in both samples following the procedure outlined in Appendix B of Lonsdale et al. (1985). This FIR value represents the total flux in an ideal 42–122μm bandpass. The median L_{FIR} for the interacting and isolated galaxies is 1.2×10^{10} $L_\odot$ and 2.2×10^9 $L_\odot$, respectively. We need some measure of the intrinsic sizes of galaxies in the two samples so that increases in infrared flux due to differences in this quantity can be factored out. The best measures we have available at present are the blue luminosities. We find that the median value of L_{FIR}/L_B for the interacting galaxies is 5.6 and for the isolated galaxies is 3.0. Thus by this measure interactions are producing an approximate doubling of the FIR flux.

An infrared color-color plot for the interacting and isolated galaxy samples is shown in Figure 1. Only those objects detected in all four *IRAS* bands are included in the plot. While there is a large amount of overlap in the range of IR colors for the two samples, the distribution of interacting systems is shifted to higher $f_\nu(60)/f_\nu(100)$ and lower $f_\nu(12)/f_\nu(25)$ values (see also Helou, 1986).

For the interacting galaxy sample there exists a high L_{FIR}/L_B tail, which is not present in the control sample, consisting of seven systems. All of these have high FIR luminosities, ranging from 1.9–3.1$\times10^{11}$ $L_\odot$, and are members of a class of objects having $f_\nu(80)/f_\nu(B)$ values only slightly lower than those of the ultra-luminous infrared galaxies described by Sanders et al. (1988). All seven of these objects also have $f_\nu(60)/f_\nu(100)$ and $f_\nu(25)/f_\nu(12)$ flux ratios that are higher than average for the interacting galaxy sample as a whole. In addition, optical spectra of the central (~2 kpc) regions of these objects (Bushouse, 1986) contain indications of intense star formation in the form of strong HII region-like emission lines and underlying continuum features characteristic of young or intermediate age stellar populations.

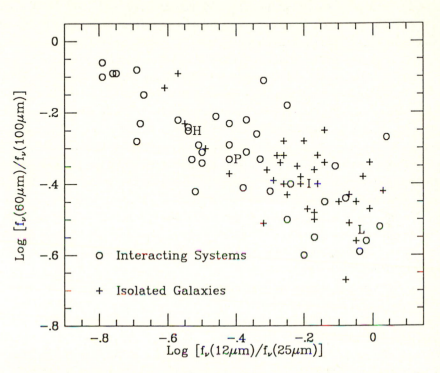

Figure 1. Infrared color-color plot for interacting systems and isolated galaxies. The average colors for various subsamples are also plotted: I = all isolated galaxies; P = all interacting pairs; H = high SFR interactors; L = low SFR interactors. The average colors of interacting systems dominated by disk star formation are nearly the same as the average for all interactors.

4. COMPARISON BETWEEN OPTICAL AND INFRARED INDICATORS OF STAR FORMATION

Using the available optical data for the interacting galaxies (Bushouse, 1986; 1987) and the isolated galaxies (Kennicutt and Kent, 1983) we can compare the classical optical indicators of star formation with the IR properties of the galaxies. There is a correlation between L_{FIR} and $L_{H\alpha}$ for both samples of galaxies. In particular the Hα flux correlates with the infrared flux in the 60μm and 100μm bands for both samples and with the flux in the 25μm band for the interacting pairs. The $L_{FIR}/L_{H\alpha}$ distribution is shifted to higher values for the interacting sample as compared to the isolated galaxies, and the median value of $L_{FIR}/L_{H\alpha}$ is 2.6 times higher. This suggests that star formation in the interacting galaxies may be occuring in more embedded environments, with consequently more localized radiation fields which results in higher levels of FIR radiation and/or suppressed Hα flux. It is generally assumed that Hα emission-line equivalent width is a measure of the ratio of recent (last 10^7 years) to more long term (averaged over last 10^9 years) star formation in a galaxy. We find no obvious correlation between this quantity and either the FIR flux or L_{FIR}/L_B for either sample.

Bushouse (1987) has found that global star formation rates (SFRs) of interacting galaxies span a large range and when interaction-induced enhancements in the SFR do occur, it usually concentrates in the near-nuclear regions of the galaxies. There are, however, some systems in which disk star formation dominates. It appears that the level of current star formation in interacting galaxies and the location of that star formation in the galaxy correlate to some extent with IR colors. In Figure 1 we note the locations of the average colors for the entire interacting and isolated galaxy samples, as well as the locations of the average colors of those interacting systems that are experiencing high and low current SFRs. [For convenience we will refer to the upper left of the diagram as the high end of the distribution and the lower right as the low end.] As Figure 1 shows, the average colors of high SFR systems are the highest of all the samples, while the low SFR systems are lower than even the average for isolated galaxies. We note that the majority of interacting systems in the high SFR sample are dominated by nuclear-region star formation. The average IR colors of interacting systems dominated by disk star formation fall very close to that of the entire interacting sample and are therefore systematically lower than the colors of the high (nuclear region) SFR sample. Thus the optical and infrared indicators here imply a difference in the character of nuclear vs. disk star formation activity.

Within the sample of interacting pairs there are a few galaxies which, on the basis of optical spectra, appear to have high nuclear region SFRs, but have low $f_\nu(60)/f_\nu(100)$ values, i.e. low dust temperatures, and low FIR fluxes as compared to the sample as a whole. Their L_{FIR}/L_B values are also below the median for both samples. Thus the optical and IR data taken together imply the existence of a group of objects which are interacting and are experiencing high SFRs, but do not show up strongly in the infrared, perhaps because they contain little dust. This implies that FIR studies will not automatically pick out all starburst systems.

5. CONCLUSIONS AND FUTURE WORK

We conclude that close interactions between disk galaxies can lead to enhanced star formation but not in all cases. The FIR luminosity is almost doubled relative to L_B when compared to non-interacting disk galaxies. The value of $L_{FIR}/L_{H\alpha}$ is increased by a factor of 2.6, which leads us to suggest that the increase in FIR flux is at least partially due to an increase in embedded star formation activity in interacting galaxies. There is a class of interacting galaxies which have very high L_{FIR}/L_B values and are experiencing very high rates of star formation in their nuclear regions. These objects form a class only slightly less luminous than the most luminous *IRAS* galaxies. There are some high SFR galaxies within the interacting sample which have lower than average FIR fluxes and dust temperatures. Taken together the optical and IR data imply a tremendous variety in star formation activity among closely interacting disk galaxies.

Future work will address several of the remaining problems in understanding star formation in interacting pairs of galaxies. We plan to compare our current results with 10μm and radio maps of these galaxies and to obtain more optical data, particularly on the disks of the interacting galaxies and the compactness of the emission regions. We also plan to supplement the *IRAS* data with higher resolution airborne observations, and to attempt to resolve some of the more widely separated pairs using deconvolving algorithms on the *IRAS* data. This will give us very valuable information about the relative contributions to the total IR flux by the individual galaxies within the pairs. Lastly, we are currently obtaining information from theoretical N-body simulations of interacting galaxies performed by Miller and Smith (1980) on the timescales for interactions and the energies available for star formation due to the internalisation of orbital energy. The experiments also give indications of where the energy is deposited and may help provide a natural explanation of why, in most cases, star formation is particularly enhanced in the nuclear regions of interacting galaxies.

References

Bushouse, H. A. 1986, Ph.D. thesis, University of Illinois at Urbana-Champaign.

Bushouse, H. A. 1987, *Ap.J.*, in press.

Helou, G. 1986, *Ap.J.(Letters)*, **311**, L33.

Kennicutt, R. C., and Kent, S. M. 1983, *A.J.*, **88**, 1094.

Lonsdale, C. J., Helou, G., Good, J. C., and Rice, W. L. 1985, Jet Propulsion Lab, preprint D-1932.

Miller, R. H., and Smith, B. F. 1980, *Ap.J.*, **235**, 421.

Sanders, D. B., Soifer, B. T., Elias, J. H., Madore, B. F., Matthews, K., Neugebauer, G., and Scoville, N. Z. 1988, *Ap.J.*, in press.

Star Formation in Galaxies 1987, ed. C. J. Persson, (U.S. Government Printing Office, Washington, D. C.).

Toomre, A., and Toomre, J. 1972, *Ap.J.*, **178**, 623.

Cosmology in the Infrared

COSMOLOGICAL BACKGROUND RADIATION IN THE INFRARED

B.J.Carr

School of Mathematical Sciences
Queen Mary College
Mile End Road
London E1 4NS.

ABSTRACT

Cosmological background radiation may already have been detected in both the near-IR and far-IR. Even if these detections are not confirmed, space experiments should soon present the possibility of detecting IR backgrounds at a level where their existence is inevitable as a result of astrophysical activity in the pregalactic or protogalactic era. These backgrounds will generally appear in the near-IR unless they have been absorbed by galactic or intergalactic dust, in which case they should have been reprocessed into the far-IR.

1. INTRODUCTION

In this talk I will first review the attempts to detect a cosmological infrared background (CIRB). Such observations are exceedingly difficult and one can usually only place upper limits on the background density in different wavebands. However, we will see that there are already some claims to have detected a CIRB and the prospects should improve dramatically within the next few years as a result of space experiments. I will next discuss the possible sources of a CIRB. In particular, I will argue that, in the absence of dust, one could expect many astrophysical sources in the cosmic "dark ages" between decoupling and galaxy formation to generate a near-IR background. I will then consider the effects of dust, arguing that one would expect most of the backgrounds to have been reprocessed into the far-IR, with a spectrum which depends only weakly on the grain parameters. Finally I will discuss the anisotropies expected in this background, stressing that these could provide a unique probe of pregalactic conditions.

2. OBSERVATIONS OF A COSMOLOGICAL INFRARED BACKGROUND

For the purposes of this talk, a cosmological background will be defined as one which originates at a cosmological redshift ($z>1$); such a background could be generated by unresolved astrophysical sources but it could also have a non-astrophysical origin. I will focus attention on wavelengths between 1μ and 1000μ, although wavelengths above 500μ would generally be regarded as submillimetre rather than infrared. In a cosmological context, the intensity of a background is conveniently measured by the energy density per logarithmic frequency interval divided by the critical density required for the Universe to recollapse ($\rho_{crit} = 2 \times 10^{-29} h^2$ g/cm^3 where h is the Hubble parameter in units of 100 km/s/Mpc). This is denoted by $\Omega_R(\lambda)$ and related to the more familiar observational quantities I_ν and λI_λ by

$$\Omega_R(\lambda) = 7 \times 10^{-7} h^{-2} \left[\frac{\lambda}{100\mu} \right] \left[\frac{I_\nu}{MJy/sr} \right] = 2 \times 10^5 h^{-2} \left[\frac{\lambda I_\lambda}{W/cm^2/sr} \right] \qquad (1)$$

For comparison, the microwave background radiation (MBR) has a total density $\Omega_R \approx 2 \times 10^{-5} h^{-2}$ and peaks at 1600μ.

The observation of any background is difficult since one needs an absolute calibration. For an IR background, the problem is compounded because atmospheric effects usually necessitate non-ground-based observations and, even then, there are a large number of *local* backgrounds which have to be subtracted before one can extract a cosmological contribution. The most important ones are zodiacal light (ZL), starlight (SL), interplanetary dust (IPD), interstellar dust (ISD), and foreground infrared galaxies (IG). Rough estimates of these local backgrounds are shown in Figure (1) but it must be appreciated that some of them are very uncertain. (For example, there is some disagreement between the six estimates of IPD emission presented at this conference.) One sees that the local backgrounds are smallest at around 4μ, 100μ and 700μ, so these are the best wavebands in which to search for a cosmological background. Although positive detections have been claimed in all of these wavebands, one must bear in mind the possibility that the local backgrounds have been underestimated.

In principle, there are two ways in which cosmological backgrounds can be differentiated from local ones. Firstly, one can use the Sunyaev-Zeldovich (1972) effect: radiation passing through the hot gas in any intervening cluster will have its spectrum distorted (Fabbri & Melchiorri 1979). The mean photon energy will be increased, leading to a deficit longward of the peak and an excess shortwards of it. If this effect is observed, it implies that the background must have originated at a higher redshift than the cluster. So far, it has only been detected for the MBR (Birkinshaw & Gull 1984). Secondly, one can use the fact that the peculiar motion of the Sun relative to the cosmological rest frame should induce a dipole anisotropy in any cosmological background. It should also modify the spectral index of the background (Ceccarelli *et al.* 1983). Again, this effect has only been detected for the MBR (Smoot *et al.* 1977, Boughn *et al.* 1981, Fabbri *et al.* 1980) but it has already been used to constrain the far-IR background.

Various *upper limits* on the background radiation density in wavebands between 1μ and 1000μ are shown in Figure (1). The points labelled "1" are the balloon results of Woody & Richards (1981); "2" is the balloon result of de Bernardis *et al.* (1984) and "3" is the dipole anisotropy limit of Ceccarelli *et al.* (1983) which can be inferred from this; the points labelled "4" are the Hauser *et al.* (1984) IRAS results; those labelled "5" are the rocket results of Matsumoto *et al.* (1984), originally claimed as a detection but now usually interpreted as an upper limit; "6" is the balloon result of Hoffman & Lemke (1978). By way of comparison, Figure (1) also shows the generally stronger limits on the background density in the optical to UV range: "7" is the Kitt Peak limit of Dube *et al.* (1979); all the rest are rocket or satellite results, the appropriate references being given in McDowell (1986). The latter constraints are relevant to my talk because I will use them to argue that some optical–UV backgrounds must have been reprocessed into the far-IR by dust.

Let us now focus on claimed *detections* of a CIRB. In the near-IR, Matsumoto *et al.* (1984) reported the detection of a 2–5μ background with $\Omega_R \approx 3 \times 10^{-5} h^{-2}$ after a rocket experiment. However, the interpretation of this was difficult because of rocket exhaust problems. A more recent rocket experiment by the same group (Matsumoto *et al.* 1987) also appears to find a background (point "8") at around 2μ but it is smaller than before ($\Omega_R \approx 3 \times 10^{-6} h^{-2}$) and so narrow that they are obliged to interpret it as a line. Whether one could expect cosmological sources (which presumably span a range of redshifts) to produce such a narrow line remains an open question.

In the far-IR, Rowan–Robinson (1986) claimed to find a background with $\Omega_R \approx 5 \times 10^{-6} h^{-2}$ at 100μ from an analysis of IRAS data (point "9"). The main problem here is knowing how to subtract interplanetary dust emission. Rowan–Robinson adopted a very specific model for the density distribution of the dust but there are uncertainties. His most recent estimates are more conservative (Rowan–Robinson 1988) and allow for the possibility that there is no CIRB at 100μ. In any case, one has a safe upper limit: $\Omega_R(100\mu) < 5 \times 10^{-6} h^{-2}$.

In the submillimetre band, Gush (1981) reported a background at 500–800μ with $\Omega_R \approx 10^{-5} h^{-2}$ after a rocket experiment. Although problems with rocket exhaust make this result questionable, so it is not included in Figure (1), it is interesting that at this conference the Nagoya–Berkeley collaboration has announced the detection of a background peaking at 700μ with $\Omega_R \approx 6 \times 10^{-6} h^{-2}$ (Matsumoto *et al.* 1988). The details of the data are discussed elsewhere in this volume but it should be emphasized that one would expect a background like this is many cosmological scenarios. The result, if confirmed, would therefore constitute one of the most exciting cosmological discoveries since the detection of the MBR itself.

The next few years should see a great improvement in our ability to detect a CIRB as a result of various space experiments. While IRAS was sensitive to backgrounds with $\Omega_R \approx 10^{-6} h^{-2}$, COBE should allow us to dig down to $\Omega_R \approx 2 \times 10^{-8} h^{-2}$ over the entire waveband 1–1000μ and down to $\Omega_R \approx 10^{-9} h^{-2}$ at particular submillimetre wavelengths (Mather 1982). If SIRTF is ever launched, it will be able to dig down to $\Omega_R \approx 10^{-8} h^{-2}$ in the waveband 2–750μ. These prospects are summarized in Figure (2). Of course, this assumes that local backgrounds can be estimated with similar accuracy.

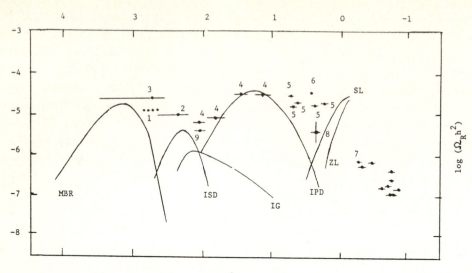

Figure (1). This summarizes the observational constraints on the background radiation density between 0.1μ and 10⁴μ, including the MBR, together with estimates of the dominant local backgrounds.

3. SOURCES OF A COSMOLOGICAL INFRARED BACKGROUND

There are many possible sources of a CIRB. The ones whose existence is most established are, of course, galaxies. These are associated with three types of IR emission: (i) the radiation from dusty disks; (ii) the "starburst" radiation coming from dust clouds; and (iii) the radiation from quasars and active galactic nuclei. A possible estimate of the associated background is indicated by the "IG" curve in Figure (1). However, it should be stressed that the form of this curve is dependent upon very uncertain evolutionary factors. One would need a very large evolutionary factor to explain the sort of background intensity claimed by Rowan-Robinson at 100μ (Hacking et al. 1987).

Here I would like to focus on pregalactic or protogalactic sources since these have been somewhat neglected in the literature. The point is that there could be several kinds of astrophysical generators of IR in the period between z=10 and z=10³: for example, primeval galaxies, Population III stars, accreting black holes, large-scale explosions, or decaying particles. If the radiation from these sources propagated to us unimpeded (unaffected by dust), it would presently reside in the near-IR to UV range. We now indicate the total energy density Ω_R and peak wavelength λ_{peak} associated with each background. The associated spectra are summarized in Figure (2). Most of them have a dilute black-body form, although this would require modification if there was enough neutral hydrogen in the background Universe to absorb photons shortward of the Lyman cut-off. In this case, most of the radiation would come out as recombination lines. If the backgrounds are reprocessed by dust, λ_{peak} will change but Ω_R will be roughly the same (as discussed in Section 4). More detailed calculation can be found in Bond et al. (1986).

Primeval Galaxies. Several arguments suggest that galaxy formation was accompanied by an initial burst of massive star formation which generated the first metals (Truran & Cameron 1971). The stars must have generated a metallicity Z of order 10^{-3} at a redshift z_G in the range 3–10 (depending on the epoch of galaxy formation) and must have had a mass M in the metal-producing range 10–100 $M_\odot$. Since these stars must also generate light, one can predict a minimum integrated background radiation density:

$$\Omega_R \simeq 2 \times 10^{-7} \left[\frac{Z_{ej}}{0.2} \right]^{-1} \left[\frac{\Omega_g}{0.1} \right] \left[\frac{1+z_G}{10} \right]^{-1} \left[\frac{M}{10^2 M_\odot} \right]^{0.5} \left[\frac{Z}{10^{-3}} \right] \qquad (2)$$

(cf. Peebles & Partridge 1967, Thorstensen & Partridge 1975, Shchekinov 1986). Here Ω_g is the initial gas density in units of the critical density (normalized to the value indicated by cosmological nucleosynthesis considerations) and Z_{ej} is the metal yield of each star. The presence of the M term in eqn (2) reflects the M–dependence of the efficiency with which nuclear burning turns the star's rest mass into radiation. The background should peak at a present wavelength

$$\lambda_{peak} \simeq 0.6 \left[\frac{1+z_G}{10} \right] \left[\frac{M}{10^2 M_\odot} \right]^{-0.3} \mu \qquad (3)$$

the M term reflecting the M–dependence of the surface temperature. If one regards M as unknown, one can use eqns (2) and (3) to express Ω_R as a function of λ_{peak} and thereby infer a background which is *minimal* in the sense that it must be attained at some wavelength. The curve "PG" in Figure (2) corresponds to a model with M=25$M_\odot$ and z_G=9. An analogous argument shows that the background associated with the stars which produce the solar metallicity is given by the curve "SM" in Figure (2). Since the appropriate normalizations are now $Z \simeq 10^{-2}$ and $z \simeq 1$, λ_{peak} is reduced by a factor of 10 and Ω_R is increased by a factor of 100. This is already in conflict with the background UV constraint, which suggests that the radiation must have been reprocessed by dust.

Population III Stars. It has been proposed that the dark matter in galactic halos is baryonic (Ashman & Carr 1987), in which case a large fraction of the Universe must have been processed through a generation of "Population III" stars which left dark remnants. Background light and nucleosynthetic constraints imply that the objects must be either jupiters or the black hole remnants of "Very Massive Objects" (VMOs) in the mass range above $M_c \simeq 200$ $M_\odot$ (Carr et al. 1984). Since VMOs have a surface temperature of 10^5K (independent of M) and radiate at the Eddington limit, one can show that the background light they generate will have

$$\Omega_R \simeq 4\times10^{-6} \left[\frac{1+z_*}{100}\right]^{-1} \left[\frac{\Omega_*}{0.1}\right] \quad , \quad \lambda_{peak} \simeq 4 \left[\frac{1+z_*}{100}\right] \mu \qquad (4)$$

where z_* is the redshift at which they burn and we have normalized the star density Ω_* to sort of value required to explain galactic halos. The curve "VMO" in Figure (2) assumes z_*=99. The background from jupiters would have a much smaller density ($\Omega_R\approx10^{-9}-10^{-8}$) but it would peak in the range 10–100μ and extend over a wider waveband. The "J" curve in Figure (2) assumes a jupiter mass of 0.085 $M_\odot$ and is based on the calculation of Karimabadi & Blitz (1984). Note that this background is non-thermal because it is associated with the *formation* of the jupiters (viz. a Hayashi phase, followed by a degenerate cooling phase) rather than their emission at the present epoch. The prospects of detecting individual jupiters directly are remote because they are so dim.

Black Hole Accretion. In order to explain quasars and active galactic nuclei, it is commonly supposed that some galaxies have giant black holes in their nuclei with a mass M of order $10^8 M_\odot$ (Rees 1978a). If the holes radiated at the Eddington limit for a "mass-doubling" time t_E, then they would generate a background with

$$\Omega_R \simeq 10^{-7} \left[\frac{\epsilon}{0.1}\right] \left[\frac{1+z_E}{10}\right]^{-1} \quad , \quad \lambda_{peak} \simeq 2 \left[\frac{1+z_E}{10}\right] \mu \qquad (5)$$

as illustrated by the curve "AGN" in Figure (2). Here ϵ is the efficiency with which the accreted material generates radiation and $z_E\approx6h^{-2/3}\epsilon^{-2/3}$ is the M-independent redshift corresponding to the epoch when the Hubble time is t_E. The wavelength estimate assumes that one has an optically thick accretion torus at a temperature of 2×10^4K, as suggested by the models of Begelman (1984). If galactic halos are the black hole remnants of VMOs, they would also accrete but the accretion would generally be sub-Eddington. In the pregalactic era, one can usually assume that they accrete at the Bondi rate from a medium with the cosmological gas density Ω_g and a temperature T$\approx10^4$K. In this case, they should produce a background with density

$$\Omega_R \simeq 7\times10^{-7} \left[\frac{\Omega_g}{0.1}\right] \left[\frac{\epsilon}{0.1}\right] \left[\frac{M}{10^6 M_\odot}\right] \left[\frac{T}{10^4 K}\right]^{-3/2} \left[\frac{1+z_*}{10}\right]^{1/2} h \, \Omega^{-1/2} \quad (6)$$

(where Ω is the total cosmological density parameter) and peak wavelength

$$\lambda_{peak} \simeq \left[\frac{\epsilon}{0.1}\right]^{-1/4} \left[\frac{\Omega_g h^2}{0.1}\right]^{-1/4} \left[\frac{T}{10^4 K}\right]^{3/8} \left[\frac{1+z_*}{10}\right]^{1/4} \mu \qquad (7)$$

(Carr *et al.* 1983). We have normalized M to $10^6 M_\odot$ since there may be dynamical evidence for halo black holes of this mass (Lacey & Ostriker 1986). The curve "HBH" in Figure (2) assumes M=$10^6 M_\odot$, ϵ=0.1 and z_*=9. The choice of z_* corresponds to the redshift at which most of the radiation is generated.

Pregalactic Explosions. It has been proposed that some features of the large–scale cosmic structure can be explained by pregalactic explosions (Ostriker & Cowie 1981, Ikeuchi 1981). One envisages each explosive seed (a star or a cluster of stars) generating a shock which sweeps up a shell of gas. In order to explain the existence of giant voids and the form of the galaxy correlation function, the shells must eventually overlap with a characteristic radius of order 10 Mpc (Saarinen et al. 1986). However, the stars which generate the explosive energy will also generate light, so one can predict a minimum background radiation density as follows. To generate density fluctuations of order 1 on a comoving scale d corresponds to a kinetic energy density (Hogan 1984)

$$\rho_{exp} c^2 \simeq m_b \; n_b \left[\frac{H(z)d}{1+z} \right]^2 \tag{8}$$

where m_b and n_b are the baryon mass and number density, z is the redshift at which the shells overlap, and H(z) is the Hubble rate at that redshift. Since stars

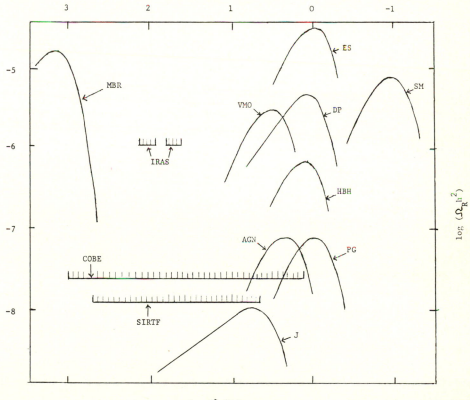

Figure (2). This summarizes the theoretical backgrounds which may have been generated by pregalactic and protogalactic events and compares them to the sensitivity of various space experiments. If the backgrounds are reprocessed by dust, Ω_R remains the same but λ moves into the far–IR.

generate η≈100 times as much radiation energy as explosive energy during the main-sequence phase which precedes the explosive phase, the background radiation density must be

$$\Omega_R \simeq 10^{-4} \left[\frac{\eta}{100} \right] \left[\frac{\Omega_g h^2}{0.1} \right] \left[\frac{d}{10 \text{Mpc}} \right]^2 \tag{9}$$

The spectrum should peak at the wavelength given by eqn (3). The curve "ES" in Figure (2) assumes η=0.3, as applies for exploding stars of mass 10$M_\odot$. This density is already in conflict with the optical to UV limits unless the radiation is reprocessed by dust.

Decaying Particles. Elementary particle relics of the Big Bang would be expected to pervade the Universe and, if their mass is sufficiently large, they could have an appreciable cosmological density. In certain models, these particles would be expected to decay radiatively on some timescale τ_d. For $\tau_d < 50y$, they would contribute to the MBR, while for $50y < \tau_d < 3 \times 10^4 y$ they would distort its spectrum (Silk & Stebbins 1983). However, for $\tau_d > 3 \times 10^4 y$, they would just generate an IR background with

$$\Omega_R \simeq 5 \times 10^{-6} \, \Omega_X \left[\frac{1+z_d}{10^5} \right]^{-1}, \qquad \lambda_{peak} \simeq 120 \left[\frac{1+z_d}{10^5} \right] \left[\frac{m_X}{\text{keV}} \right]^{-1} \mu \tag{10}$$

Here z_d is the decay redshift, Ω_X is the density parameter which would be associated with the particles had they not decayed, and m_X is the particle mass. Many models relate Ω_X, m_X and τ_d, so they are not necessarily independent. The curve "DP" in Figure (2) corresponds to $\Omega_X = 0.01$, $m_X = 1 \text{keV}$ and $z_d = 10^3$. Note that the spectrum deviates somewhat from the black-body form.

4. DUST OBSCURATION

The predicted spectra of Figure (2) apply only if the radiation propagates freely between us and the source. However, most of the backgrounds discussed in Section (3) are in the optical or UV and one might expect such backgrounds to be absorbed by intervening dust. In this case, they would be re-emitted at a longer wavelength. The dust could either be confined to galaxies (if galaxies cover the sky) or it could be uniformly spread throughout the Universe. The last situation is only likely to apply if there was a generation of pregalactic stars. We first determine the condition for dust absorption and then calculate the characteristics of the re-emitted radiation.

For simplicity we will assume that each grain has a radius r_d, that its absorption cross-section is πr_d^2 for wavelengths $\lambda < 2\pi r_d$, and that it falls off as λ^{-1} for $\lambda > 2\pi r_d$. Then we can write

$$\sigma \simeq \pi \, r_d^2 \left[1 + \left[\frac{\lambda}{2\pi r_d} \right] \right]^{-1} \tag{11}$$

In fact, the exponent of λ must increase to 2 at very long wavelengths (Draine & Lee 1984); a more general analysis is given by Bond et al. (1986). If the grains are uniformly distributed throughout the Universe and have a density Ω_d in units of the critical density, then the optical depth back to a redshift z for photons with present wavelength λ can be expressed as

$$\tau(\lambda,z) \simeq 1.3 \left[\frac{\Omega_d}{10^{-5}} \right] \left[\frac{\rho_{id}}{2} \right] \left[\frac{r_d}{0.1\mu} \right]^{-1} \left[\frac{1+z}{10} \right]^{3/2} \left[1 + \frac{5\lambda}{6\pi r_d(1+z)} \right]^{-1} h\, \Omega^{-1/2} \quad (12)$$

Here ρ_{id} is the internal grain density in g/cm^3 and we have normalized Ω_d to the dust density associated with galaxies. Thus the optical depth at short wavelengths reaches unity at a redshift

$$1+z \simeq 8 \left[\frac{\Omega_d}{10^{-5}} \right]^{-2/3} \left[\frac{r_d}{0.1\mu} \right]^{2/3} \left[\frac{\rho_{id}}{2} \right]^{2/3} \Omega^{1/3} h^{-2/3} \quad (13)$$

In practice, of course, Ω_d is itself a function of z, so it is useful to express the opaqueness condition in terms of the (Ω_d,z) space of Figure (3). This figure assumes that wavelength of the source radiation is less than $2\pi r_d$ at the time when it encounters the dust. This is usually the case. For example, the light from VMOs peaks at a wavelength of 0.04μ, so photons generated at a redshift z_* will satisfy the condition at a redshift z providing $r_d > 0.007(1+z_*)/(1+z)\mu$. The grains in our galaxy probably span a spectrum sizes between 0.01μ and 0.3μ (Mathis et al. 1977).

If one thinks of the mean cosmological dust density as following a trajectory $\Omega_d(z)$ in Figure (3), then photons from pregalactic sources will be absorbed by intervening dust providing there is some redshift between their emission and now at which the trajectory penetrates the shaded region of Figure (3). It is not clear whether this is the case. Ω_d clearly starts off below the shaded region since $\Omega_d=0$ initially; observations of distant quasars also imply that a uniform dust distribution must have $\Omega_d < 6\times10^{-5}h^{-1}$ back to z=2 (Wright 1981), so one certainly has $\tau<1$ at the present epoch. However, one could still have $\tau>1$ in some intermediate redshift range. For example, one would only need $\Omega_d>10^{-7}$ at z=300, the earliest epoch at which VMOs could complete their nuclear burning. Of course, one could only expect an intergalactic grain abundance like this if there was some pregalactic star formation.

Even if there is no intergalactic grain abundance, the dust *within* galaxies could still absorb any pregalactic radiation providing two conditions are satisfied. Firstly, the *mean* dust density (i.e. the density which would be obtained if the dust was spread uniformly throughout the Universe instead of being confined to galaxies) must be large enough for τ to exceed 1 at the redshift of galaxy formation (z_G). The contribution of galactic dust to Ω_d can be written as

$$\Omega_d \simeq 10^{-5} \left[\frac{f_d}{0.01} \right] \left[\frac{f_g}{0.1} \right] \left[\frac{\Omega_G}{0.01} \right] \quad (14)$$

where f_g is the fraction of the galaxy's mass in gas, f_d is the fraction of the gas in dust, and Ω_G is the density parameter associated with galaxies. This just corresponds to the normalization in eqn (13). Secondly, we need the galaxies to cover the sky at z_G. Otherwise, most of the background photons would be unaffected. This requires

$$1 + z_G > 11 \left[\frac{R_G}{10\text{kpc}} \right]^{2/3} \left[\frac{\rho_{iG}}{10^{-24}} \right]^{2/3} \left[\frac{\Omega_G}{0.1} \right]^{-2/3} \Omega^{1/3} h^{-1/3} \tag{15}$$

where R_G is the radius of the dust-containing part of the galaxy and ρ_{iG} is the density within a galaxy in g/cm^3. Since the redshift of galaxy formation is in the range 3 to 10, it is not clear whether these two conditions are satisfied. It is certainly possible, and Ostriker and Heisler (1984) have even proposed this as the explanation for why quasars cut off at a redshift of 4, but it is not necessarily the case. Note that we have normalized Ω_G, R_G and ρ_{iG} to values appropriate for our galaxy. In practice, these parameters will span a range of values; one would generally expect the smallest galaxies to contribute most to the covering factor.

Finally it should be stressed that there will be *in situ* dust absorption for some types of source. In particular, this will apply for sources embedded in dusty galaxies. The optical depth for an individual galaxy is

$$\tau_{gal} \simeq 1.5 \left[\frac{f_d f_g}{10^{-3}} \right] \left[\frac{r_d}{0.1\mu} \right]^{-1} \left[\frac{\rho_{iG}}{10^{-24}} \right] \left[\frac{R_G}{10\text{kpc}} \right] \left[\frac{\rho_{id}}{2} \right]^{-1} \quad [\lambda < 2\pi r_d] \tag{16}$$

For our galaxy this (angle-averaged) opacity is comparable to 1 and it may have been even larger in the past since f_g always decreases. Even if one has $\tau_{gal} < 1$, the radiation from galactic sources may still be absorbed if they are contained within clouds where the *local* dust density exceeds the average galactic value. This is evidenced, of course, by the existence of starburst galaxies. Most of the sources discussed in Section (3) are pregalactic and therefore not in this category. However, for those which are, one needs to know how much of the light is absorbed within the galaxy itself and how much is absorbed by the background Universe.

If the pregalactic radiation, whatever its source, is absorbed by dust, one can readily calculate how its spectrum is modified. If the radiation density is Ω_R, then thermal balance implies that the dust temperature T_d evolves according to

$$T_d(z) = T_c(z) \left[1 + \left[\frac{\Omega_R}{\Omega_c} \right] \left[\frac{r_d}{0.1\mu} \right]^{-1} \left[\frac{1+z}{10} \right]^{-1} \right]^{1/5} \tag{17}$$

Here $T_c(z)$ is the temperature of the MBR photons, Ω_c is the MBR density, and we have assumed $T_d \ll r_d^{-1}$, $T_c \ll r_d^{-1}$ and $T_s \gg r_d^{-1}$ in appropriate units. Thus if the radiation density is less than

$$\Omega_{crit} \simeq 2 \times 10^{-7} h^{-2} \left[\frac{r_d}{0.1\mu} \right] \left[\frac{1+z}{100} \right] \tag{18}$$

the dust temperature will just be the MBR temperature (the MBR heating alone ensuring that it never drops below this). However, if Ω_R exceeds the value given by eqn (18), the dust temperature will be somewhat larger than T_c. In this case, one expects a far-IR background with a spectrum peaking at a present wavelength

$$\lambda_{peak} \simeq 700 \left[\frac{\Omega_R h^2}{10^{-6}} \right]^{-1/5} \left[\frac{r_d}{0.1\mu} \right]^{1/5} \left[\frac{1+z}{10} \right]^{1/5} \mu \qquad (19)$$

The spectrum will not be exactly black-body: $\Omega_R(\lambda)$ will scale as λ^{-4} longward of the peak rather than λ^{-3}. There will also be spectral features associated with resonance effects, although these will tend to be smeared out by cosmological redshift effects.

What is the appropriate value of z to use in eqn (19)? Strictly speaking, one is dealing with a range of values since the reprocessed radiation comes from a shell: the outer edge of the shell corresponds to the redshift z_d at which the dust or radiation is generated (whichever is smaller) and the thickness of the shell is determined by the condition that the optical depth be unity in the waveband of the source radiation. However, the thickness is generally small, so the effective redshift is just z_d. The striking feature of eqn (19) is that λ_{peak} depends only weakly on r_d, Ω_R and z. Thus one expects all the reprocessed radiation from pregalactic sources to pile up at roughly the same wavelength. In a sense, this is unfortunate since it means that the spectrum itself contains little information about the origin of the radiation. On the other hand, it is interesting because it means that one can *predict* that a far-IR background with these characteristics ought to exist.

These considerations show that, if $\Omega_R > \Omega_{crit}$, one expects the total background spectrum to have three parts: the MBR component (peaking at 1400μ), the far-IR dust component (peaking at λ_{peak}), and the residual source component (peaking in the optical or near-IR). If $\Omega_R < \Omega_{crit}$, the dust radiation will be superposed on the MBR, so the overall spectrum will have only two peaks. However, since the dust radiation does not have a black-body spectrum, it will distort the MBR spectrum unless it is itself absorbed (Rowan-Robinson et al. 1979, Negroponte et al. 1981, Puget & Heyvaerts 1980)). If self-absorption occurs, the radiation can be thermalized, in which case the dust is actually generating part of the MBR. The condition for this is that the value of τ given by eqn (12) should exceed 1 at the wavelength given by eqn (19). This requires

$$1 + z > 65 \left[\left\{ \frac{\Omega_d}{10^{-5}} \right\} \left\{ \frac{\rho_{id}}{2} \right\} h \, \Omega^{-1/2} \right]^{-0.4} \left[\frac{r_d}{0.1\mu} \right]^{0.1} \left[\frac{\Omega_R}{\Omega_C} \right]^{-0.1} \qquad (20)$$

which corresponds to the heavily shaded part of Figure (3). In principle, provided the $\Omega_d(z)$ trajectory passes through this double-shaded region, one could hypothesize that the *entire* microwave background derives from grains (Layzer & Hively 1973, Rees 1978b, Wright 1982.) However, for a reasonable grain density, this requires the radiation to be produced at a very high redshift and the grains to be rather exotic.

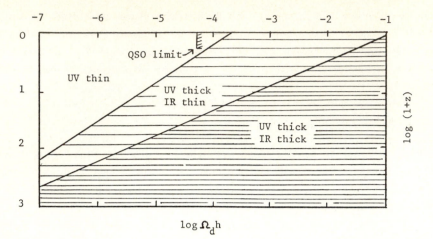

Figure (3). This shows the dust density required to absorb UV radiation at a redsdift z, thereby reprocessing it into the far-IR. In the heavily shaded region the far-IR radiation is itself absorbed, leading to thermalization. We assume $r_d = 0.1\mu$.

5. DISCUSSION

The question now arises of whether the theoretical predictions in Figure (2) could be relevant to any of the claimed detections of a CIRB. The sort of near-IR background reported by Matsumoto *et al.* (1987) could be generated by several of the models, provided one does not have too much dust absorption. Indeed, if the spectrum is cut off beyond the Lyman limit by neutral hydrogen absorption, Ly-α emission might even produce a narrow line feature (Carr *et al.* 1983, Matsumoto *et al.* 1988). If one wants to produce a far-IR background, one must clearly invoke dust. McDowell (1986) and Negroponte (1986) have already tried to explain Rowan-Robinson's 100μ background using dust and pregalactic stars and they find values of λ_{peak} which are in good agreement with the simple analytical estimate of eqn (19). Matsumoto *et al.* (1988) discuss whether their 700μ background can be explained in a similar way. Even without a detailed analysis, however, it is clear that one can explain both backgrounds providing one imposes suitable constraints on the grain size and formation redshift. Indeed the energy density involved is comparable to that expected in several of the scenarios discussed in Section (3). If the claims to have detected a CIBR are not confirmed, comparison with the background levels which will be accessible to COBE and SIRTF shows that most of the predicted backgrounds are potentially observable. Only the background associated with jupiters in galactic halos appears to be below the detectability threshold.

If there is a far-IR background, generated in the manner envisaged in Section (4), then one would expect it to display small-scale anisotropies. These would derive from inhomogenities in both the density and temperature of the dust. (The dust would generally be clumped in clouds, with T_d being higher close to the sources.)

Bond *et al.* (1986) have analysed the effect of density inhomogeneities, on the assumption that the dust clumps like galaxies. Their analysis depends on the fact that the dust radiation comes from a shell. One can show that this shell has a radius of $6000h^{-1}$Mpc (the present particle horizon size) and a comoving thickness of $900h^{-1}(z_d/10)^{-1/2}$Mpc. This thickness much exceeds the separation between galaxies, so one expects there to be many galaxies along a typical line-of-sight, implying that the fluctuations associated with the dust clumpiness will be diluted by statistical cancellations. If the galaxy correlation function is $\xi(x)=(x/x_o)^{-\gamma}$ below some correlation length x_o, then one can show that the rms intensity fluctuations for a telescope with angular resolution σ scale as σ^{-1} for $\sigma>\theta_o$ and as $\sigma^{(1-\gamma)/2}$ for $\sigma<\theta_o$. Here θ_o is the angle subtended by x_o at the emission shell, which is about 2 arcmin. The fluctuations on this scale are of order 1% to 10% (the precise value depending on how the galaxy correlation function evolves).

Whether a telescope can measure these fluctuations depends upon its angular resolution σ and its aperture size D. One clearly wants σ as small as possible since the fluctuations increase as the resolution improves. The value of D is relevant because it determines whether the fluctuations stand out above the "confusion" limit associated with unresolved foreground galaxies. COBE will have very poor angular resolution ($\sigma \approx 1^o$) and so will not be useful in this respect. However, SIRTF will have a resolution of $10"(\lambda/100\mu)$ for $2\mu<\lambda<750\mu$ and, in this case, the background fluctuations could be well above the confusion level. The situation would be even better with LDR since this would have a resolution of $0.4"(\lambda/100\mu)$ for $\lambda>50\mu$ and a very large aperture size ($D \approx 10$m). In this case, one should even be able to probe the galaxy correlation function. Unfortunately, SIRTF and LDR may not be launched for some time. Nevertheless, it is clear that studying the anisotropies in the CIRB could provide a unique probe of the Universe at large redshift.

ACKNOWLEDGEMENT

Much of this talk is based on work done in conjunction with Dick Bond and Craig Hogan. I thank them for an enjoyable collaboration.

REFERENCES

Ashman,K.M. & Carr,B.J., 1987. Preprint.

Begelman,M.C., 1984. In *Astrophysics of Active Galaxies and Quasi-Stellar Objects*, ed. J.S.Miller (Mill Valley: University Science Books), p.411.

Birkinshaw,M. & Gull,S.M., 1984. *Mon.Not.R.astr.Soc.*, 206, 359.

Bond,J.R., Carr,B.J. & Hogan,C.J., 1986. *Astrophys.J.*, 306, 428.

Boughn,S.P., Cheng, E.S. & Wilkinson, D.T., 1981. *Astrophys.J.*, 243, L113.

Carr,B.J., Bond,J.R. & Arnett,W.D., 1984. *Astrophys.J.*, 277, 445.

Carr,B.J., McDowell,J.C. & Sato,H., 1983. *Nature, 306, 666.*

Ceccarelli, C. *et al.*, 1983. *Astrophys.J.*, 275, L39.

de Bernardis,P., Masi,S., Melchiorri,B., Melchiorri,F. & Moreno,G., 1984. *Astrophys.J.*, 278, 150.

Draine,B.T. & Lee, H.M., 1984. *Astrophys.J.*, 285, 89.

Dube,R.R., Wickes,W.C. & Wilkinson,D.T., 1977. *Astrophys.J.*, 215, L51.

Fabbri,R. & Melchiorri,F., 1979. *Astr.Astrophys.*, 78, 376.

Fabbri,R., Guidi,I., Melchiorri,F. & Natale,V., 1980. *Phys.Rev.Lett.*, 44, 1563.

Gush,H.P., 1981. *Phys.Rev.Lett.*, 47, 745.

Hacking,P., Condon,J.J. & Houck,J.R., 1987. *Astrophys.J.*, 316, L15.

Hauser, M.G. *et al.*, 1984. *Astrophys.J.*, *278*, L15.

Hoffman,W. & Lemke,D., 1978. *Astr.Astrophys.*, 68, 389.

Hogan,C.J., 1984. *Astrophys.J.*, 284, L1.

Ikeuchi,S., 1981. *Pub.astr.Soc.Japan*, 33, 211.

Karimabadi,H. & Blitz,L., 1984. *Astrophys.J.*, 283, 169.

Lacey,C.G. & Ostriker,J.P., 1985. *Astrophys.J.*, *299, 633.*

Mather,J.C., 1982. *Opt.Eng.*, 21, 769.

Mathis,J.S., Rumpl,W. & Nordsieck,K.H., 1977. *Astrophys.J.*, 217, 425.

Matsumoto,T., Akiba,M. & Murakami,H., 1984. *Adv.Space Res.*, 3, 469.

Matsumoto,T., Akiba,M. & Murakami,H., 1987. Preprint.

Matsumoto,T., Hayakawa,S., Matsuo, H., Murakami,H., Sato,S., Lange,A.E. & Richards,P.L., 1988. This volume.

McDowell,J.C., 1986. *Mon.Not.R.astr.Soc.*, 223, 763.

Negroponte,J., Rowan-Robinson,M. & Silk, J., 1981. *Astrophys.J.*, 248, 58.

Negroponte,J., 1986. *Mon.Not.R.astr.Soc.*, 222, 19.

Ostriker,J.P. & Cowie,L.L., 1981. *Astrophys.J.*, 273, L127.

Ostriker,J.P. & Heisler, J., 1984. *Astrophys.J.*, 278, 1.

Peebles, P.J.E. & Partridge,R.B., 1967. *Astrophys.J.*, 148, 377.

Puget,J.L. & Heyvaerts,J., 1980. *Astr.Astrophys.*, 83, L10.

Rees,M.J., 1978a. *Phys.Scripta*, 17, 193.

Rees,M.J., 1978b. *Nature*, 275, 35.

Rowan-Robinson,M., 1986. *Mon.Not.R.astr.Soc.*, 219, 737.

Rowan-Robinsin,M., 1988. This volume.

Rowan-Robinson,M., Negroponte,J. & Silk,J., 1979. *Nature*, 281, 635.

Saarinen,S., Dekel,A. & Carr,B.J., 1987. *Nature*, 325, 598.

Shchekinov,Yu.A., 1986. *Astrophysics*, 24, 331.

Silk,J. & Stebbins, A., 1983. *Astrophys.J.*, 269, 1.

Smoot,G.F., Gorenstein,M.V. & Muller, R.A., 1977. *Phys.Rev.Lett.*, 39, 898.

Sunyaev,R.A. & Zeldovich,Ya.B., 1972. *Comm.Astr.Space Sci.*,4, 173.

Truran,J.W. & Cameron,A.G.W., 1971. *Astrophys.Space Sci.*, 14, 179.

Woody,D.P. & Richards,P.L., 1981. *Astrophys.J.*, 248, 18.

Wright,E.L., 1981. *Astrophys.J.*, 250, 1.

Wright,E.L., 1982. *Astrophys.J.*, 255, 401.

ROCKET OBSERVATION OF THE DIFFUSE INFRARED RADIATION

T. Matsumoto

Department of Astrophysics, Nagoya University
Nagoya 464, Japan

ABSTRACT

The rocket experiments were carried out to observe cosmic background radiation in both of near-infrared and submillimeter regions. In near-infrared region, a considerable amount of isotropic radiation was found at 2μm, which is possibly attributed to the cosmological origin. In submillimeter region, absolute measurement of the cosmic background radiation below 1mm was attained and a significant deviation from the blackbody spectrum of a single temperature was found. Possible origin of these two cosmic components is discussed.

1. INTRODUCTION

Infrared cosmic background is very important to study the physical processes which might have occurred between recombination of matter (z∿1000) and the galaxy formation (Carr 1987). The observation of the infrared background, however, has not been carried out well because of difficulties on experiment.

One problem is that the bright galactic and interplanetary emission obstruct the detection of the cosmic radiation in infrared region. There are, however, 2 windows which enable us to observe the cosmic background. One is a near-infrared (1-5μm) and another is a submillimeter (>300μm) region. The diffuse star light (SL) and zodiacal light (ZL) have a peak brightness below 1μm, whereas the thermal emission of the interplanetary dust (IPD) has a peak brightness above 10μm. These foreground components provide very low sky brightness in near-infrared region. Around 100μm, the thermal emission of the interstellar dust (ISD) is a dominant foreground component and well-known cosmic microwave background radiation (CMB) appears around 1mm. Submillimeter spectrum of CMB itself will provide important information, and, further, there is a possibility to detect other cosmic background radiation in submillimeter region.

Another problem is difficulties on experimental technique. A cold optics in space is absolutely necessary to observe infrared cosmic background emission. Thermal emission from the instrument itself and the bright atmospheric emission make the ground based observation impossible.

We have tried rocket experiments to observe infrared cosmic background using cold optics in near-infrared and submillimeter regions. This is a brief description of the experiments and the possible interpretation of the obtained result.

2. NEAR-INFRARED OBSERVATION

2.1. Observation

The photometer consisted of 2 parts both of which had 4° beam. One was a wide band channel, covering the standard filter bands: J(1.27μm), K(2.16μm), L(3.8μm), M(5.0μm), each of which had a lens of 14 mm dia. Another was a narrow band photometric channel consisting of two 26mm dia. Lenses and a filter wheel which rotated intermittently and covered 1-5μm region with 10% spectral resolution. A Ge detector for J band and InSb detectors for others were used. The zero level signal was confirmed by closing the cold shutter every 35 seconds and the relative response was monitored by lighting small tungstain lamps at the same time. The whole system was cooled by solid nitrogen which realized no instrumental thermal emission below 5μm.

The sounding rocket, K-9M-77, was launched on 14 Jan. 1984 at 04:30 JST (19:30 UT, 13 Jan.) from the Kagoshima Space Center of Institute of Space and Astronautical Science. At 282 sec after launch the rocket reached an apogee at an altitude of 316km. After the lid opened the survey started according to the precession of the rocket. The observed sky coverage was satisfactorily large (half cone angle ∿ 31.5°) and included the galactic and ecliptic pole regions. During the survey bright stars, such as γLeo, were observed, whose signals were consistent with the absolute responsivity measured in the laboratory. The details of the experiment is described by Matsumoto et al. (1987).

2.2. Result

The objectives of the data analysis is to look for the isotropic emission component which can not be explained by the known emission components. The observed signals are composed of the several emission sources, which were superposed on the cosmic background radiation. Separation of the observed signal to individual emission components is attained under the following procedure so that the observed data are mutualy consistent.

* Atmospheric emission is assumed to be proportional to the column density of the atmospheric molecules.

* Envoronmental emission was approximated by the exponential decay with a single time constant.

* Earthshine was estimated from the off-axis response of the teles- cope measured in the laboratory, which showed a good agreement with the observed data.

* Since a simple solar color for ZL contradicts with observed data, we assumed constant color, J-K for ZL.

* We constructed the model of the Galaxy based on the star count data which reveals that SL is represented well by the linear function of cosec (|b|) as for the observed region.

Figure 1 indicates the cosec (|b|) dependence of the sky brightness for wide band channels. The essential fea- tures are summerized as follows.

* In the J abnd, observed surface brightness at high ecliptic latitude can be ex- plained well only by SL. This indicates that the color of ZL

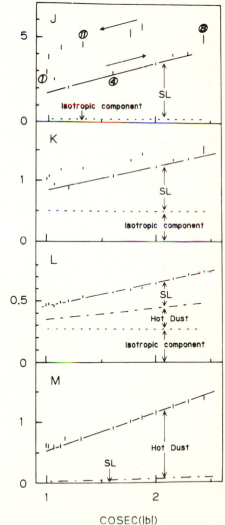

Figure 1. The observed surface brightness for the wide band channel are plotted versus cosec (|b|) for the regions where no star brighter than 3 mag at K is found. The scan direction is indicated by an arrow. The vertical bars re- present the detector noise and do not include systematic errors. The solid lines show the result of the fitting. A breakdown of the observed sur- face brightness to the star light (SL), isotropic radiation, and a new emission component (Hot dust) is also shown.

is bluer than that of the solar color, and there remains no isotropic emission in the J band.

* The K band surface brightness at high ecliptic latitude shows linear relation on cosec (|b|) with a significant offset brightness which is comparable with SL at the galactic pole. The observed spatial distribution is consistent with the J band data, if we assume constant colr for ZL which is bluer than the solar color and adopt an isotropic emission component for the K band.

* In the L and M bands, the observed surface brightness shows clear linear relation on cosec (|b|) and no dependence on the ecliptic latitude. This emission component is brighter than SL expected from the model but the origin is not certain. The L band brightness, however, can not be explained only by this emission and provides substantial amount of the isotropic emission.

Figure 2 shows the observed surface brightness at the galactic pole region, in which narrow band data and other foreground emission components are also shown. Characteristic feature of the Figure 2 is that there are humps at 2 and 3-3.5μm which are attributed to the isotropic emission. A sharp increase of the brightness above 4μm is possibly due to the galactic or inter-

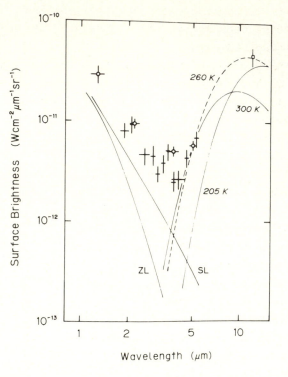

Figure 2. The spectrum of the observed surface brightness at the galactic pole region. Open circles and crosses are data of the wide band and narrow band channels, respectively. Systematic errors are included in the error bars, and horizontal bars indicate the wavelength bandwidth of filters. SL, and ZL mean star light and zodiacal light, respectively. The square represents IPD emission at 12μm towards the observed sky based on the data by IRAS (Hauser et al. 1984). The emission component around 5μm is fitted by three types of blackbody emission

planetary origin.
Figure 3 indicates the
isotropic emission
which is obtained to
be constant over the
observed sky and during
the duration of the
rocket observation.

The isotropic
emission in Figure 3
does not directry
imply extragalactic
origin, because the
analysis is based on
some assumptions and
there may be other
unknown foreground
emission. For
example, isotropic
emission at 3-3.5μm
may be environmental
emission having a
long time constant
compared with observ-
ing period, since
a similar band
feature was really
observed during the

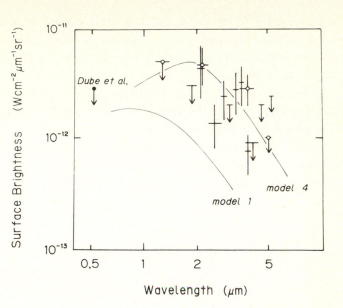

Figure 3. The spectrum of the isotropic
component. An upper limit of EBL at the
visible band (Dube et al. 1977) is also shown.
The extragalactic background light calculated
by partridge and Peebles (1967) for two
extreme cases is indicated by solid lines.
Model 1 assumes no evolution of galaxies,
while model 4 assumes that all helium were
synthesizes in stars during the early era
of the galaxy formation.

ascending phase of the rocket. On the other hand, 2μm feature was not
contaminated by the enveronmental emission and is possibly attributed
to the cosmic origin. Infrared spectrum of ZL, however, is not
established well. If ZL had a line feature at 2μm and had a quite
different spatial distribution from the optical ZL, observed 2μm
feature might be attributed to interplanetary origin. But the past
observations of ZL and comets do not show such a feature.

3. SUBMILLIMETER OBSERVATION

3.1. Observation

In order to attain the accurate measurement of the absolute sky brightness in the submillimeter region, we constructed the radiometer cooled
by liquid herium down to 1K. The radiometer consisted of the horn

antenna with 7.6° beam and the photometer which enabled us simultaneous observation in six passbands (see Table 1). Details of the
instrument are described by Lange et al. (1987), Sato et al. (1987) and
Richards (1987).

The sounding rocket, K-9M-80, which carried the radiometer was
launched on 1987 February 23, 0:00 JST (Feb.22, 15:00 UT) from the
Kagoshima Space Center of the Institute of Space and Astronautical
Science. The payload reached apogee of 317km at 287 seconds after
launch. The sky was surveyed according to the precession of the rocket
axis (30° full-angle cone) which was centered at l=203±3° and b=33±3°.

Table 1

EFFECTIVE BANDCENTERS AND BANDWIDTHS[a]

CHANNEL	λ_o μm	ν_o cm^{-1}	$\Delta\lambda/\lambda_o = \Delta\nu/\nu_o$ [b] %
1	1160	8.6	30
2	709	14.1	21
3	481	20.8	19
4	262	38.2	36
5	137	72.9	35
6	102	98.2	21

[a] Weighted by the observed I_λ as described in the text

[b] Equivalent square bandwidth

3.2. Result

During the rocket flight the instrument worked perfectly and fairly
good data were obtained for all channels. The environmental emission
was observed during the ascending phase, however, it disappeared
rapidly and became negligible by 240 sec. Temperature of the photometer and horn was cold and stable enough from 200 to 500 sec. The
absolute responsivity of each channel, as measured by the internal
calibrator, was stable enough and in excellent agreement with the responsivity calculated from laboratory measurements under similar conditions. In channel 5, atmospheric emission due to OI at 146μm may be
appreciable even at apogee, and its estimated contribution was subtracted.

Figure 4 shows the average sky brightness obtained by integrating
the signals during which the environmental and atmospheric emission
are negligible. Characteristic features of Figure 4 are as follows.

* The brightness of channel 6 is consistent with IRAS data.
* For the data of channels 4, 5 and 6 spatial modulation of the signals due to the precession phase appeared with same amplitude, which are correlated with HI column density. This indicates that the emission sources for channels 4, 5 and 6 are same and probably attributed to the ISD emission.
* Around 300μm the sky brightness is very low and the brightness of channel 4 can be explained well by extrapolating ISD emission with T ~ 20K and spectral index of 2.
* The signals of channels 1, 2 and 3 are definitely cosmological origin and indicate substantial excess at short wavelength side compared with the average blackbody temperature of 2.74K obtained from the past measurements (Smoot et al. 1987).

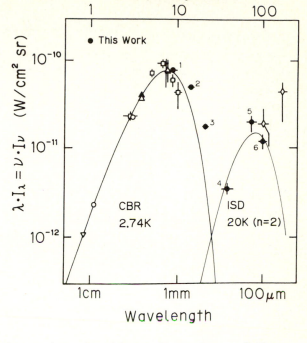

Figure 4. The observed spectrum of the astrophysical background. The fluxes obtained by the present work are shown by ● with the vertical error bars and the horizontal bars for the effective bandwidths given in Table 1. Numbers indicate channels in Table 1. The results of other measurements are shown for comparison by ○ (Peterson et al. 1985), ▢ (Meyer and Jura 1985), ▉ (Crane et al. 1986), ✕ (Smoot et al. 1985, 1987) and ▲ (Johnson and Wilkinson 1987). IRAS data at 60 and 100 μm, are shown by ◊ .

The thermodynamic temperatures for channels 1, 2 and 3 are represented in Figure 5 together with past measurements (Matsumoto et al. 1987).

4. DISCUSSIONS

In the Wien part of the spectrum of CMB, a distortion of the spectrum was expected from several theories. The most familiar one of them is an inverse Compton scattering (Zeldovitch and Sunyaev 1969). We have

tried to fit our
data with this model
so that the model
gives the highest
possible Rayleigh-
Jeans temperature,
T_{RJ}, and still lies
within the errors
in channels 1 and 2
(solid line in
Figure 5). The
model gives the
parameters, T_0=2.78K
and y=0.021. The
predicted temper-
ature for channel
3 is somewhat below
the observed temper-
ature. Even for
this case, the model
gives T_{RJ}=2.67K,
which is signifi-
cantly lower than
the average of the
past measured
temperature 2.74

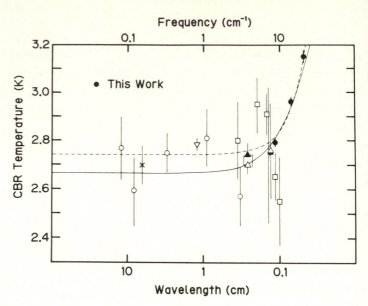

Figure 5. The equivalent blackbody temperature versus wavelength. The results of other measurements are shown for comparison by ○ (Peterson et al. 1985), □ (Meyer and Jura 1985), ■ (Crane et al. 1986), △ (Mandolesi et al. 1986), X (Smoot et al. 1985, 1987) and ▲ (Johnson and Wilkinson 1987).

± 0.002K. For the case of the relativistic electrons fitting is worse and the low brightness of channel 4 gives strong constraint (Hayakawa et al. 1987).

Another origin for a spectral distortion is redshifted dust emission in the early universe (Bond et al. 1986, McDowell 1986). In this case the excess energy was added to the CMB by the emission of dust heated by energetic sources in the early universe such as population III stars. As a simple example, we assume a spectrum of dust emission characterized by a dust temperature T_d and emissivity index n=2. The total flux and T_d are adjusted to fit the excess over the blackbody spectrum of 2.74K. The result is shown as a dashed line in Figure 5 for T_d=3.55 (1+z)K and excess radiation energy density of 10% of CMB, where z is the redshift of the emitting dust.

In both cases, energetics is a serious problem. If we assume the barionic energy when ΔX of hydrogen was converted to herium, energetics requires following relation.

$$\Delta X = 4 \times 10^{-3} (1+z) (0.1/\Omega_B) h^{-2}$$

For large z, energetics causes overproduction of herium which contradicts the standard theory of the nuclear synthesis, unless the exhausted heriums were confined in the blackholes.

The population III hypothesis explains submillimeter excess well, however, redshifted star light should be observed in the infrared or visible region (McDowell 1986). If we attribute the observed 2μm line feature to cosmological origin, it may be a redshifted Lymann α emission. In this case, redshift, z, for pop. III stars is ~17. Line band width of 2μm feature is caused by the expansion of the universe and gives ~10^8 year for the pop III era. Energetics assuming nuclear energy gives similar relation as the submillimeter excess.

$$\Delta X = 5 \times 10^{-4} (1+z) (0.1/\Omega) h^{-2}$$

Here, we took only 2μm line strength into account as a cosmological origin. Comparing above two relations, effective optical depth of $\tau \sim$ 2 is obtained. According to Bond et al. (1986), Optical depth towards the redshift z=17 is 3.0. This value is consistent with above value, since not all barionic matters were distributed in the interstellar space during and after pop. III era. Bond et al. (1986) also predicted the peak wavelength for the redshifted dust emission. If we adopt z=17 and reprocessed energy to be 10% of CMB, a peak wavelength of 650μm is obtained which is fairly in good agreement with observation.

At present, the origin of the submillimeter excess is not certain. Comptonization model is closely related with X-ray background (Guilbert and Fabian 1986), while redshifted dust emission model should be confirmed through near-infrared background. The more detailed study of the near-infrared background will be critically important.

References

Bond, J.R., Carr, B.J., & Hogan, C.J. , Astrophys. J., 306, 428.

Crane, P., Hegyi, D.J., Mandolesi, N., & Dankes, A.C., 1986, Astrophys. J., 309, 822.

Dube, R.R., Wickes, W.C., & Wilkinson, D.T., 1977, Astrophys. J., 215, L51.

Guilbert, M.G., & Fabian, A.C. 1986, M.N.R.A.S., 220, 439.

Hauser, M.G., Gilett, F.G., Low, F.J., Gautier, T.N., Beichman, C.A., Neugebauer, G., Aumann, H.H., Baud, B., Bogges, N., Emerson, J.P., Houck, J.R., Soifer, B.T., & Walker, R.G., Astrophys. J., 278, L15.

Hayakawa, S., Matsumoto, T., Matsuo, H., Murakami, H., Sato, S., Lange, A.E., & Richards, P.L., 1987, submitted to Publ. Astron. Soc. Japan.

Johnson, D.G., & Wilkinson, D., 1987, Astrophys. J., 313, L1.

Lange, A.E., Hayakawa, S., Matsumoto, T., Matsuo, H., Murakami, H., Richards, P.L., & Sato, S. 1987a, Applied Optics, 26, 401.

Mandolesi, N., Calzolari, P., Cortiglioni, S., Morigi, G., Danese, L., & De Zotti, G., 1986, Astrophys. J. 310, 561.

Matsumoto, T., Akiba, M., & Murakami, H., 1987, submitted to Ap. J.

Matsumoto, T., Hayakawa, S., Matsuo, H., Murakami, H., Sato, S., Lange, A.E., & Richards, P.L. 1987, submitted to Astrophys. J.

Meyer, D.M., & Jura, M., 1985, Astrophys. J., 297, 119.

McDowell, J.C. 1986, M.N.R.A.S., 223, 763.

Partridge, R.B., & Peebles, P.J.E. 1967, Astrophys. J., 148, 377.

Peterson, J.B., Richards, P.L., & Timusk, T., 1985, Phys. Rev. Lett., 55, 332.

Richards, P.L., 1987, in this proceedings.

Sato, S., Hayakawa, S., Lange, A.E., Matsumoto, T., Matsuo, H., Murakami, H., & Richards, P.L., 1987, Applied Optics, 26, 410.

Smoot, G.F., Bensadoun, M., Bersanelli, M., De Amici, G., Kogut, A., Levin, S., & Witebsky, C. 1987, Astrophys. J., 317, L45.

Zeldovich, YA.B., & Sunyaev, R.A., 1969, Astrophys. Space Sci., 4, 301.

SPECTRUM OF THE COSMIC MICROWAVE BACKGROUND

P. L. Richards

Department of Physics
University of California,
Berkeley, California 94720, U.S.A.

ABSTRACT

Measurements of the spectrum of the cosmic microwave background (CMB) have been actively pursued for more than twenty years in hopes of finding detailed information about the early universe. A review will be given of the technical problems associated with these background measurements in each frequency range where they have been attempted. The experimental approaches which have been used to overcome these problems will be analyzed and the most recent measurements summarized.

1. INTRODUCTION

There is a natural frequency range for measurements of the spectrum of the cosmic microwave background (CMB). This range is bounded by confusion with galactic synchrotron emission below 600 MHz (50 cm) and by confusion with galactic dust emission above 1 THz (300 μm). Measurements of the spectrum of the CMB over this range were initially stimulated by the need to verify its existence and its general blackbody character. In recent years, however, the emphasis has shifted to a search for deviations from a Planck curve, in hopes of finding detailed information about the early universe.

Much theoretical work has been done which provides a basis for these hopes. Most of this work analyzes ways in which matter could have transferred energy to the CMB. For redshifts $z > 10^6$, bremsstrahlung and radiative Compton scattering coupled the matter and radiation so tightly that a Planck spectrum is expected. For $10^6 \lesssim z \lesssim 10^5$, nonradiative Compton scattering could transfer energy to the photons, but bremsstrahlung could be effective in creating new photons only at low frequencies. Thus a Bose-Einstein spectrum with a finite chemical potential could have been created over the upper part of the frequency range. For $z \lesssim 10^5$ the electron density was too low

for Compton scattering to establish a Bose-Einstein spectrum.
Bremsstrahlung could have created a high brightness temperature at low
frequencies, while Compton scattering would reduce the temperature at
intermediate frequencies and increase it in the Wien (high frequency)
region. After recombination, re-ionization due to energy injection
from galaxy or quasar formation could have reactivated some of the
processes sketched above. Finally, the Wien region of the spectrum
could have been distorted by the redshifted emission from dust
associated with the early stages of galaxy formation.

This brief summary is sufficient to illustrate the fact that the
theory of spectral distortions is too rich to have predictive value.
Experiments are required with enough accuracy to provide unambiguous
evidence for the presence of deviations from a Planck spectrum, or at
least to constrain models in useful ways. The theories do show,
however, that such evidence in any spectral range would be extremely
interesting.

2. Low Frequency Direct Measurements

Unfortunately, it is very difficult to make accurate absolute
measurements of sky brightness. Although the detailed techniques used
vary over the wide frequency range of interest, the basic principles
of the measurement are the same. A detector is used which alternately
views a well defined patch on the sky and two or more calibration
sources or loads, to calibrate the zero level and the scale factor.
Measurement accuracies are usually limited by systematic errors, not
by detector noise.

Several necessary precautions are common to measurements at all
frequencies. The atmospheric contribution must be controlled by the
selection of an appropriate site and/or subtracted by zenith angle
scans or other appropriate techniques. The sidelobe response of the
antenna must be sufficiently small that radiation from the horizon and
other spurious signals can be avoided. The temperature of at least
one of the loads must be close to the sky temperature. Perhaps most
important, the most rigorous examination must be made of all sources
of bias and error. Appropriate allowances must be made for the
nonstatistical nature of systematic errors. The procedures used
should be reported in detail so that the community can evaluate the
results.

In addition to these general requirements, there are detailed
problems peculiar to each frequency range that must be solved or
quantified in order to make valid measurements. For frequencies below

~90 GHz (3 mm) coherent signal processing techniques are generally used. Single mode antennas and transmission lines determine the throughput $A\Omega = \nu^{-2}$. The power per unit bandwidth is then nearly independent of frequency and is a factor 10^2 smaller than for 300 K blackbody radiation. Receivers consist of a heterodyne downconverter (at the higher frequencies) followed by a transistor amplifier and a diode detector. Most of the technology required is well developed and commercially available. Since apparatus size scales with wavelength, low frequency apparatus is not generally cooled. Calibration is accomplished by viewing loads at ~3 K and ~300 K.

There has been much activity between 600 MHz and 90 GHz (50 cm and 3 mm) using these techniques from high mountain sites (Smoot et al. 1985, 1987a, Mandolesi et al. 1986, Sironi et al. 1987). The need to subtract galactic synchrotron radiation causes serious problems at the lowest frequencies. The atmospheric emission increases with frequency, but is generally small enough below 90 GHz that useful results can be obtained. The measurement techniques are relatively mature and systematic work is producing incremental improvements in the accuracy of the data. Some of the more recent experiments are reported in Table I.

A significant improvement in accuracy at 25 GHz (1.2 cm) has recently been reported by Johnson and Wilkinson (1986). Using an approach pioneered at higher frequencies, they cooled their entire apparatus to LHe temperatures to improve thermal stability and facilitate calibration, and flew it in a balloon to eliminate the atmospheric correction. There appears to be an opportunity for improved measurements from balloons over much of the 10-300 GHz (3 cm-1 mm) range as the techniques for cooling coherent receivers mature.

2.1 High Frequency Direct Measurements

For frequencies above ~90 GHz (3 mm) the single mode power per unit bandwidth in the CMB begins to fall exponentially. In the same frequency range the noise in available coherent receivers increases significantly, as does the atmospheric emission. Consequently, successful higher frequency measurements have thus far used incoherent (bolometric) detectors and spectrometers with fixed multimode throughput. When the throughput $A\Omega$ is independent of frequency, the power per unit bandwidth in the CMB increases as frequency squared at low frequencies, peaks at ~150 GHz (2 mm) and falls exponentially at higher frequencies. Because of the fall in the CMB beyond the peak,

TABLE I.   RECENT MEASUREMENTS OF THE TEMPERATURE OF THE COSMIC
MICROWAVE BACKGROUND

Reference	ν(GHz)	λ(cm)	T(K)
Sironi et al. (1987a)	0.6	50	2.45±0.7
Smoot et al. (1987a)	1.4	21	2.22±0.55
Smoot et al. (1985)	2.5	12	2.78±0.13
Smoot et al. (1987a)	3.7	8.2	2.59±0.14
Mandolesi et al. (1986)	4.8	6.3	2.70±0.07
Smoot et al. (1987a)	10	3.0	2.61±0.06
Johnson and Wilkinson (1986)	25	1.2	2.783±0.025
Smoot et al. (1985)	33	0.91	2.81±0.12
Smoot et al. (1987a)	90	0.33	2.60±0.10
Meyer and Jura (1985)	114	0.26	2.70±0.04
" " " "	227	0.13	2.76±0.20
Crane et al. (1986)	114	0.26	2.74±0.05
" " " "	227	0.13	2.75 +0.24,-0.29
Peterson et al. (1985)	86	0.35	2.80±0.16
" " " "	151	0.20	2.95 +0.11,-0.12
" " " "	203	0.15	2.92±0.10
" " " "	264	0.11	2.65 +0.09,-0.10
" " " "	299	0.10	2.55 +0.14,-0.18
Matsumoto et al. (1987)	267	0.112	2.795±0.018
" " " "	441	0.068	2.963±0.017
" " " "	651	0.096	3.146±0.022

the importance of radiation from 300 K objects such as the atmosphere,
the earth, and any warm apparatus increases exponentially with
frequency. It is therefore essential to cool all parts of the
apparatus that are in contact with the radiation. The use of broad
spectral bandwidths compounds the atmospheric emission problem, so
balloon or space platforms are required.

Because the technology required for these measurements developed
only slowly, there were many unsuccessful attempts to make
measurements above 90 GHz. The first experiment to unambiguously
observe a peak in the spectrum of the CMB (Woody and Richards 1981)
used a ⁴He cooled Winston cone antenna with apodizing horn, a ⁴He
cooled Fourier transform spectrometer, and a ³He cooled bolometric
detector. This experiment also gave evidence for a small (3 σ)
deviation from a Planck spectrum. The apparatus was later modified to

include a cold chopper, a filter wheel spectrometer and improved
calibration. The new results at five frequencies between 85 and
300 GHz (Peterson et al. 1985) are consistent with a single
temperature and suggest that the systematic errors in the Woody
Richards experiment were underestimated.

The residual atmosphere at balloon altitude causes several
problems with these measurements. Because the window must be removed,
there is a possibility of "snow" collecting in the optics. The top
end of the antenna cannot be cold, and temperature gradients in it can
change as the zenith angle is scanned. An improved version of the
Peterson et al. (1985) experiment has been flown by Bernstein et al.
(1987) but the data are not yet analyzed. There is room for radically
new approaches to measurements in this spectral range and several are
being prepared.

Experiments at frequencies higher than 300 GHz (1 mm) seem
unpromising from balloons because of the thermal problems described
above and because of the rapid increase in atmospheric emission with
frequency. The decrease in apparatus size with shorter wavelengths,
however, makes sounding rockets an attractive alternative.
Circumstances combine to make it possible to achieve the exponential
improvements in performance that are required at higher frequencies.
The atmospheric emission and the atmospheric limitations to antenna
cooling both dissappear. It is possible to make the entire optical
system cold enough that no corrections need be made for emission from
the apparatus. The flared horn antennas have been shown to have
sidelobe rejection factors as large as 10^9, despite diffraction and
scattering (Sato et al. 1987).

Lange et al. (1987) describe the construction and calibration of a
multi-channel rocket-borne photometer designed for background
measurements in six frequency bands from 267 GHz to 2.7 THz (1.1 mm to
110 µm). It includes a cold chopper, an internal calibrator, four ^{4}He
cooled bolometric detectors and two Ge:Ga photoconductive detectors.
The results of a successful flight of a similar apparatus are reported
by Matsumoto et al. (1987). Internal evidence from the flight data
shows that contamination from earthshine, atmospheric emission and
emission from rocket-borne contaminants is negligible during most of
the flight. Because of extraordinary care with the calibration before
the flight, the data are thought to be very accurate. Thermodynamic
temperatures derived from the bands at 270, 441 and 651 GHz (1.1 mm,
680 µm, and 460 µm) are shown in Table I. The flux from interstellar

dust was measured in the three highest frequency channels and used to make a small correction to the temperature reported at 460 GHz.

2.2 Optical Measurements

Observations of the optical spectrum of interstellar CN have been used by Meyer and Jura (1985) and by Crane et al. (1986) to obtain the temperature of the CMB at 114 and 227 GHz (2.6 and 1.3 mm). Upper limits are obtained from measurements of the strengths of weak optical lines, interpreted using basic theories of molecular spectroscopy. Actual measurements require, in addition, some knowledge of the local excitation in the molecular clouds being observed. Reported temperatures with relatively small errors are listed in Table I.

3. DISCUSSION

With the exception of the very recent rocket data of Matsumoto et al. (1987) the data set in Table I has evolved slowly over the last few years. Smoot et al. (1987a) give a weighted average of 2.74 ± 0.02 K for this restricted data set with a X^2 of 22 for 17 degrees of freedom. Considering the limitations of applying statistical tests to data which are dominated by non-statistical systematic errors, it appears plausible that this restricted data set is consistent with a single temperature. It has been possible to deduce interesting constraints on several cosmological models from these data (Smoot et al. 1987b).

When the rocket data are included, a substantial deviation from the Planck form is immediately apparent. The thermodynamic temperature increases systematically with frequency in the Wien region. Matsumoto et al. (1987) have fitted this complete data set to two cosmological models, one of which includes Compton scattering of the CMB from nonrelativistic electrons and the other redshifted radiation from dust in the early universe, as shown on Figure 1. Although the fit is somewhat better in the latter case, the available data cannot distinguish between these effects. Much detailed modeling will be required before the full significance of the rocket results is understood.

The fluxes measured by Matsumoto et al. (1987) reveal the presence of a deep minimum in the astrophysical backgrounds at ~1 THz (300 μm). The flux at higher frequencies is consistent with that expected from interstellar dust. Future observations in this 1 THz window may

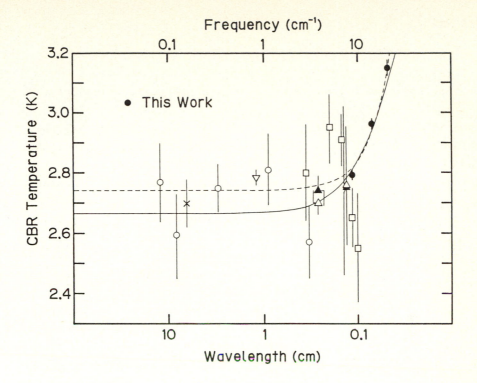

Figure 1. Recently measured thermodynamic temperatures as a function of frequency compared with model calculations for a Compton distortion (solid line) and a redshifted dust emission model (dashed line). This figure was taken from Matsumoto et al. (1987).

provide information of cosmological interest regarding early galaxy formation. This possibility will be explored in a forthcoming rocket experiment by the same group. It is also of interest for forthcoming spacecraft including COBE, IRTS, and SIRTF.

The next few years will be exciting for students of the spectrum of the CMB. New ground-based, balloon and rocket experiments are under construction. At the end of this decade, the COBE satellite will provide accurate spectral measurements and full sky coverage for frequencies above ~1 cm^{-1}.

Acknowledgements

Thanks are due to Prof. A.E. Lange for many fruitful discussions. This work was supported in part by the National Aeronautics and Space Administration Grant NSG-7205.

References

Bernstein, G.B., Richards, P.L., and Timusk, T., 1987 (to be published).

Crane, P., Heygi, D.J., Mandolesi, N., and Danks, A.C. 1986, Ap. J., 309, 822.

Johnson, D.G., and Wilkinson, D.T. 1986, Ap. J. (Letters) 313, L1.

Lange, A.E., Hayakawa, S., Matsumoto, T., Matsuo, H., Murakami, H., Richards, P.L., and Sato, S., 1987, Appl. Opt. 26, 401.

Mandolesi, N., Calzolari, P., Cortiglioni, G., Morigi, G., Danese, L., and De Zotti, G. 1986, Ap. J. 310, 561.

Matsumoto, T., Hayakawa, S., Matsuo, H., Murakami, H., Sato, S., Lange, A.E., and Richards, P.L., Ap. J. (Letters) submitted.

Meyer, D.M., and Jura, M. 1985, Ap. J. 297, 119.

Peterson, J.B., Richards, P.L., and Timusk, T. 1985, Phys. Rev. Lett. 55, 332.

Sironi, G., 1987, 13th Texas Symposium on Relativistic Astrophysics.

Smoot, G.F., De Amici, G., Friedman, S.D., Witebsky, C., Sironi, G., Bonelly, G., Mandolesi, N., Cortiglioni, S., Morigi, G., Partridge, R.B., Danese, L., and De Zotti, G., 1985, Ap. J. (Letters) 291, L23.

Smoot, G.F., Bensadoun, M., Bersanelli, M., De Amici, G., Kogut, A., Levin, S., and Witebsky, C., 1987a, Ap.J. (Letters) in press.

Smoot, G.F., Levin, S.M., Witebsky, C., De Amici, G., and Rephaeli, Y. 1987b, (to be published).

Woody, D.P., and Richards, P.L., 1981, Ap. J. 248, 18.

FROM STAR FORMATION TO GALAXY FORMATION

Joseph Silk

Department of Astronomy
University of California, Berkeley
California   94720

ABSTRACT

Current scenarios for the large–scale structure of the universe lead to rather precise predictions for the structure and evolution of primeval galaxies. These scenarios are reviewed, and models of protogalaxies are described. I discuss the origin of the first stars, and the chemical evolution of galaxies. The role of environment in affecting galaxy properties is described, and I conclude with a summary of various tests of galaxy formation theory.

1. INTRODUCTION

I am going to begin by making some rather general comments about galaxy formation. Forming galaxies are far away, and so the information to date is sketchy. However the principal message that I would like you to remember is that star formation is the key to understanding galaxy formation. Fortunately, we know enough about star formation in nearby regions that we can finally begin to deduce the rudiments of a theory of galaxy formation. I am going to discuss the following issues: The nature of a primeval galaxy, and what is a primeval galaxy?; scenarios for the large–scale structure of the universe; models of protogalaxies; how the first stars were made; chemical evolution; environment impact; and how to test the theories. I will be discussing the results both of my own work with a number of collaborators, and work by several other groups.

First of all, what is a primeval galaxy? If we wish to study galaxy formation we must define this type of object. Now one's immediate thought might be that a primeval galaxy is an object at very great redshift, but this is not necessarily so. I will give you two definitions. A primeval galaxy is an object forming stars for the very first time. One such example of a possible candidate primeval galaxy is the relatively nearby extragalactic HII region I Zw18. Another possible definition of a primeval galaxy is an object undergoing an unusually high rate of star formation, so that it is presently making most of its stars, or at least some large fraction of its stars. Again, these objects also need not be in very distant regions of the universe. It is possible, for example, that extreme starburst galaxies would meet this definition.

Merging galaxies are another example of a class of objects which may appear to exemplify triggering star formation. The very fact that an object is merging means that it is in the process of developing a new morphology, as well as undergoing chemical evolution. Finally there is a recently discovered class of objects at high redshift with strong Lyman α emission, discovered by Spinrad and collaborators. These are likely candidates for primeval galaxies, since they contain a large ($\sim 10^{10} M_\odot$ mass of gas, which is presumably being excited by a considerable number of massive stars. I am inclined to accept the latter definition, of an exceptionally high star formation rate, as being the prerequisite for a primeval galaxy. In what follows, I will first describe the basic scenarios for galaxy formation. Development of such scenarios has been one of the main goals in theoretical cosmology over the past few years.

2. COSMOLOGICAL SCENARIOS

There has been so much information that has come from studies of large–scale structure and also advances in theory that we have recently been able to make rather precise predictions, in the context of certain models, of the parameters that one might expect for protogalaxies.

2.1 The bottom–up scenario

The particular theory which I am going to describe first of all as the bottom–up theory is one to which most cosmologists subscribe. This only means that you might be lucky to find that 50 percent of them are believers, while if you polled the other 50 percent you would probably find that each cosmologist had his or her own pet theory to compete with this one. The basic input is that the universe is rather smooth at very large distances from us. We know this from the extreme isotropy of the cosmic blackbody background radiation (isotropic to better than 0.01 percent). This then leads us to a class of theories in which one has small density fluctuations which grow by gravitational instability and gradually develop into structures on different scales. One begins with a certain critical assumption concerning the initial density fluctuation amplitude on the comoving scale of a galaxy or galaxy cluster mass. The idea is that we have a well–defined spectrum predicted by the theory of the very early universe. At present the prediction of the amplitude is more of a goal than a result, but inflationary cosmology does predict that the initial spectrum should be gaussian and scale–invariant.

This spectrum amplifies in time, with no preferred scale on super horizon scales, maintaining constant curvature, until the universe becomes matter–dominated at $z_{eq} \approx 4 \times 10^4 \Omega_o h^2$. Henceforth, sub–horizon growth occurs, and this imprints a comoving scale corresponding to the horizon size, namely $L_{eq} \approx 13(\Omega_o h^2)^{-1}$ Mpc, onto the emergent fluctuation spectrum. The smaller scales first go non–linear at a well–defined epoch of the early universe. Structure starts forming on small scales and gradually bootstraps its way up to larger and larger scales which have progressively undergone less sub–horizon growth and therefore go non–linear later. Much of the motivation for this particular choice of spectrum has come from inflationary cosmology, which postulates that $\Omega_o = 1$, as well as from studies of the dark matter in galaxy halos and in galaxy clusters. Such a spectrum, with power on small scales, in which small objects form early and larger objects form later is sometimes called the cold dark matter spectrum, and reflects the fact that the dark matter is presumed to be a weakly interacting form of matter that clustered freely on all scales as the fluctuations developed. I don't really want to get into the technical details, other than to say that Figure 1 shows the prediction today for the density fluctuation spectrum in various popular scenarios. In the bottom–up theory, small structures of mass $\gtrsim 10^6 M_\odot$ formed first at $z \sim 10$, then larger and larger structures began to form.

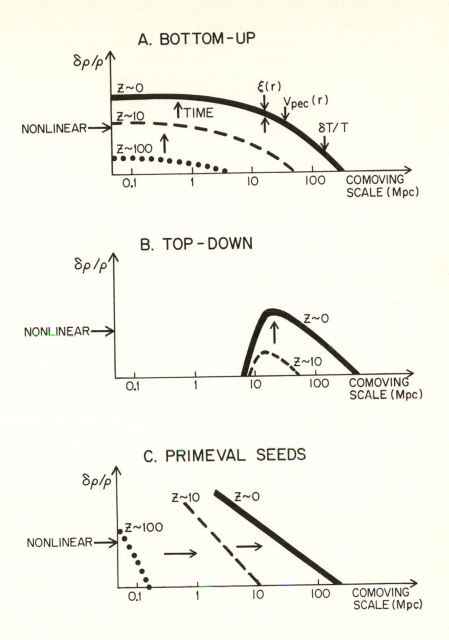

Figure 1: Primordial density fluctuation spectra

How do we test this theory? On the very large scales, we have residual fluctuations remaining around today. Linear theory is very simple to calculate. It has a specific signature, arising from the density fluctuations that drive fluctuations in the gravitational field. In particular, fluctuations in the microwave background are induced by gravitational redshifts that develop because the radiation emitted from the last scattering surface may be scattered in a local potential well or ridge. The density fluctuations produce both gravitational potential fluctuations and peculiar velocities on very large scales, which result in microwave anisotropies $\delta T/T \sim 10^{-5}$ on angular scales $\sim 1°$. Even if we average over 10 or 50 Mpc, there remains a residual low–level peculiar motion at the present epoch induced by the large–scale fluctuations. Perhaps studied in greatest detail has been the galaxy correlation function, a measure of the spatial correlations of the galaxy distribution on scales above $\sim 5h^{-1}$ Mpc. This directly probes fluctuations in the luminous matter distribution. All of these data directly constrain the linear fluctuation spectrum. While the cold dark matter spectrum is by no means a unique solution, it happens to fit reasonably well to most of the observational data that we have. Exceptions will be described below.

2.2 The Top–down Scenario

Figure 1 also shows another theory, which is often called a top–down theory. In this particular theory, one postulates that the fluctuations initially survived only on very large scales. A particular top–down theory happens to be associated with a primeval coherence length in the fluctuations that was predicted, for example, if the dark matter in the universe consists of neutrinos with a mass that happens to be of order the experimental limit on the electron neutrino mass. These neutrinos were moving with random motions at near light speed when subhorizon fluctuation growth began at the onset of matter–domination. They streamed freely over comoving scales of about 10 Mpc, so that primordial galactic scale fluctuations were suppressed but cluster scale fluctuations still grew. Eventually one has a situation in which only the surviving fluctuations on large scales developed first. These objects turned out to have masses of $\sim 10^{15} M_\odot$, collapsing at low redshift, and fragmented into galaxies.

If one compares the hot dark matter with the cold dark matter theory, one can see that they do lead to slightly different predictions for the large scale structure. The shapes of the respective curves are well known, but the respective normalizations are empirical. One doesn't yet know a priori what the normalization should be, but one appeals to the galaxy correlation function in the case of cold dark matter or to the existence of non–linear objects (quasars) at a redshift of 4 in the case of hot dark matter to specify the normalization. In principle, one can distinguish between these two theories by looking for evidence of their somewhat different effects on the microwave background radiation. There are also other probes of large–scale structure involving large–scale voids and peculiar velocity fields.

A third theory, which is even more speculative than the hot dark matter theory, involves some sort of primeval seeds. These are small scale, non–linear objects, already present near the beginning of the Big Bang, one example with an exotic name being cosmic strings. These primeval seeds act as sources of gravity, accreting ambient matter, and provide the possibility of being able to generate large–scale structure by bootstrapping from subgalactic scales all the way up to the scales of galaxies and galaxy clusters.

While we have these different possibilities for understanding the large–scale structure, there are observational arguments that make at least the top–down theory unlikely to be presently acceptable. It leads to very recent formation of galaxies and cannot account for quasars observed at $z = 4$. The primeval seed theory is rather hard to work out in detail, although some aspects look promising. This leaves us with the cold dark matter theory. This theory has been studied most carefully and meets most observational confrontations. I will discuss this theory in more detail with respect to the context of galaxy formation.

2.3 Cosmological Simulations

Computer simulations, studying the evolution of N point masses, where N $\gtrsim 10^4$, laid down in a cubical volume representative of the expanding universe, have been helpful in making meaningful comparisons of theory and observation of the galaxy distribution. In particular, comparison of cold dark matter simulations by Davis and collaborators with the observed galaxy distribution, projected so as to be a representation of galaxies on the northern sky to a depth of about 100 Mpc, suggests that one has a reasonably good representation, and this is borne out by more quantitative comparisons. On smaller scales, some recent results from high resolution simulations by the same group examine how fluctuations develop on galactic scales. For example, in a time sequence beginning at a redshift of 3, one sees structure gradually developing, with the mass–points accumulating together and gradually merging to form a potential well which resembles that of a large dark halo today with a characteristic isothermal density profile.

Other simulations with inhomogeneous initial conditions demonstrate that for a wide variety of possibilities, initial collapse results in mergers and violent relaxation of many small clumps. Only gravity is included in these simulations: incorporation of gas dynamics and star formation would modify these results. However, the density run is found to be quite similar to the $\sim r^{-3}$ light profile of an elliptical galaxy. If such mergers of clumps are presumed to occur within the $\sim r^{-2}$ dark halo generated by the cold dark matter simulations, then fitting the potential well and the light profile would seem to have met with reasonable success.

Hitherto, I have just talked about dark matter. What I want to do for the remainder of my talk is try to tell you something about the luminous component of matter and that involves talking about star formation. We would like to explain normal galaxies which typically contain stellar components distinguishable as spheroids and disks. We would like to understand why some systems have very large spheroids and some very small ones, in the family of spiral galaxies. Then we have systems which are pure spheroids, including giant elliptical galaxies and extremely low surface brightness dwarf spheroidal galaxies. And then there are the occasional more bizarre structures in the universe that we also have to try to explain.

3. PROTOGALAXY MODEL

Now the problem is of course that in order to understand the formation of the luminous cores of galaxies, and in particular the process of star formation, we have to incorporate the physics of gaseous dissipation together with the physics of gravitational collapse, a task that apparently is beyond the range of any presently available numerical simulations. So what we are going to have to do is resort to an analytic approach, and I will next discuss some of these arguments.

Common to many, but not all theories, is the following: imagine that the mass points from the computer simulations are individual clouds which are interacting gravitationally. Collisions of gas clouds are likely to be disruptive, if the relative velocity is too large compared to the internal sound speed, or else trigger star formation, whether by driving shocks and compression or by mergers of clouds. One begins with an ensemble of clouds collapsing gravitationally. We can estimate the cloud parameters, and they are inferred not to be very different in mass or density from typical giant molecular clouds in our own galaxy. While these clouds necessarily have few heavy elements at the onset of collapse, they soon acquire some heavy elements once massive stars form and evolve rapidly, and by processes of mergers, infall, and accretion, the protogalaxy gradually builds up into the systems that we see today. Let me now develop the simplest physical principles that enable one to estimate the protogalaxy parameters.

If I ask what is the surface density of the collapsing cloud or system of clouds when star formation is initiated, it is clear that one must have very strong density enhancements. The intergalactic medium is very diffuse today. To get strong density enhancements, a necessary condition is the occurrence of radiative shocks. A simple requirement for a radiative shock is that the post–shock column density must exceed a critical value that is approximately equal to the product of the following parameters: the density, the shock velocity and the post-shock cooling time. Now, because the cooling time is inversely proportional to density, the post–shock column density is simply a function of the post-shock temperature, or equivalently, the shock velocity. This critical column density is therefore just a function of the relative velocity between colliding clouds. For relative velocities characteristic of galaxy potential wells, I obtain numbers like 1 solar mass per square parsec. This is not a sufficient, but is at least a necessary, condition derived from a rather general argument in order to efficiently make stars.

Now, there are at least two other simple things that one can say, to supplement the radiative shock argument. One major prediction has come out of the theory of linear fluctuations in the expanding universe, and is due to the fact that the fluctuations are intrinsically non–spherical. Neighboring fluctuations tidally torque up one another as they grow in amplitude, moving away finally from one another as they become non–linear. One can calculate the resulting dimensionless angular momentum $\lambda \equiv J|E|^{\frac{1}{2}} G^{-1} M^{-\frac{5}{2}}$ to be ~ 0.07, which is an order of magnitude lower than seen in disk galaxies today but is a number that is very well defined in rather general classes of theories. A more illuminating way of writing this parameter is $\lambda \approx 0.3 v_{rot}/\sigma$, where v_{rot} is rotational velocity and σ velocity dispersion of a system with a de Vaucouleurs profile.

These clouds initially are far from being rotationally supported. What happens is that when these clouds start collapsing, one can imagine two possibilities. First of all, if dissipation is a dominant process, the system of clouds collapses and eventually forms a disk. This would be centrifugally supported, and one may estimate that the final surface density is enhanced by a factor λ^{-2}.

Another possibility occurs if the system does not undergo much initial dissipation, but breaks up into many stars. This is an important possible pathway of evolution, which is controlled not so much by dissipation physics as by dynamical relaxation physics. Then one has to resort to analogues of the computer simulations from which one learns that the final central density of the collapsing putative spheroid of the galaxy would depend on how cold the system was initially. The initial temperature depends on the shear in the system. In particular, one in general would expect the λ parameter to also provide a measure of the initial shear. Dynamical collapse of a cold system will lead to a higher surface density with $\Sigma \propto (T^i/W^i)^{-2} \propto \lambda^{-4}$ because the ratio of initial thermal to total energy T^i/W^i should roughly be proportional to λ^2. For a given λ, dynamical collapse results in a higher final density than dissipative collapse, at least for the central core.

We have now developed the groundwork for trying to set up some simple theoretical arguments. One would like to explain the data on galaxy surface brightnesses, based on studies by Kormendy, for example. Consider the core surface brightness as a function of absolute magnitude of galaxies. One finds that there is a huge range, but there are some intriguing patterns. Bulges, for example, have very high surface brightness, ellipticals somewhat less but are more luminous, whereas spiral disks form a distinct sequence of lower surface brightness but luminous objects. Dwarf ellipticals are the least luminous and the lowest surface brightness galaxies that are detected.

Now, theory provides most simply surface density and mass, slightly different coordinates from those of the observational astronomer. Cold dark matter provides one possible set of initial conditions, that are not meant to provide a unique description. To indicate schematically what might happen, if for example one has gas clouds which undergo dissipation, one would expect the surface density to increase by $\Sigma \propto \lambda^{-2}$ and follow a track in the (Σ, M) plane that corresponds to $M = $ constant. On the other hand, a hierarchy of merging clouds, which initially fragments into stars and violently relaxes, should follow an evolution track $\Sigma \propto M^{5/3}$ This corresponds to conserving phase–space density, an approximate property of the core of merging sub–systems. Unfortunately, one can also imagine other possibilities, for example, involving some combination of dissipation and violent relaxation. Finally, if I have a small gas–rich system that is blown apart when the first massive stars form, it can even lose mass and decrease in surface density. Figure 2, which is more of a cartoon than a theory at this point, summarizes these various evolution tracks.

The hope of the theory is that similar evolution tracks should eventually be able to account for the data. At the moment, we only have a qualitative understanding of protogalactic evolution. However the physics behind these different possibilities involving gaseous dissipation and violent relaxation, complicated by merging, accretion, and outflows, is going to be a necessary ingredient in any more realistic model.

Another way to infer the parameters of protogalaxies, in a very similar diagram but in a slightly different coordinate system, is to look at the average density as a function of the average velocity dispersion of the protogalaxy potential well. As before, there is a critical condition for cooling behind a radiative shock, which defines a minimum density at any given shock velocity. In order to form stars efficiently, the density must be higher than this critical value. Note that at low shock velocity ($\lesssim 100$ km s^{-1}), this condition becomes a lower bound on shock velocity, due to the inefficiency of low temperature cooling in a primordial gas. One finds that the observed galaxies, suitably estimating average densities of luminous matter now in old stars, lie within

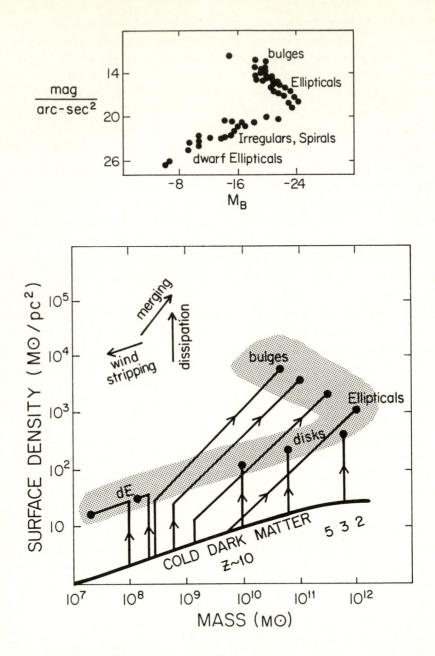

Figure 2: Evolution diagram for protogalaxies

the contours of the dissipation curve, while the largest structures, such as groups and clusters of galaxies, lie outside it. This suggests that there may be a grain of truth to this theory, which suggests that dissipation played a key role in the acquisition of the observed binding energies by galaxies but not by galaxy clusters.

4. PRIMORDIAL STAR FORMATION

I have thus far avoided any detailed description of how stars are actually made. Let me now try to tackle the process of star formation itself. Primordial stars, more or less by definition, possessed no heavy elements. Massive primordial stars are long since dead, but they left their signature behind in the halo of the galaxy. This signature can be studied in two ways. We can look at the ratios of various heavy elements in the very oldest stars of population II. These heavy elements formed in precursor massive stars, and we infer that it takes stars of appreciable masses, namely $\sim$ 5–50 $M_\odot$, given our understanding of nuclear astrophysics, to synthesize the metals observed in the oldest stars. One could not tolerate a primordial stellar population consisting, for example, exclusively of supermassive stars or of Jupiter–mass objects.

There are also one or two low mass halo stars of very low metallicity. The most extreme example is estimated to have an iron abundance of only one–millionth that of the sun and is a carbon dwarf of 0.3 $M_\odot$ in a binary system. Such low mass stars with extremely low metallicity are also presumably members of this precursor population. The inferred mass range of the primordial stars could well have been similar to that of stars forming today. And theory, as far as we can tell, basically says "why not"? Even in the absence of heavy elements to provide the cooling via fine–structure excitations of low–lying energy levels or dust emission, one can still make sufficient molecular hydrogen by H^- ion formation through some residual ionization either in shock fronts or in the very early universe to result in strong rotational cooling. This suffices to reduce cloud temperatures down to a thousand degrees or even less.

There is one intriguing property of star forming regions that may well be characteristic of the early galaxy. Let me focus for a moment on one of the best studied regions, namely the solar neighborhood. The initial stellar mass function describes the total number of stars formed as a function of stellar mass. Now the low mass stars are 10^{10} yr or older, all are observable if their mass exceeds $\sim 0.2 M_\odot$ and all we need to do is count them. However we do not know the past history of the star formation rate for stars much more massive than 1 $M_\odot$. If we assume that the star formation rate has been more or less uniform over galactic history, one would find a nearly continuous IMF. However, if far more massive stars formed in the past than are forming today, one would have bimodal star formation, with a discontinuous IMF between 1 and 2 $M_\odot$. Now what is the evidence for bimodal star formation? Consider first, the variation with time in the solar neighborhood.

4.1 Temporal evidence for bimodality

While explanations of the following phenomena are not unique to bimodal star formation, this one hypothesis could explain a number of puzzling observations. For example, the well–known G–dwarf problem: there are very few stars around us of very low metallicity, with the average metallicity in the disk being about one–tenth that of the sun. One can only understand this if the galaxy had grown with time so that one did not form many stars 10^{10} yr ago, or if the early stars rapidly enriched the ambient gas before most of the low–mass stars had formed. The metallicity–age relation for old disk stars reveals a rapid rise of metallicity during the first 10^9 yr and this can also be understood if one had enhanced Fe production in the very early galaxy.

Enhanced yields could be due either to more efficient metal production by metal–poor massive stars, or to an increase in the formation rate of such stars. The latter possibility, the bimodal hypothesis, is favoured because it can also account for some other observational puzzles. For example, in the initial stellar mass function there appears to be a kink between 1 and 2 $M_\odot$, so that there is no evidence for a continuous initial mass function. Unfortunately this feature is sensitive to the adopted mass–luminosity relation. This kink could be a memory of some different star formation rate for massive stars in the past. More importantly, there is the issue of the longevity of the gas in our galaxy, and also in other spiral galaxies. The gas would have been used up in the inner galaxy over a time much less than the age of the galaxy unless one had a preponderance of massive stars relative to the solar neighborhood. The massive stars make a difference, because low mass stars, of course, lock up the mass, being long–lived, whereas relatively short–lived massive stars will recycle the gas, ejecting enriched matter. The lock–up rate is much less with a bimodal initial mass function, and so the gas can be around for a longer time.

4.2 Spatial evidence for bimodality

Direct studies provide evidence that there is really ongoing bimodal star formation. The most dramatic case arises with starbursts, where one has a very high ratio of infrared luminosity to gas mass. This accumulating evidence suggests that one needs to truncate the initial mass function below about 2 or 3 $M_\odot$. Similar arguments apply to the inner disks of spirals, including our own galaxy. Not only is there evidence that some regions are deficient in low mass stars, but there is also evidence that in our galaxy there are cold clouds which are deficient in massive stars. Apparently some clouds form almost exclusively low mass stars whereas others form predominantly massive stars.

4.3 Theory of bimodal star formation

It is only too easy to think of theoretical reasons as to why the initial mass function may bifurcate, although no convincing analyses have been made.

One argument involves accretion onto protostar cores. The accretion rate (equal to $\sim V_s^3 G^{-1}$ for spherical accretion) and the gas reservoir determine the stellar parameters. It is not just sufficient to have a massive cloud core. One also needs a higher accretion rate to result in massive star formation, in order to suppress the convective protostellar phase. One expects convection to be important as long as the Kelvin–Helmholtz timescale is comparable to the accretion timescale. Low accretion rates allow a convective phase, which has been associated with the onset of mass outflows that limit infall. As emphasized recently by Stahler, this is especially important in stars below about 2 $M_\odot$, the Kelvin–Helmholtz timescale being inversely proportional to a high power of protostellar mass. This resulting divergence in protostar physics may lead to a bifurcation in the initial mass function. Moreover, the protostellar core accretion rate depends on the parameters of the dense core of the molecular cloud. Indeed, if the protostellar core is radiative, as for a massive protostar, the cloud evolution should determine the final mass. For example, when clouds fragment, the typical fragmentation scale of the most rapidly growing mode involves a strong sensitivity to temperature. If there is any feedback, due to heating by pre–existing massive stars or if the cloud was stirred up by considerable shear, then this would tend to favor larger fragments, that is to say, larger clumps that would presumably form more massive stars.

5. GALAXY FORMATION

5.1 Time–scale for star–formation

The time–scale for consuming the gas in a galaxy into stars is just the ratio of mass of gas divided by the rate of consumption of gas. This is a parameter one would like to know, in order to understand how galaxies formed. One can estimate the consumption rate of gas in a protogalaxy by the following simple arguments.

In order to form a round galaxy, one has to form the stars rapidly. In particular, the massive stars must form very rapidly in order to enrich the gas that forms the low mass stars, which in turn must have formed within a time–scale of order the collapse time for the protogalaxy. This could exceed the free–fall time, most plausibly if the collapse is very incoherent and there is delayed infall of a large fraction of the mass, as might occur in a merger. Most of the bulk kinetic energy will then have been preserved, and one will end up with a round galaxy. On the other hand, a system which remains gaseous for a long time will dissipate its bulk kinetic energy of collapse, and end up forming a disk.

Let us look at this from a different point of view. The timescales of order 10^9 yr for making spheroids and 10^{10} yr for forming disks can also be inferred from population synthesis arguments. These confirm that an explanation of the observed colors of galaxies also requires similar timescales. Star formation must be over within an e–folding time of a few gigayears or less in order to account for the old, red spheroidal components of galaxies.

5.2 Chemical evolution

If one really has bimodality, implying that star formation is not initially locking up a great deal of the gas, then the gas consumption time can exceed the characteristic formation time of a generation of stars by up to an order of magnitude. Typically in our galaxy at the present time, with the observed local initial stellar mass function, the gas consumption time is only slightly longer than the e–folding time for converting the gas into stars. With bimodality, the star formation time may involve only one–tenth or one–fifth of the gas consumption time. The gas is then recycled, perhaps 5 or 10 times, before finally being exhausted. These cycles need not happen continuously, but could involve a series of starbursts.

This leads to an interesting coincidence. The star formation time for forming stars in the spheroidal components of protogalaxies is inferred to be similar to the timescale observed in starbursts today, namely $\sim 10^8$ yr. This is already suggestive that the starburst phenomenon may be related to galaxy formation. The inference from the star formation timescales that massive stars must have formed early and rapidly leads to an important consequence, namely that there must have been rapid and efficient enrichment of gas in the protogalaxy. This leads to various further ramifications.

First of all, consider dwarf galaxies. One implication of the formation of massive stars in dwarf galaxies is that so much energy and momentum is transferred to the gas once it has been somewhat enriched by the first massive stars that the bulk of the gas is subsequently expelled as a wind. Low surface brightness remnants would be left behind to be identified as dwarf ellipticals. This should also produce an intergalactic medium enriched in heavy elements, and enriched moreover in the primary elements produced by the massive stars, rather than by the lower mass stars. One would expect, for example to find that intergalactic gas has an excess of oxygen and sulphur relative to iron and nitrogen. It is possible that this gas, once galaxy clustering is initiated today, cools, clumps, and is accreted by existing galaxies to provide fuel for renewed bursts of star formation, involving pre–enriched gas. One can imagine that the accumulation of this enriched gas could be a source of recent extragalactic HII region–like activity.

The fact that one has this massive star forming mode in the early phases of galaxies leads both to early enrichment and to considerable dust formation. Protogalaxies should therefore be rather dusty objects and the dust within these protogalaxies should reemit most of their radiation and provide an important contribution to the far infrared background.

In addition to the massive star mode, it is quite possible that there are systems which make exclusively low mass stars, below a few solar masses. Another outstanding problem involves trying to understand how globular clusters formed their stars. These are objects which are not very massive, but we very compact. If they are self–enriched, globular clusters must have formed stars extremely rapidly, within a crossing time ($\sim 10^6$ yr) if they had a normal initial mass function, otherwise they would have blown themselves apart due to the ionizing photon input from OB stars. However the low dispersion in metallicity within practically all globular clusters argues against self–enrichment. A possible explanation of the formation of globular clusters could involve colder pre–enriched clouds in the outer protogalaxy in which the low mass star forming mode dominated.

6. PRIMEVAL GALAXIES

What actually did primeval galaxies look like? The bottom–up scenario predicts that a protogalaxy was a collection of clumps merging together by a redshift of 1 to form a very regular looking galaxy by today. Protogalaxies are clumpy, low surface brightness objects. One interesting aspect of this scenario is that, if star formation occurred in little bits and pieces of the protogalaxy, rather than in one coherent object, the luminosity evolution would have been quite different. The requirement that in both situations the same amount of heavy elements were synthesized means that the clumpy formation model has a lower peak luminosity which occurred at a more recent epoch, than in a model with a unique formation epoch. This would lead to the expectation that because much of this activity is happening recently, there could be a considerable number of star forming, or recently post–star–forming, galaxies at low redshift.

It is likely that star formation bursts are triggered by mergers, in protogalaxies as well as in luminous IRAS systems. Once galaxy clustering commences, even more mergers between gas–rich protogalaxies will occur. The gas supply will be long–lived, because in the small units that are envisaged for the clumps ($\lesssim 10^8 M_\odot$), star formation is likely to have been very inefficient. During the initial collapse of galaxy groups and clusters, prior to virialization, one expects mergers and tidal interactions between protogalaxies to be frequent. Consequently, star formation should have been more efficient in the densest regions. Perhaps this accounts for the remarkable trend that in the denser parts of the universe, such as the cores of galaxy clusters, the early type spheroid–dominated galaxies predominate, with the disk–dominated systems being more numerous in the lower density regions.

One final point is that the bottom–up scenario allows the possibility of rare fluctuations forming very early at $z \gtrsim 10$, when Compton cooling is important. The cosmic microwave background acts like a thermostat within an ionized collapsing cloud, and should suppress supersonic motions between gas clumps to velocities below ~ 1 km s^{-1}. Such an initially cold system is expected to collapse to a high core density, especially if fragmentation occurs early and the collapse is predominantly dynamical. This allows the possibility of making compact nuclei at high redshift. The protogalaxy eventually accretes around such a nucleus, rare fluctuations tending to be highly correlated with other fluctuations. One might imagine that the nucleus is only activated much later when a large galaxy has accreted at $z \sim 1 - 2$. Intriguingly, one might expect such activity to lead to formation of quasars and radio jets. Such jets would interact with and entrain protogalactic gas, and should trigger substantial star formation in outlying regions. The rate of mass entrainment could plausibly exceed several hundred $M_\odot$ yr^{-1}.

7. TESTS

Ongoing searches for protogalaxies provide an important test of some of these ideas. Figure 3a plots the predicted number of protogalaxies per square degree versus red magnitude, and various models are shown. Limits are set by deep counts that probe well below the extragalactic background light. Other limits are set by direct spectroscopic searches, utilizing redshifted Lyman alpha, but these are very model–dependent, only setting limits on protogalaxies at $z \sim 5$. It seems that one is not so very far from detecting protogalaxies depending on which grid of models one chooses. That there is some possibility that this already has been happening, can be seen from a color–redshift diagram (Figure 3b). Although nearby galaxies are rather red, one finds a wide range in colors at a redshift of about 1. This might be evidence for enhanced massive star formation at early epochs, since a solar neighborhood IMF would not redden the colors sufficiently by today for these galaxies to be similar to nearby ellipticals.

8. CONCLUSIONS

The ultimate test of galaxy formation theory is the detection of protogalaxies. These have not yet been discovered at high redshift, but it may be that rare nearby objects, such as starburst nuclei, provide a glimpse of the formation process. The bimodality of the IMF, inferred for starbursts and for protogalaxies, leads one to expect that protogalaxies should be luminous,

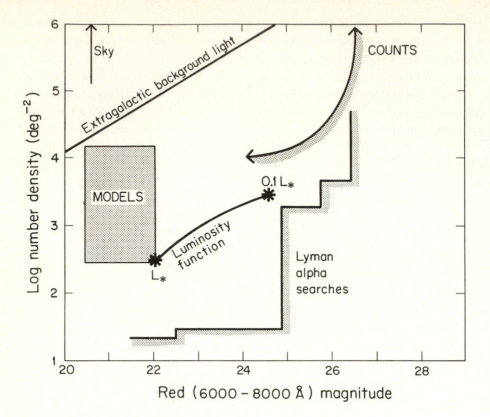

Figure 3a: Protogalaxy searches. Surface density versus red magnitude.

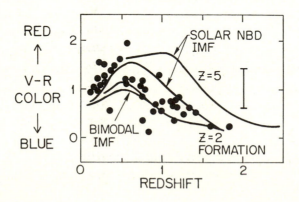

Figure 3b: Color evolution. V–R color versus redshift. From models by Wyse and Silk.

dusty objects. Resolving these objects in the far infrared is one of the greatest remaining challenges. Since even minimal evolution yields a diffuse radiation background amounting to about one percent of the cosmic microwave background in energy density, the bimodal star formation hypothesis points to a diffuse far infrared flux that could be an order of a magnitude larger.

Quasars and radio jets may provide a means of illuminating protogalaxies at high redshift if, as expected in a hierarchical formation model, galaxies form "inside–out".

ACKNOWLEDGEMENTS

I am grateful to several of my collaborators and colleagues for many discussions that have helped shape the ideas presented here. These include F. Palla, R. Pudritz, A. Szalay, and R. Wyse. I am also indebted to NSF and NASA for support.

GALAXIES AS TRACERS OF THE MASS DISTRIBUTION

G. Efstathiou

Institute of Astronomy
Madingley Road, Cambridge, CB3 0HA

ABSTRACT

We review theoretical arguments which suggest that galaxies may be more strongly clustered than the mass distribution. These ideas can be checked by testing whether the strength of galaxy clustering depends on the properties of the tracers (e.g. their luminosities, morphological type, etc.) and whether biases extend to large scales. The IRAS sample may provide strong constraints on large-scale clustering. We present preliminary results from a redshift survey of IRAS galaxies which indicate no significant correlations on scales $\geq 15h^{-1}$Mpc.

1. Introduction

Observers have traditionally assumed that bright ($\sim L^*$) galaxies are accurate tracers of the mass distribution. If this were true, then dynamical methods such as the cosmic virial theorem or Virgo infall applied to optically selected samples imply a low density universe with

$$\Omega \approx 0.2 \pm 0.1. \tag{1}$$

(Davis and Peebles 1983, Bean *et. al.* 1983, Yahil 1985).

The assumption that bright galaxies trace the mass is extremely difficult to verify observationally and should therefore be viewed with great caution. There is considerable evidence that the strength of clustering depends on the nature of the tracer. For example, Davis and Geller (1976) have shown that the two-point correlation functions of galaxies in the ($m_{pg} \leq 14.5$) Nilson (1973) catalogue depend on morphological type:

$$\xi(r) = (r_o/r)^\gamma, \quad \text{with}$$
$$\gamma_{GG} \approx 1.71, \quad r_{oGG} \approx 4.7h^{-1}\text{Mpc}, \quad \text{for all galaxies,}$$
$$\gamma_{EE} \approx 2.10, \quad r_{oEE} \approx 6.4h^{-1}\text{Mpc}, \quad \text{for ellipticals,} \tag{2}$$
$$\gamma_{SS} \approx 1.69, \quad r_{oSS} \approx 3.6h^{-1}\text{Mpc}, \quad \text{for spirals,}$$

where the amplitudes r_o have been computed from the angular correlation functions using the luminosity functions determined by Efstathiou *et. al.* (1987). There is evidence that IRAS galaxies, which are mostly late-type spirals, are more weakly clustered that optically selected galaxies (Rowan-Robinson and Needham, 1986). Correspondingly, estimates of Ω based on the IRAS dipole apparently yield higher values than equation (1), perhaps compatible with $\Omega = 1$

(see the articles by Davis and Rowan-Robinson in these proceedings and references therein). It is not at all obvious how to decide whether spirals and ellipticals should be assigned equal weight in an optical sample, or whether the IRAS sample should be regarded as a more faithful tracer of the mass fluctuations.

In this article I present theoretical arguments in support of the idea that luminous galaxies are biased tracers of the mass distribution. If the luminosities of galaxies are closely related to those of the dark haloes in which they are embedded, then gravitational clustering would automatically lead to "biases" of the kind described above. Section (2) describes results from N-body simulations of gravitiational clustering from scale-free initial conditions which show that large clumps at any time tend to be made preferentially of the largest clumps present at earlier times. In Section (3) we discuss a specific model for the formation of structure, the $\Omega = 1$ cold dark matter (CDM) model (e.g. Blumenthal *et. al.* 1984) and show that the gravitational growth of structure may lead to a strong bias in the distribution of galaxies. If these results are applicable to the real universe, the dynamical methods for estimating Ω reduce to tests of the level of bias expected in any particular model. Further constraints will have to be derived by other means, for example, by determining whether galaxies are positively or negatively correlated at large scales. The IRAS galaxy sample may well play a key role in such tests. Preliminary results are described in Section (4).

2. Gravitational Clustering from Scale-Free Initial Conditions

Gravitational clustering evolves in a self-similar fashion if the initial power-spectrum of the matter fluctuations is of power-law form $|\delta_{\mathbf{k}}|^2 \propto k^n$ and if the background cosmological model contains no characteristic lengths and timescales (Davis and Peebles 1977). Self-similar models are clearly idealizations, although they may be applicable to the real universe over restricted ranges of length and time. Here I summarize a few results from a detailed N-body study of matter dominated $\Omega = 1$ scale-free universes (Davis *et. al.* 1987). The N-body simulations contain 32768 particles; three simulations were run for each of several values of n.

Figures 1(a,b) show the evolution of the multiplicity function. The multiplicity of each particle is defined to be m if it is part of a group (identified using a "friends of friends" algorithm) with more than 2^{m-1} but no more than 2^m members. The histograms in the Figure show the fraction of particles with different values of m for groups corresponding to regions interior to contours of isodensity contrast $\delta\rho/\rho \sim 600$. The multiplicity functions at different epochs have been scaled according to the expections of self-similar evolution; this scaling does not apply for $m \sim 1$, but is clearly well obeyed over a wide mass range. The shape of the multiplicity function depends strongly on the initial fluctuation spectrum; the more negative the value of n, the broader the multiplicity function. The multiplicity functions are remarkably well described by the theory of Press and Schechter (1974), shown as the heavy lines in Figures 1(a,b).

We now turn to the origin of biases. Group catalogues were constructed for all the simulations at the last 6 output times. For each ensemble and for each time except the last, we have calculated the fraction of particles in groups of a given multiplicity which end up in massive clumps at the end of the simulation. Plots of this fraction as a function of multiplicity are shown in Figure 1(c,d). The definition of "massive clump" for each value of n was varied to ensure that such clumps contained about 1/6 of the total mass. The actual fraction contained is shown by the horizontal lines in the Figure.

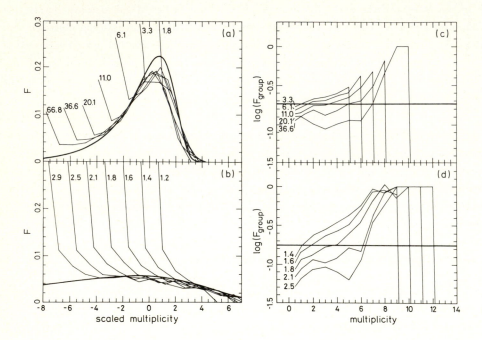

Figure 1. In (a,b) we show multiplicity functions for two values of the spectral index n. The numbers denote the cosmological scale-factor R ($R_i = 1$). In (c,d) we show the fraction of particles in groups of a given multiplicity which end up at the final epoch in massive clumps containing 1/6 of the total mass.

If the material of the massive clumps were randomly chosen from the clustering distribution at earlier times, all the curves in Figures 1(c,d) would coincide with these horizontal lines. In fact, the more massive a group at early times, the more likely it is to be incorporated in a massive clump at the end. These bias effects become stronger with decreasing n. The characteristic mass of clustering grows by a factor of 2.46 between output times, and so by a factor of ~ 100 between the first time plotted in Figures 1(c,d) and the end of the simulations. If the more massive groups at earlier times are identified as the sites of galaxy formation, "galaxies" would be overrepresented relative to the mass in the final "galaxy clusters". This is a manifestation of *natural biasing* resulting from the properties of hierarchical clustering (Frenk *et. al.* 1987, White *et al.* 1987).

3. The Galaxy Distribution in the Cold Dark Matter Model

A flat, cold dark matter dominated universe with scale invariant initial conditions can provide a remarkably good description of observed structure, but only if bright galaxies are biased tracers of the mass distribution (Davis *et. al.* 1985). In previous discussions, the required bias has been inserted "by hand" without any strong physical justification, though plausible astrophysical biasing mechanisms have been discussed (e.g. Rees 1985; Dekel and Silk 1986). Here I summarise some results on natural biasing derived from a set of N-body models in which we have simulated cubes of side 2500 km/s in a CDM universe using 262144 particles (White *et. al.* 1987). The initial amplitude of the fluctuation spectrum, and the cosmological parameters ($\Omega = 1$, $H_\circ = 50$km/s/Mpc) were chosen to match our previous work on flat CDM models (Davis *et. al.* 1985; White *et. al.* 1986).

The simulations follow the evolution of the collisionless dark matter component only, and so do not allow us to study the behavior of the dissipative gas from which galaxies must form. We have therefore included galaxy formation and merging in a way which, although plausible, remains somewhat *adhoc*. At various stages during the evolution of a a model we locate the most strongly bound particle in each dark matter halo. These are labelled "galaxies" and are assigned a circular speed $V_c = (GM(r)/r)^{1/2}$, where $M(r)$ is the mass contained is a sphere of mean overdensity 500 centred on each "galaxy". We then adopt simple algorithms to model galaxy mergers and to avoid multiple galaxy formation within each halo. The results described below are not especially sensitive to these procedures thus the details of these algorithms, which are somewhat technical, will not be repeated here (see White *et. al.* 1987). The spatial autocorrelation functions ($\xi(r)$) for "galaxies" with $V_c > 250$ km/s, $V_c > 100$ km/s and for the mass distribution are shown in Figure 2. The galaxy autocorrelation functions are steeper than that of the mass and have a larger amplitude which increases for larger rotation speeds. Over the range of separations shown in Figure 2 the mean correlation enhancement is $\sim .8$ for $V_c > 100$ km/s and ~ 5 for $V_c > 250$ km/s. For $V_c > 250$ km/s the correlations are comparable to those of observed galaxies (represented in the Figure by the dashed line of slope -1.8). This level of natural bias is evidently strong enough to reconcile the dynamics of galaxy clustering with a flat universe. The model predicts (via the Tully-Fisher (1977) relation) that the strength of galaxy clustering should depend on luminosity. There is little evidence for this in present samples, though good statistics are available for only a narrow range of luminosities.

The natural bias described above depends on the assumptions that a galaxy condenses in every halo, that the brightness of the galaxy reflects the depth of the halo potential well, and that galaxy merging is not strong enough to prevent the formation of galaxy clusters. The Tully-Fisher relation relates the luminosity of a spiral galaxy to the observed rotation velocity. Since the rotation curves of spirals are flat and at large radii are likely dominated by a dark halo component, we infer that the luminosities of spirals are indeed strongly correlated with the properties of their halos. Of course, it remains to be seen whether this observation, the small scatter in the Tully-Fisher relation, the morphological dependences summarised in equation (2) etc. can be explained with plausible astrophysics. The advantage of the argument presented above is that such uncertainities are largely bypassed.

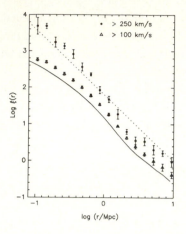

Figure 2. Two-point correlation functions in CDM simulations. The solid line shows the mass correlations, open triangles and filled circles show correlation functions for "galaxies" identified in the simulations with $V_c > 100$ km/s and 250 km/s respectively. The dashed line shows the power-law $\xi(r) = (r/10\mathrm{Mpc})^{-1.8}$ ($H_o = 50$ km/s/Mpc) which approximately describes the clustering of bright galaxies.

4. Large-Scale Clustering of IRAS Galaxies

The precise nature of the biases described above depend on the specific theory for the formation of galaxies and large-scale structure. As a consequence, we must expect that interpretations of large-scale structure in the Universe will necessarily be model dependent. Nevertheless, it is easy to think of tests which would impose stringent constraints on acceptable theories. For example, it would be extremely important if galaxies (whatever their intrinsic properties) were found to be clustered on scales $\geq 20h^{-1}\mathrm{Mpc}$. It is also important to check whether biases in the clustering properties of galaxies, such as those implied by equation (2) extend to large scales where $\xi \ll 1$.

Studies of the IRAS galaxies will certainly prove to be of interest in this regard. The IRAS Point Source Catalogue allows the delineation of a complete sample of galaxies over a wide area of sky that is deeper and more homogeneous than any available optical sample covering a comparable solid angle. It is therefore suitable for a study of the large-scale clustering of galaxies. Here I describe preliminary results results from a redshift survey of IRAS galaxies (Efstathiou, Ellis, Frenk, Kaiser, Lawrence, Rowan-Robinson and Saunders, in preparation). The flux range is $0.6 - 2$ Jy and we have sampled galaxies at a rate of 1 in every 6. Figure 3 shows the distribution of galaxies in the Northern hemisphere for which we have measured measured redshifts . The analysis described below is restricted to the complete region at $b \geq 40°$ containing about 240 galaxies. The median redshift of the sample is ~ 8500 km/s, thus it covers an extensive volume of space and is suitable for measuring $\xi(r)$ at large scales.

The sparse sampling strategy is specially designed to give small errors in ξ at large scales (Kaiser 1986). If $\xi \approx 0$ at large r, the uncertainty in the estimate for the M'th bin is

$$\delta \xi_M = \frac{1 + 4\pi f \bar{n} J_3}{\sqrt{N_p(M)}}, \quad J_3 = \int \xi r^2 \, dr, \tag{3}$$

where $\bar{n}$ is the mean density of galaxies, f is the fraction of objects sampled and $N_p(M)$ is the number of distinct galaxy pairs counted in the M'th bin. In practice, the luminosity function introduces a variable sampling rate with depth. We therefore estimate ξ in redshift space using the estimator

$$\xi = \frac{DD}{DR} - 1, \tag{4}$$

where DD is the weighted pair count of the data points and DR is the weighted pair count between data points and a set of random points distributed in redshift according to the selection function, $p(r) = \int \phi(L) \, dL$, determined from the IRAS luminosity function (Saunders et. al. in preparation). The variance in ξ at $\xi \ll 1$ is then minimised by weighting each point by

$$w = \frac{1}{1 + 4\pi f p(r) J_3}. \tag{5}$$

Adopting $4\pi J_3 \approx 13000 h^{-3} \text{Mpc}^3$ as a rough estimate, we find that $4\pi f p(r) J_3 \approx 1$ at $H_o r \approx 5000$ km/s for our survey.

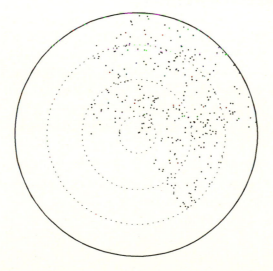

Figure 3. Distribution of IRAS galaxies in our sample with measured redshifts. The outer circle shows $b = 20°$.

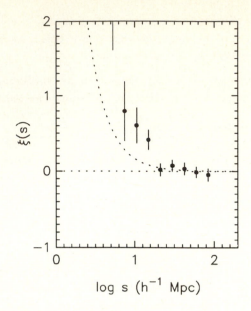

Figure 4. Two-point correlation function in redshift space for IRAS galaxies from our redshift sample. The dotted line shows $\xi(r)$ for spiral galaxies (equation 2).

The redshift space correlation function for the IRAS sample is shown in Figure 4. The error bars in this Figure have been computed from a set of Soneira-Peebles (1978) simulations. On scales $> 15h^{-1}$Mpc the correlation function is consistent with zero. This agrees with results from optical samples (Davis and Peebles 1983, Shanks *et. al.* 1983), though the IRAS data yield more accurate limits. These results are also consistent with the angular correlation function for the Lick sample (Groth and Peebles, 1977). On scale $< 10h^{-1}$Mpc, our estimate of $\xi(s)$ is higher that $\xi(r)$ for spirals, though the statistical significance of this is marginal. This could be due to peculiar motions which can distort the redshift space estimate $\xi(s)$, or perhaps a manifestation of the kind of biasing effects effects described above, since our weighting scheme favours luminous galaxies. These results will be substantially improved when our survey is completed. It will then be possible to apply more detailed tests of theories of large-scale structure.

Aknowledgements. I thank my N-body and redshift survey colleagues for allowing me to discuss our collaborative projects. This work has been supported by the SERC and by the Nuffield foundation.

REFERENCES

Bean, A.J., Efstathiou, G., Ellis, R.S., Peterson, B.A. and Shanks, T., 1983, *Mon. Not. R. astr. Soc.*, **205**, 605.

Blumenthal, G.R., Faber, S.M., Primack, J.R. and Rees, M.J., 1984, *Nature*, **311**, 517.

Davis, M. and Geller, M.J., 1976, *Astrophys. J.*, **208**, 13.

Davis, M. and Peebles, P.J.E., 1977, *Astrophys. J. Suppl.*, **34**, 425.

Davis, M. and Peebles, P.J.E., 1983, *Astrophys. J.*, **267**, 465.

Davis, M. Efstathiou, G, Frenk, C.S. and White, S.D.M., 1985, *Astrophys. J.*, **292**, 371.

Dekel, A. and Silk, J., 1986, *Astrophys. J.*, **303**, 39.

Efstathiou, G., Ellis, R.S. and Peterson, B. A., 1987, *preprint*.

Frenk, C.S., White, S.D.M., Davis, M. and Efstathiou, G., 1987, *Astrophys. J.*, in press.

Groth, E.J. and Peebles, P.J.E, 1977, *Astrophys. J.*, **217**, 385.

Kaiser, N., 1986, *Mon. Not. R. astr. Soc.*, **219**, 785.

Nilson, P., 1973, *Uppsala General Catalogue of Galaxies*.

Press, W.H. and Schechter, P., 1974, *Astrophys. J.*, **187**, 425.

Rees, M.J., 1985, *Mon. Not. R. astr. Soc.*, **213**, 75p.

Rowan-Robinson, M and Needham, G., 1986, *Mon. Not. R. astr. Soc*, **222**, 611.

Shanks, T., Bean, A.J., Efstathiou, G., Ellis, R.S., Fong, R. and Peterson, B.A., 1983, *Astrophys. J.*, **274**, 529.

Soneira, R.M. and Peebles, P.J.E., 1978, *Astron. J.*, **845**, 61.

Tully, R.B. and Fisher, J.R., 1977, *Astron. Astrophys.*, **54**, 661.

White, S.D.M., Frenk, C.S. , Davis, M. and Efstathiou, G., 1986, *Astrophys. J.*, **313**, 505.

White, S.D.M., Davis, M., Efstathiou, G. and Frenk, 1987, *preprint*.

Yahil, A., 1985, in *The Virgo Cluster of Galaxies*, eds. O.G. Richter and B. Binggeli, Munich, ESO. p359.

YOUNG GALAXIES

G. Burbidge and A. Hewitt
Center for Astrophysics and Space Sciences
University of California, San Diego

ABSTRACT

Several kinds of extragalactic systems have been suggested to be galaxies which
are very young in evolutionary terms. The evidence is reviewed, and we conclude
that it is most likely that the rare but very powerful IRAS galaxies with
luminosities $> 10^{12} L_\odot$ are young with ages less than 10^9 years.

1. INTRODUCTION

In this paper we discuss what sorts of evidence for young galaxies might be
found, since there are several kinds of observations which have lately been claimed
to provide evidence of galaxies seen at an early age.

We start by defining what we mean by young galaxies. In the conventional
evolutionary cosmologies, it is always supposed that the gaseous condensations out
of which galaxies form arise from density fluctuations at very early epochs.
Protogalaxies may be detected in the form of absorbing gas clouds. However galaxies
can only be detected directly if they form stars and emit radiation. Young galaxies
are galaxies in which only the first few generations of stars have formed and are
evolving. Since it is likely that massive stars are the first ones which form,
young galaxies can probably be detected when their ages are less than 10^9 years.

2. YOUNG GALAXIES AT HIGH REDSHIFT

Many quasi-stellar objects with redshifts of 3 or greater have been detected
(cf. Hewitt and Burbidge 1987) and many radio galaxies with redshifts of from 1 to 2
(Spinrad et al 1986) are known. Provided that the redshifts are of cosmological
origin, objects at high redshifts are being detected at much earlier epochs in the
universe, so that they may be much younger than nearby objects.

Lyα regions have been detected near to two QSOs, PKS 1614+051 (z=3.218) and
2016+112 (z=3.273) (Djorgovski et al 1985; Schneider et al 1986). The key question

is whether or not there is a continuum associated with the Lyα emission which may be due to a population of young hot stars. In further studies of the structure near PKS 1614+051, Djorgovski et al (1987) have concluded tentatively that there is such a continuum. However, Hu and Cowie (1987) cannot see a continuum and conclude that the extended object is a gas cloud photo-ionized by the QSO. This is very similar to that seen in low-redshift QSOs. Following the discovery of such features around 3C 48 (Wampler et al 1975), 4C 37.43 (Stockton 1976), and 3C 249.1, (Richstone and Oke 1977), Stockton and McKenty (1987) have shown that about 25% of 47 low redshift QSOs show such extensions when images using the [O III] 5007 line are obtained.

It appears to us that these phenomena are evidence that gas is ejected from the QSOs. There is no strong evidence that stars are forming.

In the radio galaxies there is by now a considerable body of evidence showing that gaseous emission is often found outside the nuclei and in the directions of the radio axes.

Large redshift objects such as 3C 368 (z = 1.132) (Djorgovski et al 1987) and 3C 326.1 (z = 1.82) (McCarthy et al 1987) have been studied. In 3C 326.1 through the method of imaging (broad band and Lyα) and long slit spectroscopy, it has been concluded that we are observing a very powerful H II region galaxy containing clusters of hot stars, i.e. a young galaxy. In the case of 3C 386 it is argued that the system is only about 3×10^9 years old, and that it is undergoing a merger with extensive star formation occurring.

Recently McCarthy et al (1987) have shown that there is a strong correlation between the radio and optical morphologies of powerful radio galaxies. Heckman et al (1986) found that a large fraction of the smaller redshift (less powerful) radio galaxies has peculiar optical morphology. Narrow-band imaging of these systems has shown that they also contain extended optical line-emission regions. It has been argued by Heckman et al and by Djorgovski et al (1987) that what we are seeing here are collisions or mergers between galaxies, leading to extensive star formation. These authors suggest that it is the interactions which trigger the violent activity in the center.

In our view the argument that we are seeing mergers, and that it is these mergers which give rise to violent activity in the centers is not likely to be correct. We shall come to the question of whether or not mergers are a common event later. Here we simply want to make the argument that all of the evidence points to the view that it is the violent activity in the centers which gives rise to the extended activity around those centers and not the other way around.

When evidence was first presented that violent activity often takes place in the centers of galaxies and is responsible for many types of non-thermal phenomena (Burbidge, Burbidge and Sandage 1963) it was realized that ejection was the key phenomenon. This conclusion has been completely borne out by many subsequent observations, i.e. various observations of Centaurus A and Minkowski's object, which suggest that star formation in the direction of the radio axis takes place following the ejection of matter and radiation from the center. Also the alignments of discrete objects e.g. the jet in M87 lying along the line joining M87 to M84, which has been known for more than 20 years, and has been supplemented by the observations of Arp that the X-ray sources in M87 are also aligned along the direction of the jet. Alignments such as these, and the alignment of the radio axis of Centaurus A with the other galaxies in that weak group, suggest that the activity starts in the center.

These kinds of alignments and extended activity are almost certainly what are being observed at larger distances by McCarthy et al and by Heckman et al.

Thus we conclude that these results bear out the original thesis that activity in a very small nucleus is responsible for all of the external activity, and not the other way around.

3. YOUNG GALAXIES AT LOW REDSHIFT

A number of proposals have been made concerning objects at low redshift which may be young galaxies. Obvious cases of this kind observed at optical wavelengths are so-called extragalactic H II region galaxies which are clearly being excited by young stars. In the 1960s Burbidge, Burbidge and Hoyle (1963) proposed that the remarkable system NGC 2444-45 (VV 144) was a young galaxy forming in the presence of an old system. In those days it was considered that the steady-state cosmology was a viable alternative to the evolutionary model, and thus we argued that this discovery supported the steady-state hypothesis, since in that cosmology galaxies have to be formed continuously, and the average age of a galaxy is only $(H_o^{-1})/3$ rather than an average age of $\sim(H_o^{-1})$ which is usually assumed for big bang cosmologies. Sandage (1963) immediately attempted to rebut our conclusion that NGC 2444 was a young system by measuring the continuum, and concluded that it was not distinguishable from the stellar continuum of the Large Magellanic Cloud which contains stars $\sim 10^{10}$ years old.

More recent work on H II region galaxies identified by Arp and Zwicky and studied by Sargent, Kunth and others suggests that they may be genuinely young systems. They have low metal to hydrogen ratios indicating that the enrichment of the initial gas by stellar nucleosynthesis has only recently begun.

However, we believe that the most promising candidates for genuinely young galaxies are the most luminous IRAS galaxies, those emitting radiation dominated by the far infrared flux ($\sim 100\mu$), at a level equal to or greater than $10^{12}L_\odot$.

Why is this? It is because we believe that the first stars to form, (the so-called Population III stars), are massive. If this is the case then they will generate very high luminosities, and evolve very rapidly both by mass loss and by the burning of hydrogen and helium. The most massive stars known to be evolving in this way at the present epoch have masses close to $100M_\odot$, and a prototype of such a star in our own galaxy is η Carinae. This star is both ejecting mass in the form of heavy elements, and ejecting the material which condenses into dust. If enough dust is ejected from an aggregate of massive stars the ultraviolet radiation will be absorbed and re-radiated in the far infrared.

One of us (Burbidge 1986) has made a model based on these ideas which will explain the high luminosity and the far infrared flux from the high luminosity IRAS galaxies such as Arp 220 and NGC 6240. If we suppose that the mass range of the stars formed lies between $20M_\odot$ and $120M_\odot$ it is easily shown that, for a variety of mass functions, the total mass in the stars needed to provide the luminosity

$(\sim 10^{12} L_\odot)$ is $10^7 - 10^8 M_\odot$, and $10 - 100$ generations of such stars, each with a lifetime $\sim 10^6$ years, will be required to produce enough dust $(\sim 10^8 M_\odot)$ to give rise to the far infrared flux. Thus it seems perfectly reasonable to suppose that such galaxies are young, with ages less than 10^9 years. There are several observations that can be made to test this hypothesis. First, there should be no evidence for the presence of old stars (with ages of 10^{10} years), and second, we might expect that the chemical abundances both in gas and dust will be highly anomalous, in the sense that the abundances of the heavy elements with respect to hydrogen will be higher than they are in well-evolved galaxies where all of the elements made in the Pop. III stars have been well-mixed and diluted throughout the whole of the galaxy.

We end with a final remark on the subject of mergers. In the model just described, as in others, the onset of massive star formation is attributed to tidal interactions with some outside galaxy. A merger is not necessarily required. There is a tendency in this field to call almost every irregular system which is a powerful IRAS source a merger, though the evidence for such an event is practically always absent. Tidal interactions of the kind orginally investigated by Toomre and Toomre may very well give rise to enhanced star formation. Also the case for a few bright peculiar galaxies being the results of mergers between previously separate galaxies has been made extremely strongly, and extremely persuasively by Schweizer.

However the fact remains that for most of the distant IRAS galaxies, and the very faint radio galaxies, the kind of evidence required to demonstrate a merger is not available. Increased activity, nuclear activity and the like are not enough to demonstrate that a merger has taken place. There are many other possible explanations. And sometimes we should not be afraid to say "I do not know."

REFERENCES

Burbidge, E. M., Burbidge, G. and Hoyle, F., 1963, Ap. J. 138, 873.
Burbidge, G., 1986, Pub. Astron. Soc. Pacific 98, 1252.
Burbidge, G., Burbidge, E. M. and Sandage, A. R., 1963, Rev. Mod. Phys. 35, 947.
Djorgovski, S., Spinrad, H., Pedelty, J., Rudnick, L., and Stockton, A.,
     1987, A. J. 93, 1307.
Djorgovski, S., Strauss, M. A., Perley, R. A., Spinrad, H. and McCarthy,
     P., 1987, A. J. 93, 1318.
Heckman, T., Smith, E., Van Breugel, W., Balick, B., Miley, G. K.,
     Bothun, G., Illingworth, G. and Baum, S., 1986, Ap. J. 311, 526.

Hewitt, A. and Burbidge, G., 1987, Ap. J. Suppl. 63, 1.
Hu, E. and Cowie, L., 1987, Ap. J. (Letters) 317, L7.
McCarthy, P., Spinrad, H., Djorgovski, S., Strauss, M., Van Breugel, W.
    and Liebert, J., 1987, Ap. J. (Letters) 319, L39.
Richstone, D. O. and Oke, J. B., 1977, Ap. J. 213, 8.
Sandage, A., 1963, Ap. J. 138, 863.
Schneider, D. P., Gunn, J. E., Turner, E. L., Lawrence, C. R., Hewitt,
    J. N., Schmidt, M. and Burke, B. F., 1986, A. J. 91, 991.
Spinrad, H., Djorgovski, S., Marr, J. and Aguilar, L., 1985,
    Pub. Astron. Soc. Pac. 97, 932.
Stockton, A., 1976, Ap. J. (Letters) 205, L113.
Stockton, A. and McKenty, J., 1987, Ap. J. 316, 584.
Wampler, E. J., Robinson, L. B., Burbidge, E. M. and Baldwin, J. A.,
    1975, Ap. J. (Letters) 198, L49.

INFRARED AND OPTICAL OBSERVATIONS OF DISTANT RADIO GALAXIES

J.S. Dunlop and M.S. Longair

Department of Astronomy, University of Edinburgh,
Royal Observatory, Blackford Hill,
Edinburgh EH9 3HJ, U.K.

ABSTRACT

Infrared and optical observations of 4 complete samples of radio galaxies are reviewed in the light of the most recent observations. Evidence is found for evolution of the stellar populations of the galaxies in all the samples. The most recent data on a 0.1 Jy sample of radio galaxies selected at 2.7 GHz are used to show that (i) the optical–infrared colours indicate the presence of a very old stellar population; (ii) all radio galaxies exhibit some star formation activity after the initial formation epoch; (iii) there is a significant dispersion in the amount of star formation activity at any epoch.

1. INTRODUCTION

Radio galaxies are the only objects which can at present be observed in reasonable numbers at redshifts greater than one and in which the light is the integrated emission of stars. As such they are at the moment the only tools which we can use to explore directly how the stellar populations and other properties of galaxies have changed with cosmic epoch. The obvious problems to be addressed are: (i) how typical are their properties of galaxies in general? (ii) does the fact that the galaxy is a strong radio source influence the properties of its stellar population in some direct or indirect way? We have been attempting to address some of these problems by studying well–defined complete samples of radio sources. Our collaborators in this project include John Peacock, Jeremy Allington–Smith, Hy Spinrad, Bruno Guiderdoni and Brigitte Rocca–Volmerange.

2. HISTORY OF THE PRESENT PROGRAMME

The story began about 7 years ago when it was appreciated that, in order to understand the evolutionary history of the radio source population, it was essential to embark upon a major endeavour to identify complete samples of radio sources and then to obtain for the identifications optical and infrared photometry and optical spectra from which redshifts could be determined. It was known from the work already completed on the brightest radio sources that many of the identifications would be with very faint objects and that it would be a very major undertaking to attempt to obtain the optical spectra of objects which could be as faint as 24 or 25 magnitude.

This programme has been much more successful than we would have predicted thanks to a number of pieces of pure astronomical luck. First, sensitive optical and infrared detectors became available at exactly the right time to enable these difficult programmes to be undertaken in a reasonable time. Second, mainly through the heroic efforts of Hy Spinrad and his colleagues, it has been found possible to measure the spectra and redshifts of many of the very faintest radio galaxies because they turn out to have extremely strong narrow emission lines in their spectra. It is important to emphasise that these strong narrow emission line spectra have a very different origin from the strong broad and narrow emission line spectra observed in Seyfert galaxies and quasars. In these cases, the strong emission lines originate close to the nucleus and are excited by the non-thermal nuclear continuum radiation. In contrast, the strong narrow emission lines observed in the spectra of the radio galaxies with redshifts greater than one seem to be associated with diffuse regions of ionised gas which can extend to distances of up to about 100 kpc from the nucleus of the galaxy. It is likely that these regions are similar to standard HII regions in which the excitation of the gas clouds is associated with energy sources such as young stars distributed throughout the gas. These topics are discussed in the publications of Spinrad and his colleagues (see e.g. Spinrad 1988).

The complete samples of radio sources which have been studied are as follows:

COMPLETE RADIO GALAXY AND RADIO QUASAR SURVEYS

LOW FREQUENCY (178 and 408 MHz)	HIGH FREQUENCY (2.7 GHz)
$S_{178} \geqslant 10$ Jy	$S_{2.7} \geqslant 2$ Jy
173 radio sources (3CR sample)	233 radio sources (2.7 GHz all-sky survey)
Laing, Riley and Longair (1983)	Wall and Peacock (1985)
$1 \leqslant S_{408} \leqslant 2$ Jy	$S_{2.7} \geqslant 0.1$ Jy
59 radio sources (1 Jy sample)	178 radio sources (0.1 Jy sample)
Allington-Smith (1982)	Downes, Peacock, Savage and Carrie (1986)

It is important to recall that one of the prime motivations for studying these samples was to understand more about the astrophysical evolution of the radio source population. The importance of using both low and high frequency samples is to investigate the properties of both the flat and steep spectrum radio sources. This simple spectral division results in two very different classes of radio source, the flat spectrum objects being compact sources associated with the nuclear regions of the radio galaxies whereas the steep spectrum sources are extended radio sources, the radio emission in general originating from regions very much larger than the parent galaxy. As is well known, the flat spectrum sources are poorly represented in the low frequency samples but are much more common at

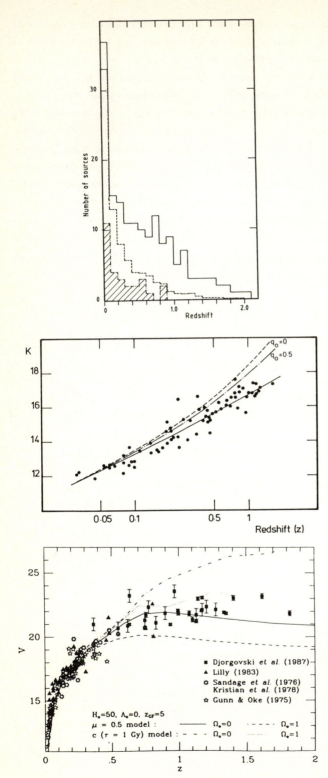

Figure 1. Comparison of various redshift distributions for radio sources in the 3CR sample. The hatched histogram shows the redshift distribution of sources with $S_{178} \geqslant$ 25 Jy. The dashed histogram shows the redshift distribution predicted at 10 Jy from that observed with $S_{178} \geqslant$ 25 Jy assuming a uniform Friedmann model with Ω = 2 (q_O = 1). The redshift distribution observed for sources with $S_{178} \geqslant 10$ Jy is shown by the solid line. The predicted numbers of sources agree well with the observed numbers in the low redshift bins, $z \leqslant 0.2$, but there is a large excess at all larger redshifts.

Figure 2. The infrared K magnitude − redshift diagram for 3CR radio galaxies (Lilly and Longair 1984). The dashed lines show the relations expected for standard giant elliptical galaxies observed through a fixed aperture in world models having Ω = 1 (q_O = $\frac{1}{2}$) and Ω ≃ 0 (q_O = 0). The solid line is a best fit to the data, consisting of Friedmann world models with Ω = 0 − 1 and including evolution of the stellar population of the galaxy with cosmic epoch.

Figure 3. The V magnitude − redshift relation for 3CR galaxies and other selected samples. The points at redshifts greater than 0.5 are almost exclusively 3CR galaxies. Details of the parameters of the Bruzual models used to fit the data are included on the diagram (Spinrad and Djorgovski 1987).

high frequencies. Of particular interest are the surveys at 1 Jy at 408 MHz and 0.1 Jy at 2.7 GHz because it is at these flux densities that the greatest divergence between the observed numbers of sources and the expectations of the uniform world models is found.

3CR Survey   The complete sample of 173 sources defined by Laing, Riley and Longair (1983) has been the most intensively studied of all samples. It consists of all the bright sources in the northern hemisphere in directions away from the Galactic plane at low frequencies. Thanks to the efforts of many workers, the identification and redshift content of this sample is now essentially complete. These data can be used to demonstrate how radio astronomers have been assisted by the effects of cosmological evolution in finding objects with large redshifts. In Figure 1, the observed distribution of redshifts for radio galaxies having $S_{178} \gtrsim 25$ Jy is shown. From this, it is a straightforward task to work out, for any of the standard world models, the redshift distribution which would be expected at 10 Jy and this is shown by the dashed line. This can be compared with the observed redshift distribution at 10 Jy shown by the solid line. It can be seen that vastly larger numbers of radio sources with redshifts greater than 1 are observed than would be expected. It is this enormous excess which is attributed to the effects of cosmological evolution of the properties of the radio source population.

The 2.7 GHz All-sky Survey   This survey is the high frequency counterpart of the 3CR survey. There is considerable overlap between it and the 3CR survey in the northern hemisphere for steep spectrum sources. One of the important aspects of this survey is that it contains many very luminous flat-spectrum quasars which, in conjunction with similar sources in the deeper 0.1 Jy survey, enable strong constraints to be set on the high redshift behaviour of this class of object.

The 1-Jy and 0.1 Jy Samples   The 1-Jy sample was the first of the deeper surveys to be studied in detail. It was selected at a flux density at which the excess of faint radio sources approaches its maximum value. The 0.1 Jy sample performs the same role as the 1-Jy sample but is considerably larger and now includes both the steep and flat spectrum sources. It is the results of this last survey which form the principal new material of this paper.

3. INFRARED AND OPTICAL OBSERVATIONS OF THE 3CR AND 1-Jy SAMPLES

Most of the identification work has been carried out in the optical waveband although it was realised in the early 1980s that the infrared waveband, $1 - 2.2\mu m$, had a number of important advantages for this type of study. First, if the spectral energy distribution of a giant elliptical galaxy is redshifted to redshifts of 1 or more, the galaxy becomes relatively a much stronger emitter in the $1 - 2.2$ μm waveband rather than in the optical. This is because the spectral energy distribution of a giant elliptical galaxy at zero redshift peaks at about 1 μm. It is partly for this reason that it has become possible to identify distant galaxies as easily in the near infrared as in the optical waveband. The second important point is the fact that the stars which contribute most of the light of the giant elliptical galaxy in the near infrared waveband are derived from the old stellar population of the galaxy. The light is dominated by the emission of stars on the red giant branch whose progenitors are roughly solar mass stars which have completed their evolution on the main sequence.

3CR Survey These ideas were first tested in detail on the 3CR sample. Lilly and Longair (1984) analysed the infrared and optical properties of a complete sub–sample of radio galaxies from the 3CR sample. Similar results have been found by Eisenhardt and Lebofsky (1987). By selecting only the radio galaxies in which there was no contamination from non–thermal nuclear emission, they showed that the infrared colours of the galaxies were entirely consistent with the (H–K) v redshift and (J–K) v redshift relations expected if the spectral energy distribution of a giant elliptical galaxy is redshifted. Interestingly, once account was taken of the observational uncertainties in each point, the intrinsic scatter in (H–K) colour was the same at all redshifts.

Perhaps the most striking result was the K magnitude – redshift relation (K–z relation) for 3CR radio galaxies (Figure 2). It can be seen that the dispersion in absolute K magnitudes remains the same at low (z $\leqslant$ 0.3) and high (z $\geqslant$ 1) redshifts. However the diagram is interpreted, it is clear that there is some systematic behaviour in the evolutionary history of these galaxies with redshift. Since the diagram is produced from a complete sample of radio sources selected according to radio selection criteria, it has the advantage that the Malmquist bias which dogs optical studies of the redshift–magnitude relation is eliminated. The lines on the diagram show various predicted relations once account is taken of the fact that the observations are made through a fixed aperture. Standard world models with Ω = 0 to 1 are a poor fit to the data and only if Ω were as large as about 7 would it be possible to achieve an adequate fit. Our preferred model is indicated by the solid line which takes into account the evolution of the stellar populations of the galaxies. This analysis can be undertaken in a reasonably model–independent way because by far the dominant contribution to the stellar evolution is simply the fact that when a galaxy was younger, the rate of evolution of stars off the main sequence was greater than it is today. Since we are only concerned with a relatively narrow range of stellar masses over look–back times to about one third the present age of the Universe and since the giant branches are not expected to be very different for stars of about 1 M$_\odot$ the correction for the effects of stellar evolution are not very sensitive to the precise model adopted. All the models predict that the galaxies should be about a magnitude brighter at a redshift of 1 relative to their present luminosities. The best fitting line shown in Figure 2 involves using a world model with Ω ~ 0 – 1 and a simple correction for the stellar evolution of the luminosity of the galaxy. There are several new points to be made about the analysis of the 3CR data

(i) It is interesting to compare Figure 2 with the corresponding diagram using optical magnitudes (Figure 3 – from Spinrad and Djorgovski 1987). There are two important differences (cf Figures 3 and 4 of Spinrad and Djorgovski 1987). First, there is a significantly greater dispersion in the optical absolute magnitudes of the radio galaxies at redshifts greater than z ~ 0.5 as compared with their K magnitudes. Second, the predicted V magnitude–redshift relation is much more sensitive to the model for the evolution of the stellar population of the galaxies than the K–z relation. In both cases, the root cause of the problem is the fact that the visual luminosities of the galaxies are much more sensitive to the presence of young stars in the stellar population. This makes the determination of the underlying evolution of the galaxies and the dynamics of the Universe much more difficult in the optical as compared with the infrared waveband.

(ii)   One of the factors which could cause a bias in the K−z relation is if there were a correlation between radio and optical luminosities of the members of the 3CR sample.   Yates, Miller and Peacock (1986) have carried out such an analysis for the radio galaxies in the 3CR sample.   A weak positive effect has been found although it is rather sensitive to the presence or absence of a few data points.   Our interpretation of this result is that the influence of any correlation between radio luminosity and the absolute magnitude of the galaxies is not as strong as the effects of the stellar evolution of the stellar content of the galaxies.

(iii)   A consequence of the greater dispersion in the optical as compared with the infrared Hubble diagram is that there must be a dispersion in the optical−infrared colour−redshift diagrams.   In the analysis of Lilly and Longair (1984), there were few points available to compare with the models but even then the results were suggestive (Figure 4).   At redshifts greater than 0.5, the dispersion in (r−K) colours increases markedly.   A comparison was made between the observations and various models.   The NE model results from simply redshifting the energy distribution of a standard giant elliptical galaxy.   The *passive evolution (or C−model)* model is more physical in that it incorporates the evolution of the stellar populations of a model galaxy in which all the stars form in an initial burst and the subsequent evolution simply reflects the life and death of these stars.   The model is constrained to produce the correct spectral energy distribution at zero redshift.   The third model, the μ−model, corresponds to Bruzual's model with $\mu = 0.5$ in which there is an exponentially decreasing rate of star formation.   As a result, there is ongoing star formation at redshifts z ∼ 1 which accounts for the fact that the model galaxy is "bluer" (i.e. smaller r−K) than in the other cases.   It is significant that all the points lie on or below the C model which can be regarded as a reasonable null hypothesis.   The dispersion below that line probably reflects different amounts of star formation going on in these galaxies and the dispersion is greater than the uncertainties in the estimates of the optical−infrared colours.   This was one point in which the analyses of Spinrad and ourselves has differed.   He found that he could represent his optical data with a single μ−model whereas, when the infrared data are taken into account, a real dispersion is found.

(iv)   One of our reasons for believing in the reality of this dispersion to the blue of the C−model was the fact that there seemed to be a weak correlation between the "blue−excess" in the galaxy, i.e. its colour excess with respect to the C−model, and the strength of the [OII] lines.   This made intuitive good sense in that both could be associated with star formation within the galaxies.   This picture is consistent with the beautiful new results of Spinrad, Djorgowski and their colleagues (e.g. Djorgovski et al 1987, McCarthy et al 1987b) who have shown that the narrow emission lines originate from very extended regions which can exceed the size of the galaxy itself and which are likely to be excited by young stars.

The 1−Jy Survey   It was fully expected that the 1 Jy sample would be biassed towards larger redshifts than the 3CR sample and this was reflected in the fact that the 1−Jy galaxies turned out to be on average about a factor of 5 fainter as infrared emitters than 3CR radio galaxies (Figure 5). Over 50% of the 1−Jy galaxies now have redshifts but naturally the easier, brighter objects have been the prime targets for spectroscopy. Nonetheless, a number of faint large redshift galaxies have been measured and the new points all lie within the envelope of points in Figure 2. These results give us confidence that the K−z relation for powerful radio galaxies is very well defined out to K = 17.5 and can be used to make estimates of redshifts. We note, however, that the 3 1−Jy galaxies with K > 17.5 for which redshifts have been measured all lie close to the upper envelope of the 3CR K−z diagram. This might suggest that the very faintest 1 Jy galaxies (i.e. K ~ 18) may not be extreme high redshift counterparts of the 3CR galaxies but simply intrinsically fainter galaxies at z ~ 1.5.

An intriguing part of the analysis of these data was the extrapolation of the K−z relation to large redshifts using the passive evolution models which give a good representation of the infrared data. These extrapolations clearly depend upon the age of the Universe (and hence on H_0) and also upon the epoch when the galaxies first formed. The calculations which were presented by Lilly, Longair and Allington−Smith (1985) are no more than indicative of what might be learned from this type of data. We know that the radio galaxies must lie along the observed K−z relation at redshifts 0 < z < 1.5. It is therefore possible to constrain the evolutionary behaviour at larger redshifts depending solely upon the redshift at which the galaxy formed. The results of some of these calculations are shown in Figure 6. If the galaxies formed at a relatively low redshift (z_F = 3.5 is shown in Figure 6), the galaxy is expected to be very bright through the redshift interval 1.5 < z < 3 before it joins the observed redshift−magnitude relation. In fact, the models suggest that if the formation redshift were low, the radio galaxies would never be much fainter than K = 18. On the other hand, if the epoch of formation were large (z_F = 20 in Figure 6), the apparent brightness of the galaxy decays at large redshifts and they would be expected to be as faint as K = 19 at a redshift of 2.5. The observed magnitude distribution is displayed on the ordinate of Figure 6 in which it can be seen that there is a group of galaxies at magnitude 18 and a tail of fainter objects which extends to K = 19. It is not clear whether or not these are galaxies at very large redshifts or, as the few available redshifts suggest, simply a low luminosity tail of the radio galaxy luminosity function. The nature of these galaxies is of the greatest interest.

In view of the incompleteness in the redshifts of the 1−Jy sample, we have plotted the (optical−infrared) colour v K magnitude diagram in Figure 7 with an estimated redshift scale along the upper abscissa based upon the mean redshift−magnitude diagram of Figure 2. The solid line is the expectation of the C−model and again it can be seen that there is a significant scatter of points to the blue side of this relation. The crosses on this diagram show the location of the 3CR points and it is clear that roughly the same distribution of "blue−excess" is observed in the 1−Jy sample. Of particular interest is the clump of points with (r−K) ~ 6 and K ~ 18. Interpreted literally, these are galaxies which appear to be undergoing passive evolution at a redshift of about 1.5 or possibly greater. More measurements of these galaxies are very important to obtain better colours and, if possible, redshifts. These points are difficult to explain in terms of μ−models with μ ~ 0.5.

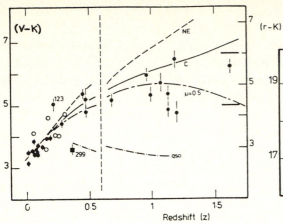

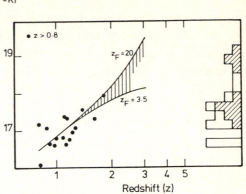

Figure 4. The variation of the optical – infrared colours (V–K and r–K) for 3CR radio galaxies as a function of redshift. The meanings of the various predicted relations are explained in the text. NE – redshifting the spectrum of a standard giant elliptical galaxy; C – passive evolution of the stellar population of the galaxy; $\mu = 0.5$ – Bruzual model of galaxy evolution with exponentially decreasing star formation rate (Lilly and Longair 1984).

Figure 6. The large redshift K–z relation for radio galaxies from Figure 2. The solid lines show extrapolations of the observed relation for passively evolving galaxies with epochs of formation of 3.5 (lower line) and 20 (upper bound). The histogram of observed K magnitudes for sources in the 1 Jy sample are shown on the ordinate (Lilly et al., 1985).

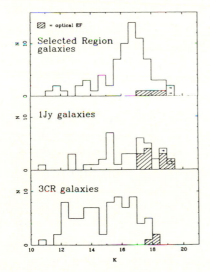

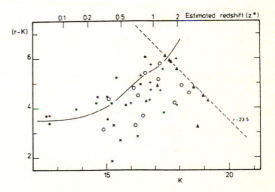

Figure 5. The apparent K magnitude distributions for radio galaxies in the following samples: (a) the 0.1 Jy sample; (b) the 1 Jy sample; (c) the 3CR sample. Sources which were optically empty fields are shown.

Figure 7. The (r–K) – K relation for radio sources in the 1 Jy sample. The solid line shows the relation expected for a passively evolving galaxy. The sources indicated by triangles are lower limits to the values of r–K. The crosses show the locations of 3CR galaxies on this diagram. The line marked r = 23.5 corresponds to the limit of the optical identification survey (Lilly et al 1985).

4. THE 0.1 Jy SAMPLE

The optical and infrared observations of the radio sources in the 0.1 Jy sample have been carried out over the last $2\frac{1}{2}$ years. Of the 178 sources in the sample, complete infrared and optical photometry is currently available for about two thirds of the sample which forms an unbiased complete sub-sample of the data. These results are presented in a preliminary form here.

One of the most important results is that it is possible to identify essentially all the objects in the sample in the infrared waveband. This is illustrated in Figure 5 which shows that the peak in the magnitude distribution for these identifications occurs well above the limit to which observations could be made. The apparent K magnitude distribution is similar to that of the 1 Jy sample suggesting that they are spanning a similar range of redshifts.

Even more surprising was the fact that it has proved possible to identify essentially all the objects in the optical B and R bands as well. These results are presented as a set of colour-redshift diagrams, the redshifts being estimated redshifts on the basis of the K-z relation of Figure 2. Figures 8 (a) and (b) show the basic data as well as two reference models. In our new work, we have used the models of galaxy evolution of Guiderdoni and Rocca-Volmerange (1987). These models are similar to those of Bruzual but incorporate a number of refinements and improvements in the computational procedures. Specifically, they include new data on the the ultraviolet spectra of different stellar populations and utilise stellar spectra of higher resolution. Other improvements include a mass-dependent giant branch and the introduction of post-giant branch evolutionary stages.

For the reference models, we have taken the stellar energy distributions which result from evolving what Rocca-Volmerange and Guiderdoni (1987) describe as "UV-hot" and "UV-cold" elliptical galaxy models to the present day. These spectra provide excellent fits to the observed range of optical and ultraviolet spectra of nearby elliptical galaxies, those with a strong ultraviolet excess being termed the UV-hot ellipticals and those with only a weak ultraviolet continuum the UV-cold ellipticals. From the point of view of the evolution of the galaxy models, the UV-hot ellipticals correspond closely with the end point of evolution of a μ-model with $\mu = 0.3$ while the UV-cold elliptical corresponds to the end point of evolution of a μ-model with $\mu = 0.6$.

Figure 8(a) illustrates the expected (R-K) - z relations if the UV-hot and UV-cold elliptical galaxy spectra at the present day are redshifted. These illustrate clearly why it was exciting that the galaxies were not only detectable in the K waveband but also in the R waveband. Both the UV-hot and UV-cold ellipticals would have been far too red in (R-K) to be detected at 24th magnitude in the optical waveband. The reference model is not particularly physical since it involves simply redshifting a standard spectrum. However, an interesting point arises when the optical colours (B-R) of the galaxies are plotted against redshift (Figure 8(b)). The UV-cold elliptical galaxy is too red to explain the observations but the UV-hot elliptical galaxy spectrum passes more or less through the centre of the optical colours of the galaxies. This can account for the fact that some authors have

335

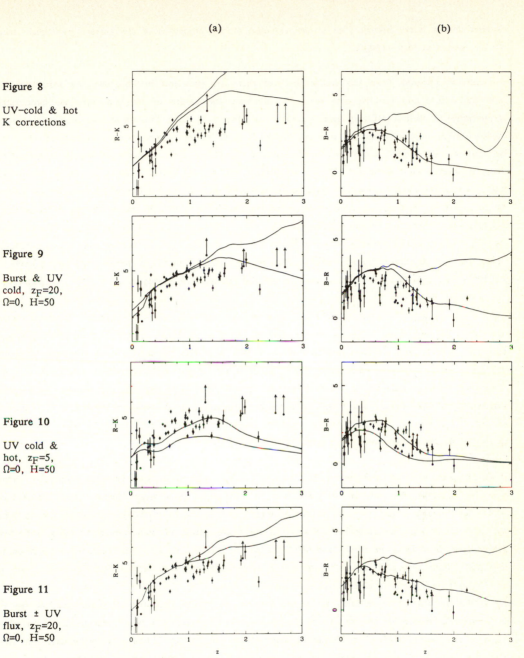

(a)   (b)

Figure 8

UV–cold & hot
K corrections

Figure 9

Burst & UV
cold, z_F=20,
Ω=0, H=50

Figure 10

UV cold &
hot, z_F=5,
Ω=0, H=50

Figure 11

Burst ± UV
flux, z_F=20,
Ω=0, H=50

Figures 8 to 11. Comparison between the observed (R–K) v redshift and (B–R) v redshift diagrams with the predictions of different models (for details see text). Figure 8: redshifting UV–hot and UV–cold spectra; Figure 9: Burst model and UV–cold evolving model; Figure 10: Young UV–hot and UV–cold evolving models; Figure 11: burst model plus continuous star formation.

claimed that it is not necessary to invoke evolution of the properties of the galaxies to explain their optical colours at high redshift.

It is already apparent from this comparison that there are problems in simultaneously accounting for both the optical and infrared data sets. To illustrate the nature of the problems, we show some comparisons of various evolving galaxy models with the data. An interesting comparison is that of models similar to those used previously in Figure 4. The closest correspondence is obtained for what we term a burst model, similar to the C-model of Figure 4, and the UV-cold model corresponding to a μ-model with μ = 0.6 (Figure 9(a) and (b)). We have used the galaxy models of Rocca-Volmerange and Guiderdoni who kindly provided us with the results of their computer codes. As in Figure 4, these loci provide a reasonable description of the upper bound to the distribution of points in the (R-K) v z diagram although there are a few points which lie above this locus. One important point is that, because the model chosen has a long age, it can account for the reddest points on this diagram. In the case of the (B-R) colours, the burst model does not provide a satisfactory fit and the cold model scarcely makes the galaxies blue enough.

Figures 10 (a) and (b) show the predictions of models involving stronger evolution at more recent epochs. The epoch of formation of the galaxies has been brought forward to a redshift of 5 and both the UV-cold and UV-hot models compared with both sets of data. It can be seen that the galaxies never become red enough in (R-K) although they span nicely the range of (B-R) colours. The basic problem with Figure 10(a) is that the stellar populations are not old enough. This is a feature which has already been noted by a number of authors - to obtain red galaxies at a redshift of one, it is necessary for the simple evolution models to have enough time for the bulk of the population to be well-evolved.

It is too early to know how severe this problem is but there are many possible solutions. If we maintain the assumption that the stars are born with the same initial mass function at all epochs, we could, for example, argue that the underlying models correspond to behaviour similar to the burst or old UV-cold models superimposed upon which there are bursts of star formation which make the galaxy very much bluer in (R-K) and (B-R). In any case, there must be considerable variation in the evolutionary history of the radio galaxies to account for the considerable dispersion in colours at any redshift. Another possible variant is shown in Figure 11 (a) and (b) in which we show the expected distribution if, in addition to the initial burst, we assume that there is a constant low rate of star formation throughout the history of the galaxies, the rate adopted being that needed to account for the UV properties of giant elliptical galaxies at the present day. These models indicate how the arbitrary inclusion of star formation can help to account simultaneously for the redness of the (R-K) v redshift diagram and the relative blueness in the (B-R) v redshift diagram.

Our preliminary conclusions from this work are therefore as follows. First, we require the bulk of the stellar population to be rather old in order to fit the red envelope of the (R-K) diagram (i.e. to achieve R-K ~ 6 at z = 1). In the simple models of galaxy evolution, this means present day ages of ~ 17 Gy and hence Ω ~ 0 and H_o ~ 50 km s^{-1} Mpc^{-1}. It should of course be noted that this timescale is ultimately tied to the globular cluster ages and hence this is not an independent argument for long cosmological timescales. Second, the (B–R) v redshift diagram illustrates that **ALL** the radio galaxies have undergone some star–formation activity after the initial burst, since the C–model which fits the upper envelope in the (R–K) – redshift diagram is then too red to fit the upper envelope of the B–R diagram. The old UV–cold model with age $\sim$ 17 Gy provides a reasonable description of the upper envelope in both diagrams. Third, we need to account for the scatter blueward of the red envelope in both the R–K and B–R diagrams. This could either be the consequence of a range of star formation histories or a range of formation redshifts.

5. IMPLICATIONS FOR THE EVOLUTION OF THE RADIO SOURCE POPULATION

The success in identifying completely the 0.1 Jy sample has important implications for the determination of the evolution functions which describe how the comoving space density of radio sources of different radio luminosities have changed with cosmic epoch. For most of the quasar candidates in the sample with flat radio spectra, redshifts have been determined. For the radio galaxies, the programme is at a much earlier stage, particularly for the very faint identifications. However, the tight K–z relation for the radio galaxies enables us to make reasonable estimates of their redshifts and these data have been included in the most recent estimates of the evolution functions for both the steep and flat spectrum source populations. The procedures adopted were the same as those described by Peacock (1985) but incorporating the new data on the 0.1 Jy survey and the new identification and redshift data on the brighter surveys. Figure 12 shows the variation with cosmic epoch of the comoving space density of radio sources with radio luminosities 10^{27} and 10^{26} W Hz^{-1} sr^{-1} for the case of an Einstein–de Sitter world model ($\Omega = 1$). The pairs of lines indicate the statistical uncertainties in the determination of the comoving space densities. The evolution is found to be strongest for the most luminous sources and similar evolution functions are found for both the steep and flat spectrum sources. Both classes of source exhibit a cut–off at look–back times of about 75% of the present age of the Universe. This result had been established previously for the flat spectrum radio sources (Dunlop et al., 1986) but the result is now found for the steep spectrum sources as well. The most important new pieces of data which contribute to this result are the

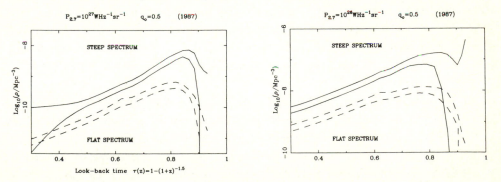

Figure 12. Examples of the evolution functions for radio sources with steep and flat radio spectra following the procedures of Peacock (1985). In each case, the pairs of lines indicate the range of likely comoving source densities as a function of look–back time.

infrared identifications of all the sources in the 0.1 Jy sample. The latter result is important in that it indicates that both classes of radio source exhibit a cut–off in their evolutionary behaviour with cosmic epoch at large redshifts. In the case of the new data we have now identified with galaxies objects at about the epoch of maximum high–energy astrophysical activity, making them of special interest for spectroscopic studies.

6. ASTROPHYSICAL CONSIDERATIONS

A wholly new dimension has been added to these studies by the recent work of Spinrad and his collaborators on the nature of the diffuse narrow emission line regions found in the radio galaxies at large redshifts (Spinrad 1988). There are now three independent pieces of information about the evolution of the properties of the radio galaxies used in these studies.

(i) The probability of radio source activity changes rapidly with cosmic epoch. As illustrated in Figures 12, there is between a factor of 10 and 100 increase in the probability of strong radio source activity between redshifts of about 0.5 and 2. Notice that we do not attribute any specific physical origin for this evolution which could be attributed to the changing luminosities of the sources or to changes in the probability with which they occur with cosmic epoch or to some combination of these and other factors.

(ii) There are significant changes in the infrared and optical luminosities of the stellar components of the radio galaxies with cosmic epoch. The evolution of the infrared luminosity of the galaxy is consistent with the decay of the underlying old population of stars but superimposed upon this there must be star formation activity which produces the much bluer optical–infrared colours observed at large redshifts in all the samples studied.

(iii) There is a very much larger probability of finding extended strong narrow emission line regions in large redshift radio galaxies than in their counterparts at low redshifts. Indeed, in some cases, it appears that a significant fraction of the galaxy population may be forming within these diffuse emission line regions (Spinrad 1988). In many of the cases studied, the extended region coincides roughly with the optical image of the galaxy suggesting that star formation is taking place throughout the body of the galaxy. Spinrad and his colleagues show that the excitation of these large emission regions can be reasonably accounted for by enhanced star formation rates within the galaxy.

The key questions are whether or not these different types of evolutionary phenomena are related and what their significance is for the evolution of galaxies in general. A number of scenarios can be envisaged. For example, a simple picture would be one in which the burst of star formation throughout the galaxy results in the diffuse emission line region and the consequent enhancement of the optical relative to the infrared stellar luminosity of the galaxy. Some of the gas liberated finds its way down into the nucleus of the galaxy where it acts as the source of fuel for the black hole in which, by means as yet poorly understood, the radio source activity is generated. Many sources for the enhanced rate of star formation in the past could be envisaged – mergers between galaxies, late formation of a significant fraction of the stars in the galaxy, galaxy formation at late epochs, etc. This picture would result in a direct causal relation between the rate of star formation in these galaxies and the resulting properties of the radio source.

A major spanner may have been thrown into this picture by the recent work of McCarthy et al (1987a) who show that there is a correlation between the axis of the double radio sources and the major axis of the extended emission line regions. The full significance of this result has yet to be understood but it raises the possibility that exactly the opposite of the process described in the last paragraph may occur, namely that the radio jets compress the gas through which they pass and this stimulates star formation activity in the disturbed regions. This work raises a whole host of new possibilities and undoubtedly will be the subject of much further study. The data discussed here provide the basis for a deeper understanding of the relation between the evolutionary properties of the host galaxies of strong radio sources and the occurence of radio source activity within them.

References

Allington-Smith, J.R., 1982. *Mon. Not. R. astr. Soc.,* **199**, 611.

Bruzual, G.A., 1983. *Astrophys. J.,* **273**, 105.

Djorgovski, S., Strauss, M.A., Perley, R.A., Spinrad, H. & McCarthy, P., 1987. *Astr. J.,* **93**, 1318.

Downes, A.J.B., Peacock, J.A., Savage, A. & Carris, D.R., 1986. *Mon. Not. R. astr. Soc.,* **218**, 31.

Dunlop, J.S., Downes, A.J.B., Peacock, J.A., Savage, A., Lilly, S.J., Watson, F.G. & Longair, M.S., 1986. *Nature,* **319**, 564.

Eisenhardt, P.R.M. & Lebofsky, M.J., 1987. *Astrophys. J.,* **316**, 70.

Guiderdoni, B. & Rocca-Volmerange, B., 1987. *Astr. Astrophys.,* in press.

Laing, R.A., Riley, J.M. & Longair, M.S., 1983. *Mon. Not. R. astr. Soc.,* **204**, 151.

Lilly, S.J. & Longair, M.S., 1984. *Mon. Not. R. astr. Soc.,* **211**, 833.

Lilly, S.J., Longair, M.S. & Allington-Smith, J.R., 1985. *Mon. Not. R. astr. Soc.,* **215**, 37.

Longair, M.S., 1988. *V.L. Ginzburg Festschrift,* in press.

McCarthy, P.J., van Breugel, W., Spinrad, H. & Djorgovski, S., 1987a. *Astrophys. J. Lett.* in press.

McCarthy, P.J., Spinrad, H., Djorgovski, S., Strauss, M.A., van Breugel, W. & Liebert, J., 1987b. *Astrophys. J. Lett.* in press.

Peacock, J.A., 1985. *Mon. Not. R. astr. Soc.,* **217**, 601.

Rocca-Volmerange, B. & Guiderdoni, B., 1987. *Astr. Astrophys.,* **175**, 15.

Spinrad, H. & Djorgovski, S., 1987. *Proc. IAU Symposium No. 124, 'Observational Cosmology',* A. Hewitt et al (eds.), D. Reidel, p. 129.

Spinrad, H., 1988. *Proc. 3rd IAP Astrophysics Meeting, 'High Redshift and Primeval Galaxies',* in press.

Wall, J.V. & Peacock, J.A., 1985. *Mon. Not. R. astr. Soc.,* **216**, 173.

Yates, M.G., Miller, L. & Peacock, J.A., 1986. *Mon. Not. R. astr. Soc.,* **221**, 311.

A VERY DEEP IRAS SURVEY AT THE NORTH ECLIPTIC POLE

J. R. Houck, P. B. Hacking

Astronomy Department, Cornell University,
Ithaca, NY 14853

and

J. J. Condon

NRAO
Edgemont Road
Charlottesville, VA 22901

ABSTRACT

The data from approximately 20 hours observation of the 4- to 6-square degree field surrounding the North Ecliptic Pole have been combined to produce a very deep infrared survey at the four IRAS bands. Scans from both pointed and survey observations were included in the data analysis. At 12 and 25 microns the deep survey is limited by detector noise and is approximately 50 times deeper than the IRAS Point Source Catalog, PSC. At 60 microns the problems of source confusion and galactic cirrus combine to limit the deep survey to approximately 12 times deeper than the PSC. These problems are so severe at 100 microns that we only quote flux values for locations corresponding to sources selected at 60 microns. In all, 47 sources were detected at 12 microns, 37 at 25 microns and 99 at 60 microns. Here we describe the data analysis procedures and discuss the significance of the 12- and 60-micron source count results.

1. INTRODUCTION

The sensitivity limit of the IRAS all-sky survey was set by detector noise and the requirement that each source be observed on several different scans. Most of the sky was scanned six times resulting in an average of 12 detections per source. By combining the data together before selecting the sources the sensitivity limit can be lowered significantly. This is being done in the IRAS Faint Source Survey to be published in 1988. By increasing the number of scans of a given region the sensitivity limit can be further lowered. Because of the IRAS orbital geometry the areas surrounding ecliptic poles were scanned on nearly every orbit. In addition the location of the IRAS secondary flux standard, NGC 6543, is very near the North Ecliptic Pole so large number of calibration scans were made of the region. The Large Magellanic Cloud is very near the South Ecliptic Pole so that data set is not appropriate for a deep survey of the type discussed here.

2. DATA

The North Ecliptic Pole is located at 18 hr $+66°$ ($\ell = 97°$ and b $= 30°$), 10-arc minutes from the position of NGC 6543. The deep survey at 12 and 25 microns covers the 4.3-square degrees surrounding the pole. At 60 and 100 microns the survey covers 6.3-square degrees. In all, 488 survey scans and 838 calibration scans have been combined to produce a very deep survey at the North Ecliptic Pole. Individual survey scans were 1/2-degree wide and crossed the field at virtually every location along a wide range of position angles. The calibration scans were typically 1/2 degree wide by 1- to 1 3/4-degrees long and roughly centered on NGC 6543. They too were aligned along a wide range of position angles. Because the calibration scans were too short to cross the entire survey field, there is uneven coverage of the field. It is therefore important to keep track of the coverage or equivalently the noise in the maps in constructing number count diagrams. Two sets of maps were generated by separately combining the intensity mode and point source filtered data. The intensity mode maps for 60 and 100 microns show strong extended emission arising from dust in our own Galaxy, infrared cirrus. Source extraction was performed on the point-source-filtered maps. A point source was selected at 12 and 25 microns if it had an amplitude greater than five times the noise in the surrounding region of the map. At 60 microns, because of source confusion and the effects of cirrus, it was further required that the selected source have a detected flux $\geq$ 50 mJy and that its appearance on the 60-micron point-source-filtered-contour map closely resembled the contours for other nearby point sources. The results of the point source selection are summarized in Table I. The details of the data analysis, source selection and the complete source list are presented by Hacking and Houck (1987).

TABLE I

λ Microns	Total Number*	Stars	Galaxies	Unidentified
12	46	41	5	0
25	36	18	17	1
60	98	0	$\sim$ 90	$\sim$ 8

*In addition to NGC 6543.

2.1. 12-Micron Sources

The source selection process resulted in the detection of 47 sources. Of these 41 are clearly associated with stars visible on the POSS prints, five are associated with galaxies and the remaining object is the planetary nebula NGC 6543. Most of the detected stars are K-type giants without significant circumstellar shells. Stars with thick shells which dominate the sources observed at high flux values are absent from the faint survey because their implied distances, given their high luminosities, would place them well outside the Galaxy. We have compared the number of detected stars at 12 microns with the predicted number counts based on an optical model of the galaxy (Bahcall and Soneira 1980) and the transformation from optical magnitudes and colors to the 12-micron magnitude given by Waters, Coté and Aumann (1985). The resulting comparison is shown in Figure 1. The dotted line is a power law fit,

$$n_{S \geq S_o} = \{109^{+70}_{-40}\}S_o^{-0.74 \pm 0.15} \, \text{stars/sq. deg} .$$

where S is measured in mJy. The slope of the power law is consistent with the model; however, the magnitude of the observed counts is 30% higher than predicted by the model, a difference which is marginally significant. The difference can be removed by increasing the number of stars in the model by 30 percent or by increasing the scale height for giants from the assumed 250 pc to 270 pc.

2.2. 60-Micron Sources

Most and perhaps all of the 60-micron sources are galaxies. Approximately 80% of the sources have obvious counterparts on the POSS prints. We have secured five-minute CCD images of all the fields using the Palomar 60- inch telescope and the fraction of objects that are clearly extragalactic rises to 90 or 95%. The remaining few sources do not show obvious candidates but

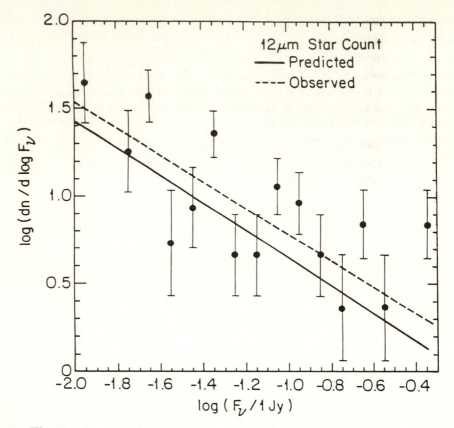

Figure 1. The 12-micron number counts for stars is shown. The prediction of the model of Bachall and Soneira, transformed by the optical to 12-micron transformation of Waters, Coté and Aumann is shown by the solid line.

often show a relatively dense cluster of up to 25 faint galaxies. The number counts are shown in Figure 2. Number count results at higher flux levels from the PSC are also shown. The curves were derived using the local luminosity function from the PSC data (Soifer *et al.* 1985). A very similar luminosity function has been independently derived by several other authors (see Soifer et al. 1987 and references therein). The luminosity function was first converted into a local visibility function and then into number counts correcting for source motion and including a luminosity dependent K correction term (see Hacking *et al.* 1987). The curve labeled "No Evolution" assumes that the local luminosity function is appropriate for all distances represented in the survey. The curve labeled "Evolution" assumes that the luminosity function evolves in a fashion similar to that found for radio sources (Condon 1984). The deep counts appear to fall above the no evolution curve but not as high as the evolution curve. The excess in the numbers at low flux levels, if real, could be due to evolution or the presence of a statistically significant fluctuation in the form of a cluster of galaxies at some modest redshift. Although there is no obvious evidence on the POSS prints for the existence of a cluster, it may reveal itself as a spike in the redshift distribution.

3. ACKNOWLEDGEMENTS

We are very grateful for the assistance we have received from the staff at IPAC and for many helpful conversations with Jim Cordes, Martin Harwit, Martha Haynes and Tom Soifer. This work was supported by NASA through IPAC.

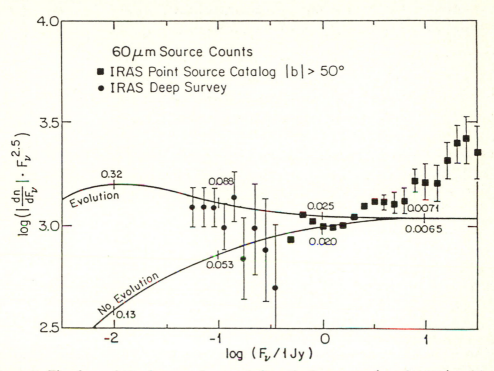

Figure 2. The observed 60-micron number counts from the deep survey (round points) and from the IRAS Point Source Catalog (square points). The curves show the expected number counts based on the local luminosity function and either no evolution or evolution of the type invoked in the analysis of radio number counts. The numbers along the curves indicate the median redshift at that flux level.

References

Bahcall, John J. and Soneira, Raymond M., 1980, *Ap. J. (Supplement)*, **44**, 73.
Condon, J.J., 1984, *Ap. J.*, **287**, 461.
Hacking, Perry and Houck, J.R., 1986, *Ap. J. (Supplement)*, **63**, 311.
Hacking, Perry, Condon, J.J., Houck, J. R., 1987, *Ap. J.(Letters)*, **316**, L15.
Soifer, B.T., Sanders, D.B., Neugebauer, G. *et al.*, 1986, *Ap. J. (Letters)*, **303**, L41.
Soifer, B.T., Houck, J.R. and Neugebauer, G., 1987, *Ann. Rev. Astr. and Astrophysics*, to be published.
Waters, L.B.F.M., Coté, J. and Aumann, H.H., 1986, *Astronomy and Astrophysics*, **172**, 225.

COSMOLOGICAL EVOLUTION OF STARBURST GALAXIES
AND IRAS COUNTS AT 60 μm

A. Franceschini   and   L. Danese
Dipartimento di Astronomia,
Vicolo dell' Osservatorio 5,
I-35122 Padova, Italy

G. De Zotti
Osservatorio Astronomico,
Vicolo dell' Osservatorio 5,
I-35122 Padova, Italy

and C. Xu
International School for Advanced Studies,
Strada Costiera 11,
I-34014 Trieste, Italy

ABSTRACT

Cosmological evolution of actively star-forming galaxies on timescales $\simeq 20 \div 25\%$ of the Hubble time has been recently called for by Danese *et al.* (1987) to explain the radio source counts at mJy and sub-mJy levels. Owing to the tight correlation between far-IR and radio emissions from these sources, the IRAS survey data provide a direct test of such models. Hints of evolution consistent with the model predictions already emerge from the redshift distribution of galaxies brighter than 0.5 Jy at 60 μm. The model also provides a good fit to the recent deep IRAS counts, which are consistently higher than expected in the absence of evolution. The predicted contribution of galaxies to the 100 μm background amounts to only a few percent of the intensity tentatively estimated by Rowan-Robinson (1986).

1. INTRODUCTION

The IRAS survey has presented a new view of the universe, emphasizing objects endowed with intense star formation activity. Thus the far-IR survey data are particularly well suited to study the cosmological evolution of bursting galaxies and to explore the early phases of galaxy formation.

In view of the tight correlation between far-IR and radio emission of disk galaxies (de Jong *et al.*, 1985), it is particularly useful to analyse IRAS data in conjunction with those provided by recent VLA surveys. As shown by Danese *et al.* (1987), Starburst/Interacting (S/I) galaxies evolving on time scales of order of $20 \div 25\%$ of the Hubble time can account for the upturn of deep radio source counts at mJy and sub-mJy levels both at 1.4 GHz and at 5 GHz (Windhorst, van Heerde & Katgert 1984; Fomalont *et al.* 1984; Partridge, Hilldrup & Ratner 1986), as well as for their identification statistics and their preliminary redshift distributions (Windhorst *et al.* 1985; Windhorst, Dressel & Koo 1987).

We explore here the implications of the models by Danese *et al.* on IRAS counts and related statistics. We stress that no free parameters come in: the models fitted to the radio data are directly translated into the far-IR band exploiting the observed distributions of far-IR to radio luminosity ratios.

2. BRIGHT FAR-IR SAMPLES AND LOCAL LUMINOSITY FUNCTIONS

To understand the evolutionary properties of far-IR selected galaxies we need *i)* to distinguish among those classes of sources in which basically different astrophysical processes are operating – since their evolution is also likely to be different – and *ii)* to determine reliable local Luminosity Functions (LFs) for each class. To this end we have exploited the rich information content of a complete sample of 1671 galaxies selected from the UGC catalogue, with almost complete redshift data, observed at 1.4 GHz by Dressel & Condon (1978) and cross-correlated with the IRAS Point Source Catalogue (see Franceschini et al., 1987 for details). Separate 60 μm luminosity functions have been derived for the following classes of sources: *a)* E + S0 galaxies; *b)* Spirals + Irregulars; *c)* Seyferts; *d)* S/I galaxies (including non-Seyfert Markarians and galaxies classified in the UGC as peculiars, distorted, interacting). As shown by Fig. 1, the global LF at 60 μm, sum of contributions from the various classes, is generally in very good agreement with that of IRAS selected galaxies (Soifer *et al.* 1987; Lawrence *et al.* 1986).

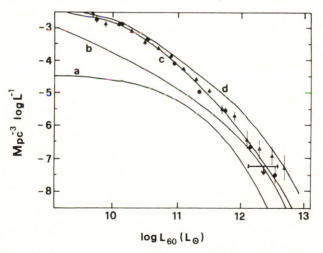

Figure 1. Comparison of 60 μm luminosity functions. Filled circles are from Soifer *et al.* (1987), open triangles from Lawrence *et al.* (1986). The upper limit at high luminosities comes from an analysis of the UGC sample. Curves *a* and *b* are the local LFs of Seyfert and Starburst/Interacting galaxies, respectively. Curves *c* and *d* are the LFs at z=0 and z=0.25. The latter is approximately the mean redshift of the highest luminosity sources in the sample of Lawrence *et al.* (1986); note that the associated LF matches their space density.

On the other hand, the UGC sample is not deep enough to allow a direct evaluation of the space densities of galaxies with $L_{60} > 10^{12} L_\odot$. It yields, however, an upper limit, consistent with the estimates by Soifer *et al.* (1987) but below those by Lawrence *et al.* (1986). As noted by Soifer *et al.*, taking into account that their sample is shallower (maximum redshift ~ 0.08) than that of Lawrence *et al.*, the difference between the two luminosity functions may be ascribed to evolution. Our results, based on an independent, optically selected sample, strengthen their conclusion (cf. Fig. 1).

3. THE DEEP IRAS SURVEY

The model used to interpret the deep radio source counts assumes that the LFs of E + S0 and Spiral + Irregular galaxies do not change with cosmic time, while S/I and Seyfert galaxies undergo luminosity evolution: $\phi[L(z), z] = \phi[L(0), 0]$, $L(z) = L(0)e^{\kappa \cdot \tau(z)}$, $\tau(z)$ being the look-back time and κ^{-1} the evolution time scale in units of the Hubble time (fits to the radio data yield $\kappa = 4.3$ for a density parameter $\Omega = 1$, $\kappa = 5.1$ for $\Omega = 0$). This simple recipe, in combination with the observed distribution of far-IR to radio luminosity ratios and with the observationally determined far-IR K-corrections, allows us to reproduce all available IRAS survey data.

In Fig. 2, the predicted counts at 60 μm (curve a) are compared with those from the deep IRAS survey by Hacking & Houck (1987); curve (c) in the same figure shows the contribution of evolving S/I galaxies. In the absence of evolution the counts of the latter sources would be given by curve (d) and the total log N - log S by curve (b). The χ^2 test rejects this possibility with a high confidence ($> 99\%$): the reduced χ_ν^2 for the five faintest bins is $\simeq 5$ (see also Hacking *et al.*, 1987).

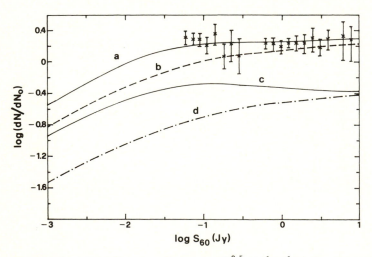

Figure 2. Differential counts at 60 μm normalized to 600 $S_{60}^{-2.5}$ $Jy^{-1}sr^{-1}$. Data points at $S_{60} < 0.7$ Jy are from Hacking & Houck (1987), those at brighter flux levels by Rowan-Robinson *et al.* (1986). Curves (a) and (b) are the predicted total counts calculated for the cases of evolution and no-evolution of the Starburst/Interacting galaxies. The separate contributions of the latter class are shown as curves (c) and (d) for the cases of evolution and no-evolution, respectively.

Although, in the present scenario, the evolution timescale of S/I galaxies is comparable to those of QSOs and of powerful radio sources, the predicted counts are never as steep as are observed to be in other wavebands: actually, they are slightly flatter than euclidean for $S_{60} > 0.1$ Jy and decline significantly at fainter fluxes. Two factors contribute to keep the log N - log S quite flat:

i) non-evolving Spirals and Irregulars outnumber S/I galaxies: the surface density of the latter approches that of the former only at $S_{60} < 0.1$ Jy, where also the counts of S/I galaxies start converging.

ii) The very steep K-correction flattens down effectively the log N - log S of S/I galaxies themselves. Note, in this respect, that source counts at $\sim 300 \ \mu m$ should be particularly effective in emphasizing the cosmological evolution of this population. Such counts could be provided by the Far Infrared and Submillimeter Space Telescope (FIRST) under study by ESA.

The contributions to the 60 μm counts of normal and radio loud E + S0 galaxies and of QSOs (both radio loud and radio quiet) are always small.

Primeval galaxies re-radiating in the far-IR the bulk of energy released in the production of a solar metallicity at $z \leq 2.5$ should show up in the counts at $S_{60} < 0.02$ Jy and could then be detected by the Infrared Space Observatory (ISO), to be launched be ESA in 1993.

In the present framework the 60 μm counts of galaxies are expected to start converging shortly below 0.1 Jy. Correspondingly, the 60 μm diffuse background component due to galaxies is predicted to be quite modest: $\simeq 2 \ 10^4 \ Jy \ sr^{-1}$ from both S/I and Spiral + Irregular galaxies, $\simeq 0.5 \ 10^4 \ Jy \ sr^{-1}$ from Seyferts and, possibly, $\simeq 1 \ 10^4 \ Jy \ sr^{-1}$ from primeval galaxies (ellipticals and S0s yield a negligible contribution). A conservative extrapolation to 100 μm gives $\simeq 0.2 \ MJy \ sr^{-1}$, i.e. only a few percent of the diffuse background intensity tentatively estimated by Rowan-Robinson (1986).

REFERENCES

Danese, L., De Zotti, G., Franceschini, A., and Toffolatti, L. 1987, *Ap. J. Lett.*, **318**, L15.

de Jong, T., Klein U., Wielebinski R., and Wunderlich, E. 1985, *Astr. Ap. (Letters)*, **147**, L6.

Dressel, L.L., and Condon, J.J., 1978, *Ap. J. Suppl.*, **36**, 53.

Fomalont, E.B., Kellermann, K.I., Wall, J.V., and Weistrop, D., 1984, *Science*, **255**, 23.

Franceschini, A., Danese, L., De Zotti, G., and Toffolatti, L. 1987, submitted to M.N.R.A.S.

Hacking, P., and Houck, J.R. 1987, *Ap. J. Suppl.*, **63**, 311.

Hacking, P., Condon, J.J., and Houck, J.R. 1987, *Ap. J. (Letters)*, **316**, L15.

Kron, R.G., Koo, D.C., and Windhorst, R.A. 1985, *Astr. Ap.*, **146**, 38.

Lawrence, A., Walker, D., Rowan-Robinson, M., Leech, K.J., and Penston, M.V. 1986, *M.N.R.A.S.*, **219**, 687.

Partridge, R.B., Hilldrup, K.C., and Ratner, M.I., 1986, *Astr. Ap.*, **308**, 46.

Rowan-Robinson, M. 1986, *M.N.R.A.S.*, **219**, 737.

Rowan-Robinson, M., Walker, D., Chester, T., Soifer, T., Fairclough, J. 1986, *M.N.R.A.S.*, **219**, 273.

Soifer, B.T. et al. 1987, IRAS Preprint.

Windhorst, R.A., van Heerde, G.M., and Kargert, P. 1984, *Astr. Ap. Suppl.*, **58**, 1.

Windhorst, R.A., Miley, G.K., Owen, F.N., Kron, R.G., and Koo, D.C., 1985, *Ap. J.*, **289**, 494.

Windhorst, R.A., Dressler, A., and Koo, D.C. 1987, in *IAU Symp. N° 124, Observational Cosmology*, eds. G. Burbridge and L.Z. Fang (Dordrecht:Reidel), Reidel Publ. Co.

THE IRAS DIPOLE

Michael Rowan-Robinson

Astronomy Unit

Queen Mary College,

Mile End Road,

London E1 4NS

ABSTRACT

The IRAS dipole is calculated for a new complete and reliable 60μm galaxy catalogue compiled from the IRAS Point Source, Small Scale Structure, and Large Galaxy Catalogs. The catalogue covers 81.8 % of the sky and includes 17,710 sources. The dipole direction and amplitude agree well with those of the earlier study by Yahil *et al* (1986). If IRAS galaxies trace the overall matter distribution, $\Omega_0 \sim 1$.

1. INTRODUCTION

The IRAS all-sky survey at 60μm has given us the first unambiguous demonstration of the cause of the dipole anisotropy in the cosmic microwave background radiation (CBR). Yahil, Walker and Rowan-Robinson (1986, YWR) have analyzed the surface brightness distribution of IRAS 60μm sources in spherical harmonics and found that the direction of the dipole component agrees well with that of the CBR dipole. With a simple analytical approximation to the 60μm luminosity function of Lawrence *et al* (1986), YWR calculate the gravitational acceleration acting on the Local Group due to matter distributed like IRAS galaxies within 200(50/H_0) Mpc. Hence they derive a value for the cosmological density parameter due to such matter, $\Omega = 1.0 \pm 0.2$.

The failure of earlier optical studies to derive a convincing value for the velocity of the Local Group of galaxies with respect to the cosmological frame probably lies in the well-known problems of maintaining a homogeneous calibration of galaxy magnitudes over the whole sky and of allowing for the effects of interstellar extinction. In a novel approach, Lahav (1987) has constructed an all-sky diameter-limited catalogue and, using the square of the angular diameter in place of flux, has carried out an analysis similar to that of YWR. His dipole direction differs by only 21° from the IRAS dipole, though by 47° from the CBR dipole. A more careful analysis by Lahav, Rowan-Robinson and Lynden-Bell (1987), correcting for systematic differences between the ESO and UGC catalogues and with improved correction for interstellar extinction, yields a dipole direction in good agreement with the CBR.

The analysis of YWR was subject to a number of limitations : (1) The mask used to exclude contamination by interstellar dust emission (infrared "cirrus") was severe, excluding 53 % of the sky. Meiksin and Davis (1986) excluded a much smaller fraction of the sky (24 %) in their analysis of the dipole component of the number-density of IRAS sources. This led to significant contamination by cirrus, especially in their faintest quartile (Rowan-Robinson 1987), though their dipole direction is close to that of YWR. (2) Version 1 of the IRAS Point Source Catalog overestimates the fluxes of sources which are weak or of low flux-quality (sources not detected on all occasions observed). Although this is not a serious problem for the high flux-quality sources above the completeness limit used by YWR, it is obviously desirable to take advantage of the release of Version 2 of the PSC, in which this problem is fixed (Chester *et al* 1987). (3) The fluxes of sources extended with respect to the IRAS beam are underestimated in the PSC and in some cases such sources were not detected by the point-source recognizer. This could lead to an underestimation of the contribution of nearby galaxies to the gravitational acceleration experienced by the Local group. (4) The approximation to the luminosity function used by YWR was not the best fit to the data of Lawrence *et al* (1986). This problem has been discussed by Villumsen and Strauss (1987) and Lahav, Rowan-Robinson and Lynden-Bell (1987). (5) The sample of YWR contained a small number of sources due to stars with circumstellar dust, whose colours satisfied the condition $S_{25} < 3S_{60}$. The latter condition was extremely effective in eliminating most stellar sources.

All these problems are fixed in the present study. No correction has been applied for photon-induced detector responsivity enhancement ("hysteresis"). The magnitude of this effect for 60μm point-sources is small except within 5 to 10° of the Galactic plane (Whitlock 1987). Similarly, no correction has been applied for the residual effects of the South Atlantic Anomaly or the Polar Horns of the radiation belts, the effects of which are believed to be small (YWR, Rowan-Robinson 1987).

2. A COMPLETE AND RELIABLE ALL-SKY 60μm GALAXY CATALOGUE

The goal of this study is to create a 60μm galaxy catalogue which covers a large fraction of the sky and which, within the area covered, is highly complete and reliable. Specifically the aim is that the fraction of true IRAS galaxies brighter than the completeness limit which are excluded, and the fraction of sources included which are not galaxies, should be less than 1 %.

The excluded area obviously includes the 4 % of the sky which failed to achieve two hours-confirmed coverages. In addition I have excluded areas flagged as of high source-density in any band. There are two reasons for this. Within such areas we found it necessary, during the preparation of the IRAS Point Source Catalog, to introduce additional constraints for acceptance of a source into the Catalog (IRAS Explanatory Supplement 1984), with the result that the completeness limit in such regions varies in a complex way with the environment. The high figures for completeness and reliability quoted for the IRAS PSC (Rowan-Robinson *et al* 1984, Chester *et al* 1987), and on which the possibility of cosmological studies depends, apply only outside regions of high source-density. The second reason is that the majority of 60μm sources in such areas are clearly Galactic. The total area of the sky remaining in the study is 10.28 sr (81.8 % of the sky).

The excluded area includes half a dozen or so 1° × 1° bins in which the high source-density flag has been set in the 100μm band due to a high density of galaxies. While the exclusion of these will have a negligible effect on the calculation of the IRAS dipole, their exclusion may introduce a slight bias into clustering studies. I am looking for ways of including such bins on an objective basis but the remaining conditions applied below would allow through large numbers of Galactic sources, so for the moment these 30 or so galaxies are excluded.

The catalog is then prepared from Version 2 of the Point Source Catalog, the Small Scale Structure Catalog (Helou and Walker 1985), and the IRAS Large Galaxy Catalog (Rice *et al* 1987), as follows:

(1) Point Source Catalog. Sources which do not have counterparts in the SSS Catalog are included provided they are detected at 60μm (flux-quality flag > 1), have $S_{60} \geq 0.5$ Jy, and satisfy :

(a) $\log(S_{60}/S_{25}) > -0.5$. If the source is not detected at 25μm, the condition is imposed with the quoted upper limit in place of S_{25}. No catalogued galaxies violate this condition, while the overwhelming majority of stars do.

(b) If the source is detected at 25μm, then $\log(S_{25}/S_{12}) < 1.0$, unless the source is identified with a catalogued galaxy. If the source is not detected at 12μm, the condition is imposed with the quoted upper limit in place of S_{12}. This condition is needed to exclude planetary nebulae. Only two catalogued galaxies violate it, Arp 220 and NGC 4418 (both obviously interesting and unusual objects). There are no other sources violating the condition at $|b| > 10°$.

(c) If the source is detected at 12μm, then $\log(S_{60}/S_{12}) > 0.0$. This excludes a few additional stellar sources, but no catalogued galaxies.

(d) $\log(S_{100}/S_{60}) < 0.6$, unless the source is identified with a catalogued galaxy. If the source is not detected at 100μm, the condition is imposed with the quoted upper limit in place of S_{100}. This condition excludes cirrus sources very effectively. However many nearby, low luminosity galaxies have cool $60 - 100\mu$m colours similar to the cirrus in our Galaxy and it is clearly important to include the contribution of these normal galaxies. Most galaxies with $\log(S_{100}/S_{60}) > 0.6$ have $\log(L_{60}/L_{\odot}) < 9.0$ (Rowan-Robinson *et al* 1987) and so can only be in this sample if their distance $D < 30(50/H_0)$ Mpc. From consideration of the range of 60μm to blue luminosities for such galaxies (Rowan-Robinson *et al* 1987), I estimate that few such galaxies will have $m_{pg} > 15$, so that virtually no cool, normal galaxies will fail to be in an optical galaxy catalogue and therefore have been excluded by the above condition.

Finally, PSC sources are excluded if their sole identification is with a star, a planetary nebula or a star cluster. PSC (and SSS) sources are excluded if they are associated with Local Group galaxies.

(2) SSS Catalog. If sources in the PSC are flagged as confirmed small extended sources and have corresponding entries in the SSS Catalog then the SSS fluxes are used, provided these are greater than the PSC fluxes and provided there is an association with a catalogued galaxy. Sources which appear in the SSS Catalog and which have an association with a catalogued galaxy are also included even if there is no corresponding PSC entry. However SSS Catalog sources with no galaxy association are not included. An investigation of galaxies in the SSS Catalog shows that very few lie at distances greater than $30(50/H_0)$ Mpc. If we assume that a source must be at least $1'$ in diameter to be flagged as extended at 60μm, this corresponds to few galaxies having far infrared diameters > 10 kpc. Since no cases are known of galaxies with infrared diameters greater than their optical diameters, and since all galaxies with optical diameters $> 1'$ should be in existing galaxy catalogues, I conclude that very few of the unidentified $60\mu m$ sources in the SSS Catalog are galaxies.

A sample of sources in the PSC which are flagged as confirmed small extended sources but which do not appear in the SSS Catalog have been examined in coadded raw IRAS data by John Crawford using the IPMAF line COADD facility. In all cases the PSC flux was found to be a good estimate of the total flux and the overwhelming majority are bona fide point sources. I conclude that for most such sources the PSC flux is an accurate representation of the source flux. Several unidentified PSC sources flagged as extended but not appearing in the SSS Catalog fell in the samples studied by Lawrence et al (1986) and by the QMC-Cambridge-Durham redshift survey (A.Lawrence, M.Rowan-Robinson, G.Efstathiou, N.Kaiser, R.Ellis, C.Frenk). In all such cases the sources were found to be identified with faint galaxies. Table 1 gives the numbers of SSS galaxies in my sample as a function of SSS flux down to the SSS Catalog $60\mu m$ flux-limit of 1 Jy. I estimate that within the coverage area of my sample, 20 galaxies are extended with respect to the IRAS beam and have total fluxes in the range $0.6 - 1$ Jy, and so may either have been omitted or have had their fluxes underestimated.

Table 1: Flux distribution of SSS Catalog galaxies in present sample

log S(Jy) =	0.0	0.1	0.2	0.3	0.4	0.5	0.6	0.7	0.8	0.9	1.0	1.1	1.2	1.3	1.4	1.5	1.6
number		8	11	20	34	39	32	25	10	22	14	13	6	5	6	1	5

(3) Galaxies in the IRAS Large Galaxy Catalog. Rice et al (1987) have estimated fluxes from coadded raw IRAS data for 83 galaxies with optical diameters $> 8'$. Where their $60\mu m$ fluxes are greater than 0.5 Jy, I have used the LGC fluxes in preference to either PSC or SSS fluxes. Fig 1 shows a comparison of LGC and SSS Catalog fluxes for 21 galaxies in common. Apart from the discrepant case of NGC 4945, where the SSS flux lies well below both the LGC and PSC fluxes, the fluxes are in remarkably good agreement and the mean value of $\log(S_{LGC}/S_{SSS})$ is 0.01 ± 0.03. For an individual source the dispersion in this quantity is 0.12, so that the SSS fluxes appear to be accurate to $\pm 30\%$ for extended sources. Although it is desirable to obtain accurate fluxes by coaddition of raw data for the remaining extended sources in my sample, it seems unlikely that this will lead to any significant modification of the IRAS dipole.

The complete sample consists of 17,710 $60\mu m$ sources brighter than 0.5 Jy, at a source-density of 0.525 per sq deg. 14,341 have $S_{60} \geq 0.6$ Jy, the completeness limit, i.e. a source-density of 0.425 per sq deg. Differential source-counts for different ranges of Galactic latitude show that there is (i) incompleteness at $0.5 - 0.6$ Jy, at all Galactic latitudes, (ii) contamination by Galactic sources at $|b| < 5°, S_{60} > 2$ Jy, and (iii) incompleteness at $|b| < 5°$ in the range $0.6 - 1$ Jy. I estimate that ~ 240 spurious Galactic sources brighter than 2 Jy are present in the sample at $|b| < 5°$, and that ~ 140 sources are missing at $|b| < 5°, 0.6 < S_{60} < 1$ Jy. Study of Sky Survey plates may help to identify which sources at $|b| < 5°$ are genuine galaxies, though high interstellar extinction may limit the effectiveness of this approach.

Table 2 shows the distribution of the $60\mu m$ correlation coefficient with a point-source template, CC, as a function of the cirrus flag, CIRR1, for sample sources which are not in the SSS or LGC Catalogs. There is no evidence for degradation of the $60\mu m$ data in cirrus regions. Only 347

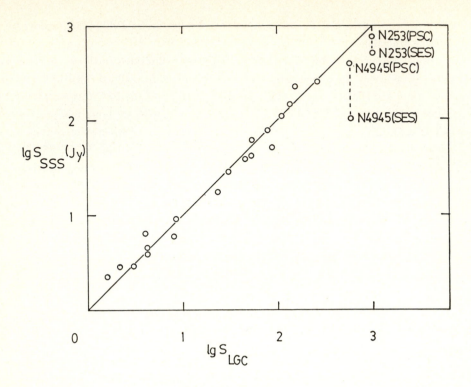

Fig. 1. Comparison of fluxes from IRAS Small Scale Structure and Large Galaxy Catalogs for 23 galaxies in common.

Table 2: Distribution of CC as a function of CIRR1

CIRR1	total number	CC=	1.00	0.99	0.98	0.97	0.96	0.95	0.94	0.93	0.92	0.91
0	7986	%=	35.8	40.6	16.7	5.4	1.2	0.2	0.1	0.	0.	0.
1	3755		35.3	41.2	15.3	5.8	1.8	0.4	0.1	0.	0.	0.
2	1770		36.1	40.5	16.1	5.6	1.3	0.2	0.	0.1	0.1	0.
3	1097		34.6	41.3	15.8	5.9	1.1	0.6	0.2	0.4	0.1	0.
4	782		33.9	40.4	18.4	5.4	1.2	0.6	0.	0.1	0.	0.
5	531		39.7	37.9	15.8	4.5	1.3	0.6	0.2	0.	0.	0.
6	430		41.6	38.4	14.7	3.0	1.6	0.2	0.2	0.2	0.	0.
7	310		34.5	36.8	19.0	6.1	2.3	0.6	0.3	0.	0.3	0.
8	224		38.4	38.4	14.7	2.7	2.7	1.8	0.4	0.4	0.	0.4
9	524		35.1	39.3	14.9	7.1	1.7	1.3	0 .2	0.2	0.2	0.

sources have $CC < 0.97$, indicating noisy or extended data, and of these 241 have $CIRR1 \leq 1$ and so are not in regions of bad cirrus. Figs 2 a,b,c show the distribution of the 17710 catalogue sources on the sky. Figs 3 a,b,c show the distribution of the masked areas.

3. SURFACE BRIGHTNESS DIPOLE

YWR have shown that the dipole moment of the surface brightness distribution, $\underline{\sigma}(S)$, of a population of sources with a luminosity function, $\phi(L)$, in which the total space density of sources follows the total mass-density, is related to the peculiar gravitational acceleration acting on the Local Group due to matter out to radius $r, \underline{G}(r)$, by

$$4\pi S\underline{\sigma}(S) = \int_0^\infty L^2 \phi(L) \frac{dG}{dr} dr \tag{1}$$

where

$$\underline{G}(r) = \frac{3}{4\pi} \int_0^\infty D(\underline{r}') \frac{r'}{r'^3} d^3 r'$$

and $D(\underline{r})$ is the relative total mass-density at r , i.e.

$$D(\underline{r}) = \rho(\underline{r})/\rho_0$$

where ρ_0 is the total mean mass-density.

To take account of the masked areas, the spherical harmonic components are calculated over the unmasked areas (YWR). The orthogonality matrix for the spherical harmonic components ceases to be diagonal. The true spherical harmonic components, on the assumption that the sky behind the mask is similar to that in the unmasked area, are then calculated by matrix inversion.

Figure 4 shows the 3 components of the $60\mu m$ surface brightness dipole as a function of flux-density, together with their uncertainties. Table 3 gives the mean dipole amplitude and direction, with statistical uncertainties, calculated by weighting the values of the components in each flux-bin by the inverse square of the uncertainties. The first line in Table 3 shows the dipole calculated for $0.63 < S_{60} < 31.6$ Jy, and areas with $|b| < 5°$ added to the mask, which is my best estimate of the surface brightness dipole for the present sample. Other lines in Table 3 show the effect of varying the extent of the mask and flux range used: the variation of the dipole amplitude and direction with these assumptions is small. Going from the YWR data to that of the new galaxy catalogue (Version 2 of the PSC, correction for extended sources) , with the same $n = 1$ mask as used by YWR, changes the dipole direction by only $2.0°$. Extension of the sky coverage from 47 % to 79 % changes the dipole direction by only $7.8°$. The latest microwave background dipole, for an assumed solar motion of 300 km s^{-1} towards $(l,b) = (90,0)$, corresponds to 600 km s^{-1} towards $(l,b) = (268, 27)$ (Lubin and Villela 1986). The IRAS dipole differs from this by only $21°$. Table 4 gives the quadrupole components corresponding to the solution in the first line of Table 3. A significant quadrupole is seen and the magnitude of the components is similar to the dipole components. This reflects the fact that the actual distribution of matter is far from dipolar.

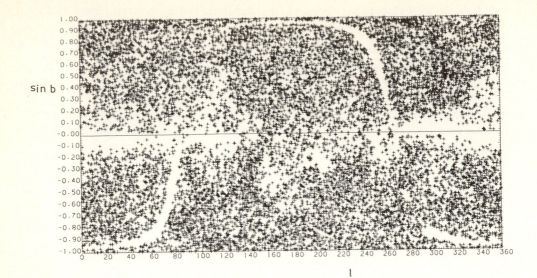

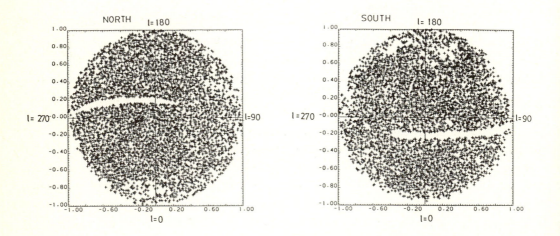

Fig. 2. Distribution of 17,710 IRAS 60 μm sources in the galaxy catalogue described in the text. (a) sin b vs l, (b) North Galactic Hemisphere, (c) South Galactic Hemisphere

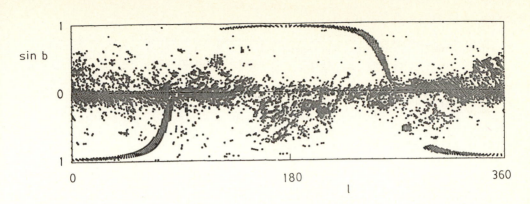

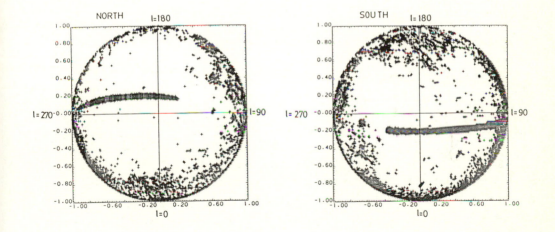

Fig. 3. Distribution of masked bins for same areas as in Fig. 2. The two tongues extending from the Galactic plane are the coverage gaps.

Table 3. The IRAS 60 μm dipole

sample	mask[a]	area (sr)	flux-range (Jy)	no. of sources	dipole amplitude $|4\pi S\underline{\sigma}|$	l (o)	b (o)	θ^b_{CBR} (o)
present	**basic** $+|b|<5°$	**9.94**	**0.63 - 31.6**	**1244**	**1342 Jy** ±148	**248.2** ±9.6	**39.5** ±9.6	**20.7**
”	”	”	0.63 - 31.6[c]	12448	1425	252.1	45.7	22.5
”	”	”	0.63 - 200[c]	12495	1323	233.8	48.9	34.3
”	”	”	0.63 - 63	12480	1347	246.3	41.2	22.8
”	”	”	0.5 - 31.6	16694	1430	239.3	40.6	27.3
”	”	”	1.0 - 31.6	6126	1468	247.9	44.1	23.6
”	”	”	2.0 - 31.6	2285	1656	245.5	48.7	27.8
”	basic only	10.28	0.63 - 31.6	12950	1320	241.2	40.6	26.0
”	basic $+|b|<10°$	9.28	0.63 - 31.6	11715	1408	249.2	37.0	18.1
”	basic $+|b|<20°$	7.63	0.63 - 31.6	9698	1375	256.9	37.7	14.2
”	$CIRR1 > 1$	5.94	0.63 - 31.6	7089	1513	250.6	31.9	16.0
”	$CIRR1 > 1$	5.94	0.5 - 31.6	9521	1518	248.8	38.1	19.6
YWR	$CIRR1 > 1$	5.94	0.5 - 31.6	9903	1550	248	40	21.0

[a]basic mask is coverage gaps + high source density bins
[b]CBR direction taken as $(l,b) = (268, 27)$
[c]straight average of flux bins (rest are error-weighted averages)

Table 4. Spherical Harmonic components

Component	magnitude $(0.63 - 31.6\ Jy)$	$(1 - 31.6\ Jy)$
1	8561 ± 92	6944
$\cos b \cos l$	-384 ± 157	-389
$\cos b \sin l$	-961 ± 153	-982
$\sin b$	854 ± 143	1044
$1.5 \sin^2 b - 0.5$	557 ± 217	656
$\sin 2b \cos l$	615 ± 159	430
$\sin 2b \sin l$	-1092 ± 154	-864
$\cos^2 b \cos 2l$	632 ± 181	359
$\cos^2 b \sin 2l$	-1636 ± 177	-1832

4. CALCULATION OF THE GRAVITATIONAL ACCELERATION

Lahav, Rowan-Robinson and Lynden-Bell (1987) have discussed the inversion of the integral eqn (1) to determine the peculiar gravitational acceleration acting on the Local Group, $\underline{G}(r)$, generated within a distance r. The analysis of YWR is equivalent to the assumptions (i)$\phi(L) = CL^{-2}$ for $L_1 < L < L_2$, and (ii) the dipole is generated within the distance range (r_1, r_2). Then (1) gives

$$4\pi S\underline{\sigma}(S) = C\underline{G}(r_2)$$

independent of S, for

$$\frac{L_1}{4\pi r_2^2} < S < \frac{L_1}{4\pi r_1^2}$$

since $\underline{G}(r_1) = 0$
A simple approach is to analyze the relative density, $D(\underline{r})$, in terms of spherical harmonics

$$D(\underline{r}) = 1 + \underline{a}(r).\hat{\underline{r}} + \cdots \tag{2}$$

in which case $d\underline{G}/dr = \underline{a}(r)$, independent of higher order terms in (2). The simplest model for $\underline{a}(r)$ is a shell model (Villumsen and Strauss 1987) in which

$$\underline{a}(r) = A\hat{\underline{n}} \text{ for } r_1 < r < r_2$$
$$= 0 \text{ otherwise} \tag{3}$$

where A is the density amplitude and $\hat{\underline{n}}$ is a fixed vector, in which case $\underline{G} = A(r_2 - r_1)\hat{\underline{n}}$ for $r \geq r_2$. A fit to the data of Fig 4 of the form (3), using the luminosity function calculated by Lawrence et al (1986, column (5) of their Table 2) yields a mimimum chi-square fit with

$$A(r_2 - r_1) = 2020^{+530}_{-280} \text{ km s}^{-1}, \qquad l = 251.1°, \qquad b = 46.3°,$$

$$r_1 = 250^{+100}_{-250} \text{ km s}^{-1}, \qquad r_2 = 2000^{+1150}_{-1200} \text{ km s}^{-1}, \tag{4}$$

which is illustrated in Fig 4. Distances have been given in km s^{-1} (i.e. r means $H_0 r$) since the Hubble constant plays no part in this analysis.

5. EVALUATION OF THE DENSITY PARAMETER, Ω_0

Peebles (1980) has shown that in linear theory $\underline{G}$ is related to the peculiar velocity of the Local Group , $\underline{u}$, by

$$\underline{u} = \Omega_0^{0.6}\underline{G}/3 \tag{5}$$

where $\underline{G}$ is measured in units of km s^{-1}.
Since the misalignment of the IRAS and CBR dipoles is negligible, we can substitute for $\underline{G}$ from (4) and use the CBR dipole value for u of 600 km s^{-1}, which gives

$$\Omega_0 = 0.83^{+0.13}_{-0.17} \tag{6}$$

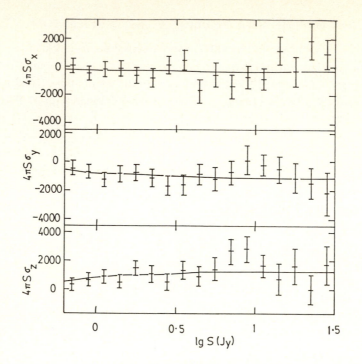

Fig. 4. Components of the 60 μm surface brightness dipole $4\pi S \underline{\sigma}(S)$. x is towards the Galactic Centre and z is towards the Galactic pole. The solid curve is the model give by eqn (4).

However we do not expect that linear theory will give us a precise estimate for Ω_0. YWR estimated that this method may underestimate Ω_0 by 15 %. Villumsen and Strauss (1987) estimate that this non-linear correction may lie in the range 3 to 24 %. Vittorio and Juszkiewicz (1987) conclude that for a particular model for initial density fluctuations (which is not, incidentally, consistent with the distribution of IRAS galaxies), $\underline{G}$ is dominated by galaxies with distances $< 10(50/H_0)$ Mpc and that Ω_0 is indeterminate. The first of these conclusions is certainly inconsistent with the results of this paper and of Strauss and Davis (1987). Probably the non-linear correction to (6) can only be estimated by an n-body simulation that mimics the distribution of IRAS galaxies.

6. DISCUSSION

We now compare our results with those of some other recent studies. Harmon, Lahav and Meurs (1987) have discussed the IRAS dipole for a colour-selected galaxy sample based on Version 2 of the IRAS Point Source Catalog. Their basic sample consists of 10554 galaxies with $S_{60} > 0.7$ Jy covering an area similar to that of the present study. For comparison, my catalogue includes 11,500 sources brighter than 0.7 Jy, so the Harmon et al colour conditions have rejected about 10 % of the sources accepted in the present study. The galaxies rejected are mainly those with $\log(S_{100}/S_{60}) > 0.6$ and / or $\log(S_{60}/S_{25}) > 1.0$, i.e. cool normal galaxies. Harmon et al have made no correction for extended sources. These two factors combine to underestimate the contribution of nearby, normal galaxies. Because no discrimination against planetary nebulae is employed, a number of these will be present in the Harmon *et al* sample, especially at lower Galactic latitudes. Planetary nebulae probably account for the band of excess source-density seen at $|b| < 5°$ in their Fig 1.

These factors may explain why their flux-weighted dipole, calculated for a fairly conservative mask excluding 34.2 % of the sky, differs in direction so markedly from those found by YWR and the present study. They find $(l, b) = (228.1, 22.3)$. The fact that dropping the flux-weighting (i.e. by calculating the dipole of the number-density distribution instead of the surface brightness) brings the dipole close to that of the CBR does not seem a good reason for abandoning the whole principle of the approach of YWR and the present paper, by which the gravitational acceleration, $\underline{G}$, and Ω_0 are estimated.

Strauss and Davis (1987) have carried out a redshift survey for IRAS 60μm sources with $S_{60}/S_{12} > 3, |b| > 10°$ and $S_{60} > 1.936$ Jy. They find that the bulk of the dipole is generated at velocities between 1000 and 3500 km s^{-1}. The values for r_1, r_2, derived in eqn (4) above are consistent with this. The dipole amplitude and direction derived by Strauss and Davis using PSC fluxes are also consistent with the values of eqn (4). The procedure of doubling the fluxes for all sources flagged as confirmed SSS sources to correct for extension however, seems dubious. Most of the sources flagged as confirmed SSS, but not in the SSS Catalog, are not truly extended and were flagged because of the generous threshold adopted for the SSS recognizer.

The dipole direction derived here, $(l, b) = (248.2, 39.5)$, differs by 20.7° from the latest CBR dipole, $(268, 27)$. Is this difference significant ? It is not clear that any further analysis of IRAS data will lead to a significant shift in the flux-weighted dipole or the derived value of $\underline{G}$. The substantial revisions made here resulted in a surprisingly small shift in dipole direction and amplitude from the YWR value, only 1°. The estimated direction of the CBR dipole has in fact shifted by a larger amount, 8°, over the same period of time. The difference of 20.7° between the acceleration and velocity vectors is not statistically very significant, but there is no a priori reason why these dipoles should be perfectly aligned, since the matter contributing to the acceleration vector today has changed its relative positions due to peculiar motions over the period of time that the velocity was generated. The comparison of the IRAS and optical dipoles is the subject of a separate paper by Lahav, Rowan-Robinson, and Lynden-Bell (1987) and I do not propose to discuss this subject here. Suffice it to say that the dipole directions now agree quite well and that the significant differences in the amplitude of the derived gravitational accelerations are probably attributable to bias in the galaxy-forming process. Either early-type galaxies are biassed towards high-density regions (in which case $\Omega_0 \sim 1$) or spirals are biassed to low-density regions (in which case $\Omega_0 \sim 0.1$).

7. FUTURE WORK

There is little prospect of extending the sky coverage using IRAS data much beyond the 80 % or so achieved here. Some regions flagged as high source-density at 100μm only may be usable at 60μm with further work. But in most high source-density regions, confusion results in the survey completeness limit being much higher than in the present sample. It would be valuable to reanalyze the IRAS raw data for regions outside the mask used in the present paper which are at $|b| < 5°$, correcting for the effects of hysteresis and the time-delay in the noise-estimator. Identification programmes may help to eliminate the remaining Galactic sources in the catalogue.

It will also be desirable to obtain accurate fluxes by coaddition of the raw data for all sources flagged as extended or which have poor correlation coefficient at 60μm. Despite the crudity of the SSS recognizer, however, the SSS fluxes seem on average to be remarkably accurate.

N-body simulations are needed to estimate the non-linear correction to the estimate of Ω given here. Studies of biased galaxy-formation scenarios may help to explain the difference in amplitude of the gravitational acceleration estimated from optical and far infrared samples.

REFERENCES

Chester, T., Beichmann C., Conrow, T., 1987. *Revised IRAS Explanatory Supplement, Ch. XII.*

Harmon, R.T., Lahav, O., and Meurs, E.J.A., 1987. *MNRAS,* **228,** 5p.

Helou, G., and Walker, D.W., 1986. *IRAS Small Scale Structure Catalog.* (JPL D-2988).

Lahav, O., 1987. *MNRAS,* **225,** 213.

Lahav, O., Rowan-Robinson, M., and Lynden-Bell, D., 1987. *MNRAS,* (submitted).

Lawrence, A., Walker, D., Rowan-Robinson, M., Leech, K.J., and Penston, M.V., 1986. *MNRAS,* **219,** 687.

Meiksin, A., and Davis, M., 1986. *AJ,* **91,** 191.

Peebles, P.J.E., 1980. *"Large-scale Structure of the Universe"*(Princeton University Press).

Rice, W., *et al* , 1987. *IRAS Large Galaxy Catalog.*

Rowan-Robinson M. *et al* , 1984. *IRAS Introductory Supplement, Ch VIII.*

Rowan-Robinson, M., 1987. *IAU Symposium No. 124 "Observational Cosmology",* ed. A.Hewitt *et al* (Reidel) p.229.

Rowan-Robinson, M., Walker, D., and Helou, G., 1987. *MNRAS* (in press).

Strauss, M.A., and Davis, M., 1987. This volume.

Villumsen, J.V., and Strauss, M.A., 1987. *ApJ* (in press).

Vittorio, N., and Juszkiewicz, R., 1987. In *"Nearly Normal Galaxies"*ed. S.M.Faber (Springer-Verlag) p.451.

Whitlock, S., 1987. Third IRAS conference, Queen Mary College, London.

Yahil, A., Walker, D., and Rowan-Robinson, M., 1986. *ApJ,* **301,** L1.

A Redshift Survey of IRAS Galaxies

Michael A. Strauss and Marc Davis

Astronomy and Physics Departments
University of California, Berkeley
Berkeley, California, 94720

ABSTRACT

We have completed a redshift survey of approximately 2200 *IRAS* galaxies, flux limited at 60μ. The survey covers 76% of the sky and has a characteristic depth of $\approx$ 6000 km s^{-1}, making it ideal for large scale structure studies requiring whole sky coverage. We have calculated the gravitational acceleration on us due to the inhomogeneous distribution of galaxies in the sample by summing the dipole acceleration in successive shells centered on us. The acceleration converges at $\approx$ 4000 km s^{-1} and we derive density estimates in the range $0.4 < \Omega < 0.9$. We discuss the various biases of the sample in detail: the paucity of elliptical galaxies, the problem of extended sources, and hysteresis, and suggest ways to accommodate them and thus decrease the uncertainty in Ω. Finally, we discuss the use of the survey to make predictions for the peculiar velocity flowfield in space.

1. INTRODUCTION AND SAMPLE SELECTION

The *IRAS* database offers a unique opportunity to study the large-scale structure of the distribution of galaxies over large solid angle. It is estimated that the Point Source Catalog (hereafter PSC) contains some 20,000 galaxies (Soifer *et al.* 1987) that have been selected in a uniform way over the sky. Moreover, because galactic extinction is negligible in the *IRAS* bands, galaxies can be selected very close to the plane, at least until confusion makes the PSC unreliable. Meiksin and Davis (1986) and Yahil, Walker and Rowan-Robinson (1986) were the first to make galaxy catalogs from the PSC, and showed that the angular distribution of galaxies has a dipole moment that points very close to the peculiar velocity of the Local Group, as inferred from the dipole anisotropy of the Cosmic Microwave Background. This result has been confirmed now with two more careful analyses of the *IRAS* database, by Rowan-Robinson (1987, this conference), and Harmon, Lahav and Meurs (1987) (see also Villumsen and Strauss 1987). The implication is that the *IRAS* galaxies at least approximately trace the matter that gives rise to our peculiar motion.

On large scales where linear perturbation theory applies, an object's peculiar velocity is directly proportional to its peculiar gravity where the constant of proportionality depends solely on Ω, the cosmological density parameter. A redshift survey of an unbiased subsample of the *IRAS* galaxies offers the possibility of measuring Ω on the scale of the material giving rise to our peculiar velocity. The analysis of the peculiar gravity can determine the coherence length of expected peculiar velocities if the radius of convergence of our acceleration can be determined.

Finally the survey can be used to construct the peculiar gravity field out to considerable distance.

We have extracted a sample of 2407 objects from the PSC (first version) based on the Meiksin and Davis (1986) criteria: $f_{60}/f_{12} > 3$, $|b| > 10°$, and $f_{60} > 1.936$ Jy. The galactic latitude limit is imposed to avoid excessive contamination from galactic sources, to avoid problems of confusion at 60μ, and to avoid problems of hysteresis after plane crossings (see, e.g., the *IRAS* Explanatory Supplement 1984). In addition, a few high-latitude star-forming regions, such as Orion and Ophiuchus, are excluded. The flux limit corresponds to the brightest quartile of the original Meiksin and Davis sample. The resulting sky coverage is 9.55 steradians, 76% of the sky.

In collaboration with John Huchra, Amos Yahil, and John Tonry, we have obtained optical spectra of the $\approx 60\%$ of these objects that did not have redshifts in the literature, using telescopes at Cerro Tololo, Lick Observatory, and Mount Hopkins. Approximately 8% of the objects turn out not to be galaxies, but rather planetary nebulae, T Tauri stars, cirrus clumps, and other members of the galactic zoo.

We have included in our observing lists an additional 498 objects satisfying our color criterion, but with $f_{60} < 1.936$ Jy, and flagged as extended, weeks-confirmed, in bands 1, 2 or 3. As we discuss below, these objects have point source fluxes that are underestimates of their true flux, and thus these objects may make it into our catalog. The IPAC people are currently ADDSCANing these sources for us, and we hope to have results by the end of the summer.

Finally, we are starting to observe galaxies in the unconfused regions of the excluded zones and in the region $5° < |b| < 10°$ in an attempt to extend the survey to still greater sky coverage.

2. THE SPACE DISTRIBUTION AND OUR PECULIAR GRAVITY

In Figure 1, we present the distribution on the sky of the 2176 extragalactic objects in the survey as a function of redshift. Note how the level of anisotropy decreases markedly as a function of distance. All the well known clusters within 5000 km/s are identifiable in the distribution, but they are not as conspicuous as in an optically-selected catalog, because the *IRAS* galaxies are more loosely distributed than optical galaxies. A redshift-space correlation analysis of the *IRAS* galaxies shows that $\xi(s)$ is virtually identical to that of late type galaxies in the CfA survey for $s > 2h^{-1}$ Mpc. We find $\xi(s) = (s/s_0)^{-\gamma}$ with $s_0 = 5h^{-1}$ Mpc and $\gamma = 1.8$ to be consistent with the data. On smaller scales the *IRAS* galaxies appear to be less clustered than typical spirals, but here the statistics are poor. These results are not surprising because the majority of *IRAS* galaxies are a dilute and apparently random subsample of all late type galaxies (Soifer *et al.* 1987).

Peebles (1980) gives the relation between peculiar velocity of an observer and the dipole moment of the matter distribution around him when linear growing modes predominate:

$$\mathbf{V} = \frac{H_0 \Omega^{0.6}}{4\pi} \int d^3 \mathbf{r} \delta(\mathbf{r}) \frac{\hat{\mathbf{r}}}{r^2} \tag{1}$$

where $\delta(\mathbf{r})$ is the matter overdensity. In practice, we compute

$$\mathbf{V} \Omega^{-0.6} = \frac{H_0}{4\pi n_1} \sum_i \frac{1}{\phi(r_i)} \frac{\hat{\mathbf{r}}_i}{r_i^2} \tag{2}$$

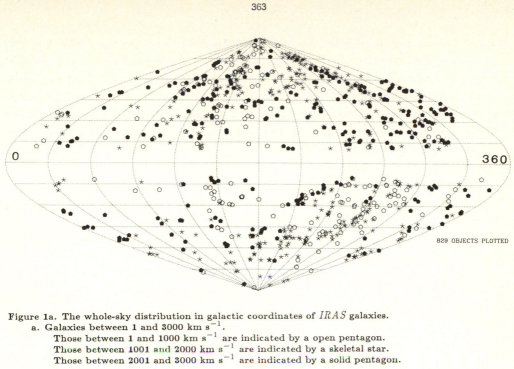

Figure 1a. The whole-sky distribution in galactic coordinates of *IRAS* galaxies.
a. Galaxies between 1 and 3000 km s^{-1}.
Those between 1 and 1000 km s^{-1} are indicated by a open pentagon.
Those between 1001 and 2000 km s^{-1} are indicated by a skeletal star.
Those between 2001 and 3000 km s^{-1} are indicated by a solid pentagon.

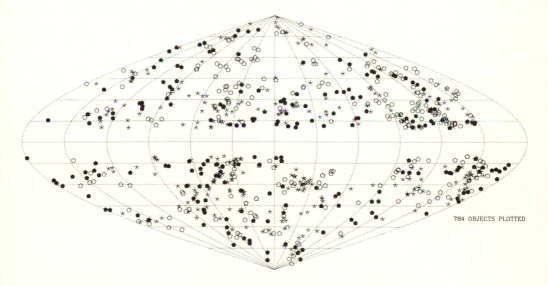

Figure 1b. Galaxies between 3001 and 6000 km s^{-1}.
Those between 3001 and 4000 km s^{-1} are indicated by a open pentagon.
Those between 4001 and 5000 km s^{-1} are indicated by a skeletal star.
Those between 5001 and 6000 km s^{-1} are indicated by a solid pentagon.

where the sum is over all galaxies sufficiently luminous to be observable to a redshift of 4000 km s^{-1}, n_1 is the mean density of these galaxies, and $\phi(r)$ is the selection function, which is just the proportion of the luminosity function sampled at a given depth in the magnitude-limited sample (see Davis and Huchra 1982), and $\mathbf{r}$ is calculated assuming pure Hubble flow, after correcting redshifts for Virgocentric infall with a non-linear spherical model (Meiksin 1985). Our infall velocity was taken to be 250 km s^{-1}, although our results are rather insensitive to this. The absolute magnitude limit imposed implies $\phi(v < 4000 \text{ km s}^{-1}) = 1$ and we find $\phi(8000 \text{ km s}^{-1}) \approx 0.1$, which is as distant as we dare go in deriving results of cosmological significance. Finally, we must add a small correction to (2) to remove the dipole contribution of the excluded zones. It is important to note that our procedure is not a flux weighted sum; instead the sum is weighted by a function of the distance of each galaxy to correct for the fact that our sample is not volume limited. We assume implicitly that the *IRAS* galaxies roughly trace the mass and that their luminosity distribution is approximately independent of environment.

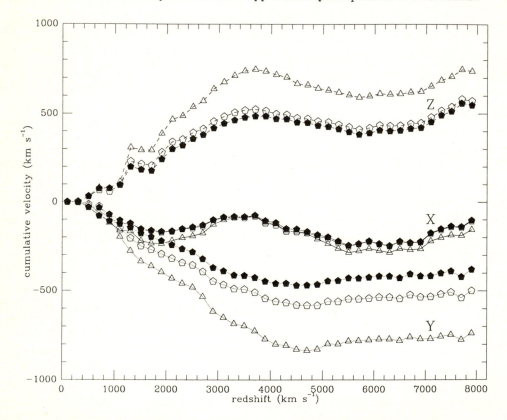

Figure 2. The three components of the cumulative peculiar gravity. The X component is the solid curve, Y is the dotted curve, and Z is the dashed curve. The solid pentagons use the PSC flux. The open pentagons double the apparent flux for objects flagged as extended. The open triangle curves also double count those objects with the most neighbors. Note that the curves have very similar shapes, but different amplitudes.

In Figure 2 we plot the three components of what we will call the "peculiar acceleration", $V\Omega^{-0.6}$, as given by Eqn. (2), as a function of the limiting distance of the summation (solid pentagons). The z-axis points to the North Galactic Pole and the x-axis points to the Galactic Center. Our peculiar acceleration converges quite well at a distance of 4000 km s^{-1}, with the bulk of the remaining fluctuations being consistent with a random walk induced by Poisson shot noise of the limited number of galaxies per shell.

The peculiar acceleration points in the direction $l = 255°$, $b = 54°$, only 26° from the microwave dipole direction. Assuming $|\mathbf{V}| = 600$ km/s, as implied from the observed microwave dipole anisotropy, the inferred value for Ω is 0.83. However before we quote this as a conclusion, we must understand any and all sources of systematic error in the method. The first of these is the problem of extended sources. Any galaxy with an angular size greater than $\approx 100''$ at 60μ will be resolved by *IRAS*, and thus the 60μ flux listed in the PSC will be an underestimate of the true flux. In a first attempt to correct for this problem, we looked for matches in the SES Catalog of all sources flagged in the PSC I as extended in one of the first three bands. Unfortunately, only 1/3 of these sources had matches. As an alternative, we are having the IPAC people ADDSCAN each of these 900-odd sources. Until we have results from them, we will use the simple kludge of multiplying the Point Source flux of each extended source within a redshift of 5000 km s^{-1} by a factor two, a number based on our admittedly limited experience with ADDSCANing.

There are additional complications with the extended sources. There are certainly sources in our sample that will violate our color criterion once accurate fluxes are known for them. Similarly there must be numerous sources *not* in our sample which belong, rejected only because the inaccurate PSC fluxes do not satisfy the color criterion. Finally, a large number of the sources flagged as extended turn out to be quite distant galaxies ($v > 5000$ km s^{-1}), with angular size $\ll 100''$. These are extended because they are viewed through a foreground of Galactic cirrus, because they are in a group of infrared-bright galaxies, or because of some glitch in the *IRAS* processing. In any case, it is not yet clear to us how to treat these objects. The open pentagons in Figure 2 show the peculiar acceleration with the additional factor of two in flux for the extended sources. The curves have essentially the same shape as before, but the more numerous foreground galaxies increase the peculiar acceleration, resulting in a reduction of the inferred Ω to 0.67. The direction of the calculated peculiar acceleration is now 20° from the microwave vector and the convergence radius is approximately the same. Clearly, it is very important to properly account for extended sources!

Another bias in our sample is the fact that very few early type galaxies are included. The *IRAS* sample definitely gives low weight to cluster centers. Clusters such as Coma are completely invisible in the *IRAS* distribution, and the maps of Figure 1 show weaker concentration toward cluster centers than in equivalent optically selected samples. We have addressed this problem in two ways. First we considered an analysis of the peculiar gravity only in the 1.83 steradians of sky covered by the original CfA survey. The comparison of the z-component of the optical to *IRAS* "gravity" in this region of sky is shown in Figure 3. The two are remarkably similar, suggesting that the *IRAS* galaxies trace the large-scale distribution of optically-selected galaxies very well, although the agreement is not perfect. A test of the robustness of our results is to add the missing cluster cores by hand, as it were. Early type galaxies comprise typically 20% of the objects in optically selected magnitude-limited samples and are more clustered than late types. Therefore we simply double-counted the 25% of the *IRAS* galaxies that had the most companions within $7h^{-1}$ Mpc (5 or more companions). This again results in a larger peculiar gravity (open triangles in Fig. 2) and reduces Ω to 0.39. The convergence radius remains nearly unchanged, but the convergence is more pronounced. The gravity vector moves to $l = 258°$, $b = 44°$, only 16° from the microwave vector. This shows that indeed our results are sensitive to the addition of early-type galaxies. We are currently investigating other ways to quantify this systematic effect.

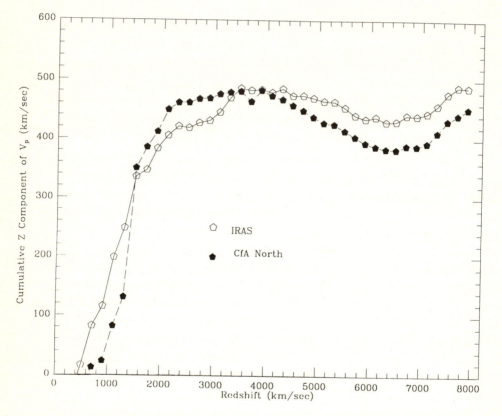

Figure 3. A comparison of the z-component of the peculiar gravity of the CfA survey (solid symbols) and the $IRAS$ survey restricted to the same area of the sky (open symbols).

3. THE PECULIAR VELOCITY FLOW FIELD

The completeness of sky coverage of the *IRAS* redshift survey allows us to extend the computation of the peculiar acceleration to observers at other points in space. For instance, we could put an imaginary observer at the center of the Virgo Cluster, and ask what her expected peculiar velocity is, due to the *IRAS* galaxies, by carrying out the sum in Eqn. (2). We have carried out this analysis for a grid of observers distributed over the sky, and have made maps of the expected peculiar velocity field. As above, we doubled the fluxes of sources flagged as extended. We added 424 random sources in the excluded zones, using the same selection function, to generate a catalog of sources with 100% sky coverage. Thus we are assuming that there are no large over- or under-densities lurking behind the Galactic Plane or Ophiuchus. The analysis was applied on a grid of points in latitude and longitude, on spherical shells centered on us. Figures 4a and 4b show the results, on shells at 2000 and 4000 km s^{-1}, respectively. Beware of the Mercator projection of these plots; they can be misleading!

Figure 4a is on the far side of Virgo, and thus we see strong infall with V_r negative, centered at $l = 290°$, $b = +70°$. We also see the effect of the Centaurus and Hydra clusters at 3000 km s^{-1} in the outflow with large shear centered at $l = 300°$, $b = +30°$.

Figure 4b is a slice just in front of Perseus, and the flow field is very strong there. The radial field in this region is dominated by the backwards pull of Hydra and Centaurus. The transverse fields in this region are rather jumbled, however; perhaps this is telling us something.

4. THE WORK THAT LIES AHEAD

Clearly, we are far from finished with this project. We have described above our first crude attempts to deal with two of the more blatant systematic effects of our sample, namely the problems of extended sources and of the paucity of early-type galaxies. The problem of hysteresis has not yet been addressed. We have recently discovered a strong anti-correlation between the Point Source Correlation Coefficient (CC) (*IRAS* Explanatory Supplement 1984) and the ratio of ADDSCANed to PSC flux; there is a large number of point sources with poor CC's, but not flagged as extended. Should all of these be ADDSCANed as well? We should also remake our sample based on the new version of the PSC; undoubtedly the galaxy list will change slightly.

The relatively short radius of convergence of the peculiar gravity implies that we do not expect large amplitude bulk flows with coherence length larger than this. The apparent disagreement of our peculiar gravity results with the velocity field measurements (Rubin, 1987; Collins *et al.* 1986; Dressler *et al.* 1987; Lynden-Bell *et al.* 1987) is very intriguing and must be investigated further. We have assumed pure Hubble flow throughout this discussion, while our results in Figures 4 clearly show this to be an inadequate assumption. Perhaps we can use these results to correct our observed redshifts for peculiar motions, and iterate. Over the next year we hope to reconcile the reported large scale velocity fields to the *IRAS* gravity field.

Current models of the nature of the dark matter pervading the Universe give definite predictions for the nature of the flowfield on different scales. The convergence radius inferred for our data is consistent with the currently popular Cold Dark Matter model, for instance.

Finally, we point out that we have perhaps the world's largest collection of spectroscopic data on the *IRAS* galaxies. We have just begun to study their emission-line properties, and hope to do detailed comparison with optically-selected samples of galaxies. A great deal of work lies ahead in digesting the mountains of data we have. The known systematic errors in the catalog appear to be treatable, and in a few month's time we hope to reduce the uncertainty in Ω and to better understand the limitations of the data base.

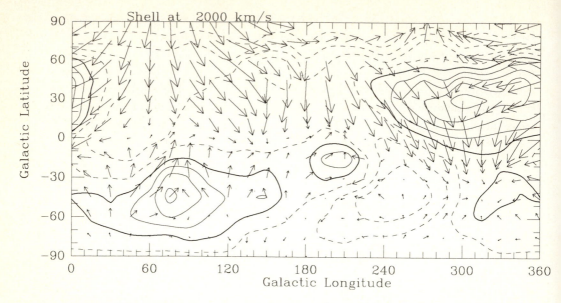

Figure 4a. Peculiar velocities on a shell at 2000 km s^{-1}. The contours are of constant peculiar acceleration (i.e., $V\Omega^{-0.6}$) in the radial direction; solid contours are positive, dashed contours are negative, and the heavy contour marks the $V_r = 0$ contour. The contours are at intervals of 200 km/s. The transverse components are represented by the arrows; the length of the arrow indicates the magnitude of V_{trans}.

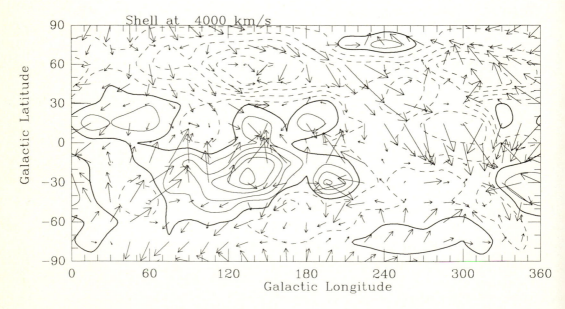

Figure 4b. Peculiar velocities at 4000 km s^{-1}.

We thank our collaborators for allowing us to use as-yet unpublished data. Thanks to Avery Meiksin, for encouragement and constructive criticism at all stages of the project. This work is supported by the NSF and NASA/IPAC. MAS acknowledges the support of an NSF Graduate Fellowship.

References

Collins, C. A., Joseph, R. D., and Robertson, N. A. 1986, *Nature* **320**, 506.

Davis, M., and Huchra, J. 1982, *Ap. J.* **254**, 437.

Dressler, A., Faber, S. M., Burstein, D., Davies, R. L., Lynden-Bell, D., Terlevich, R. J., and Wegner, G. 1987, *Ap. J. (Letters)* **313**, L37.

Harmon, R.T., Lahav, O., and Meurs, E.J.A. 1987, *M.N.R.A.S.*

IRAS Catalogs and Atlases, Explanatory Supplement 1985, edited by C. A. Beichman, G. Neugebauer, H. J. Habing, P. E. Clegg, and T. J. Chester (Washington D.C.: U.S. Government Printing Office).

Lynden-Bell, D., Faber, S. M., Burstein, D., Davies, R. L. Dressler, A. J., Terlevich, R. J., and Wegner, G., 1987, *Ap. J.*

Meiksin, A. 1985, Private Communication.

Meiksin, A., and Davis, M. 1986, *A. J.* **91**, 191.

Peebles, P. J. E. 1980, *The Large-Scale Structure of the Universe* (Princeton: Princeton University Press), §14.

Rubin, V. 1987, in "Large Scale Structure in the Universe", IAU 130.

Soifer, B. T., Houck, J. R., and Neugebauer, G. 1987, Ann. Rev. Astron. Astrophys., in press.

Villumsen, J. V., and Strauss, M. A. 1987, *Ap. J.* **322**.

Yahil, A., Walker, D., and Rowan-Robinson, M. 1986, *Ap. J. (Letters)* **301**, L1.

THE IMPACT OF INFRARED ASTRONOMY ON THE DISTANCE SCALE

L. Gouguenheim

Observatoire de Paris, section de Meudon
92195 Meudon Cedex, France
and Université Paris Sud
91405 Orsay Cedex, France

ABSTRACT

There is a dramatic impact of infra-red astronomy on the cepheid distance criterion: not only the effects of absorption and reddening are much lowered in H-band, but also those of chemical composition variations and temperature differences across the instability strip.

The impact on the Tully-Fisher distance criterion is not so obvious. It is shown that the Malmquist bias and the cluster incompleteness bias, which have been underestimated or even ignored, are the predominant effects.

1. INTRODUCTION

An accurate determination of the extragalactic distance scale is needed in order to obtain:

(1) a good determination of the expansion rate of the universe: the Hubble constant is one of the fundamental cosmological parameters and when compared to independent determinations of the age of the universe, it puts some constraints on the cosmological constant.

(2) informations for deviations from uniform Hubble flow For instance, (i) the mass concentration in Virgo cluster exerts a gravitational influence on the motion of the Local group, (ii) the observation of the dipole anisotropy in the microwave background leads to important implications regarding the large-scale mass distribution in the universe.

The accuracy needed is twofold. Firstly, we need a good linearity of the distance scale (particularly important for studying the velocity field) and secondly a good calibration which gives the accuracy on H_0.

At the time being, the various estimates of H_0 are in the range $50 - 100$ km s^{-1}Mpc^{-1}. It is widely assumed that the distance determinations suffer mostly from an uncertainty on the zero point and that possible non-linearities are small enough so that the individual distance determinations allow a good study of the kinematics of the nearby universe.

The extragalactic distance scale is built on (at least) two steps, the so-called primary and secondary distance indicators.

The **primary calibration** relies on individual stars which are well studied in our Galaxy and recognized in nearby ones. It is assumed that similar stars – whose similarity is recognized from directly observable parameters – have the same intrinsic luminosity wherever they are observed. The calibration of the intrinsic luminosity relies on stellar distances within our Galaxy.

These primary indicators have generally sound physical basis, related to our good knowledge of stellar properties and evolution, but they fail at large distances (except for the supernovae), when individual stars are no more observable.

A second kind of indicators is thus needed: they involve global properties of galaxies. They are calibrated from the properties of the previous sample whose distances have been obtained from primary calibration. Their physical basis are not so well understood, due to the limited knowledge of galaxies evolution, but they have a larger range, up to 100 Mpc.

There have been mainly two different approaches: Sandage and Tammann have choosen at each step what seemed to be the best indicator, putting "all the eggs in the same basket", while de Vaucouleurs had preferred to spread the risks, using at each step the largest number of (rather) independent indicators. An extensive study of the various primary and secondary distance indicators, including their physical basis, their accuracy and a comparison of their advantages, is given by Rowan-Robinson (1986).

2. PRIMARY CALIBRATION

2.1. The various calibrators

(1) The **RR Lyrae** variable stars are a good indicator, but due to their low luminosity, they are observable only in the Magellanic Clouds and, recently (Pritchett and van den Bergh, 1987a), in M31. They provide a good check of the other indicators.

(2) The **novae** are observable at larger distances, up to the Virgo cluster (Pritchett and van den Bergh, 1987b) and the theoretical understanding of their phenomenon is well improved. However, they show a large range in their observed properties.

(3) The **supernovae** have the greatest potential, because they are observable at very large distances. In practice there remains much difficulties considering both their theoretical understanding and the calibration of their absolute magnitudes. There are two different approaches. The first one involves the applicaton of the Baade-Wesselink method

to type II supernovae. The main problem relies on the deviation from the black body radiation. The second approach relies on type I supernovae taken as standard candles. The criterion must be restricted to early type galaxies in order to avoid difficult extinction problems. However, their origin is still under debate and it appears that a significant fraction of them are of a peculiar nature, with different luminosity characteristics.

(4) The **cepheid** variable stars are the most promising of the primary indicators: (i) they are rather luminous, (ii) the period-luminosity-colour relation is on secure theoretical basis and (iii) considerable improvements are coming from near IR photometry.

2.2 The impact of IR astronomy on the cepheid distance criterion

Much of the uncertainty in broad-band optical studies relies on the difficulty to account for the effects of interstellar extinction and chemical composition variations among cepheids in different galaxies. The main advantages in the near IR compared to the visible are the following:

(1) the effects of **absorption** and **reddening** are much lowered: the attenuation of starlight in the H-band is lower that in B-band by a factor of 6

(2) U B V magnitudes and colours are relatively sensitive to **chemical composition variations,** due to the importance of stellar absorption lines in the blue part of the spectrum. These effects are drastically reduced at shorter wavelengths where the density of metallic absorption lines is low.

(3) **Temperature differences** across the instability strip give rise to a **finite magnitude width** in the (P,L) relation, at constant P. A three parameter (P,L,C) relation is thus needed to characterize the properties of an individual cepheid. It is not easy in the blue to disentangle this effect from differential extinction. Because the monochromatic flux is less sensitive to the temperature at longer wavelength, the width of the (P,L) relation at H is only one third of the width at B. In addition the amplitude of the lignt curve of an individual cepheid, which is due primarily to temperature variations is reduced in the IR; thus, **random** IR observations are competitive with **time-averaged** B-band photometry.

The r.m.s. dispersion of the (P,L) relation among a sample of about 40 LMC cepheids, when using either random B, time-average B, random H or time-average H magnitudes is 0.65, 0.46, 0.27 and 0.15 respectively (Mc Gonegal et al., 1982). Welch et al. , (1985), using time-averaged H magnitudes obtained σ = 0.19 for a sample of galactic, LMC and SMC cepheids.

The technical limitation comes from the lack of spatial resolution. Due to the spatial switching method, some IR emission could arise from optically faint IR sources in each of the $\simeq 5"$ beams.

An alternative method uses CCD detectors in BVRI band photometry. The background is substracted by using profile fitting techniques, thus removing the problem of contamination due to the chopping. However, the extinction has to be corrected, but it is lower at I (Freedman et al., 1985).

Future projects and improvements are:

(1) H-band observations with panoramic detectors,

(2) detection of new cepheids in nearby galaxies (see for example the discovery of cepheids in M101 by Cook et al., 1986),

(3) a good calibration of the (P,L) relation from galactic cepheids. The calibration relies presently on the technique of cluster main sequence fitting and thus on the distance of the Hyades. Cepheid parallaxes are expected from the ESO satellite HIPPARCOS.

3. SECONDARY CALIBRATION

The following relations are examples of secondary calibrators:

$$-M = a \log V_m + b \qquad (1)$$

$$-M = a \log \sigma_v + b \qquad (2)$$

$$-M = a \wedge_c + b \qquad (3)$$

They give the absolute magnitude of a galaxy as a function of the maximum circular velocity V_m in disk galaxies [Tully-Fisher relation (1)], of the internal velocity dispersion σ_v in bulge galaxies [Faber-Jackson relation (2)] or the luminosity index $\wedge_c$ [de Vaucouleurs relation (3)], which applies for disk galaxies. The first two relations are the expression of a mass luminosity relation and all of them have the general shape

$$-M = a p + b \qquad (4)$$

where p is a directly observable parameter.

It is generally considered that the Tully-Fisher (1977), hereafter TF, relation is the most accurate, with a dispersion $0.4 - 0.5$ mag at given $p = \log V_m$, and the following discussion will concentrate on it.

3.1 Calibration of the TF relation.

The **maximum circular velocity V_m** is deduced from the width of the 21-cm line observed in a galaxy seen as a point source by the radio telescope, after correcting for inclination effects and for non-circular motions which tend to widen the line (Bottinelli et al., 1983). This last correction is controverted by some authors.

Two main **system of magnitudes** have been used: the B_T^0 system of de Vaucouleurs et al. (1976) and the $H_{-.5}$ system of Aaronson et al. (1979). The B_T^0 magnitudes have the advantage of being total, but they are sensitive to extinction effects; on the contrary the $H_{-.5}$ magnitudes are not sensitive to the extinction effects, but they are measured within an aperture which is one third of the **blue** photometric diameter a_{25}. This has two consequences: firstly, they do not measure the same fraction of the total light, depending on the bulge to disk ratio, and secondly they involve B-band diameters, which are subject to light extinction. For these reasons, I-band studies are being developed as an alternative (Bothun and Mould, 1986).

The **calibration** of the TF relation needs the determination of 2 parameters, the slope a, using either local calibrators, or cluster data or kinematic distances, and the zero point, b, which is determined from local calibrators.

3.2. Problems under discussion

(1) **The slope** depends on the system of magnitudes and also, to a smaller extent, to the system of line-widths, corrected or not for non-circular motions. It is in the range of 6 at B and 10 at H.. This steeper value in H-band weakens the accuracy of the method, because it reflects more widely on the magnitude the errors on $\log V_m$.

(2) It has been suggested that the slope and/or the zero point could depend on the **galaxy type**. The promising method of "sosie" galaxies, introduced by Paturel (1984), overcomes these problems

(3) Because we are dealing with magnitude limited samples, these samples are subject to a **Malmquist bias,** which has been generally underestimated

4. MALMQUIST AND CLUSTER INCOMPLETENESS BIAS IN TF RELATION.

4.1 Field galaxies.

4.1.1. The Method.

The bias arising when determining distances from a magnitude limited sample using relation (4) has been studied by Teerikorpi (1984). Its main properties can be understood from fig. 1.

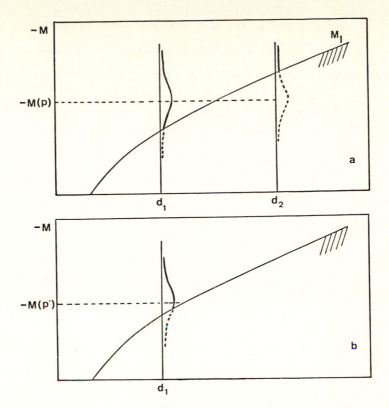

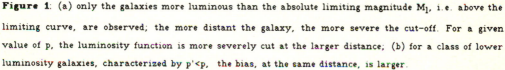

Figure 1: (a) only the galaxies more luminous than the absolute limiting magnitude M_l, i.e. above the limiting curve, are observed; the more distant the galaxy, the more severe the cut-off. For a given value of p, the luminosity function is more severely cut at the larger distance; (b) for a class of lower luminosity galaxies, characterized by p'<p, the bias, at the same distance, is larger.

We consider first a class of galaxies characterized by se same value of $p = \log V_m$. Through the TF relation their mean absolute magnitue M_p is known ($-M_p = a\,p + b$) and the individual magnitudes are distributed around M_p. We assume this distribution to be gaussian, with dispersion σ_{Mp}. If the sample is cut at the apparent limiting magnitude m_l, it results an absolute limiting magnitude at distance d, $M_l(d) = m_l - 5 \log d - 25$. When d increases $M_l(d)$ becomes brighter, the mean absolute magnitude <M> of the sample selected becomes more luminous and the bias $\Delta M_d = -(<M>-M(d))$ increases. For a larger m_l the limiting absolute magnitude is less luminous and the bias is smaller.

If 2 different samples of galaxies, characterized by p and p' (p'<p) are considered, it is a easily seen that the bias is stronger for the p' class, which is the less luminous. Thus:

– at given m_l and given p, the bias increases with the distance

– at given m_l and distance, the bias is stronger for smaller p

– at given p and distance, the bias is smaller for larger m_l.

The shapes of the bias curves are shown in fig. 2

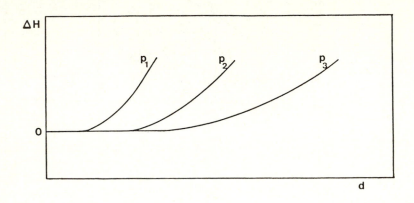

Figure 2: 3 bias curves, obtained for 3 different values of parameter p, with $p_1 < p_2 < p_3$, are represented; the other parameters, m_l and σ_{Mp} are the same.

– For a given class of galaxies (p) the bias is negligeable up to a threshold and then it increases with distance

– The threshold depends on p: it is larger for larger p

– When considering a sample of galaxies with all p, it results a cloud of points, distributed around the various curves and the remaining plateau is the smallest one. As a consequence, the bias is not conspicuous.

In order to overcome these problems, Bottinelli et al. (1986a) have introduced the concept of **normalized distance**:

$$d' = d \times dex[-0.2a(2.7 - \log V_m]$$

where d is the kinematic distance, which is an unbiased estimate of the distance. At same d', all galaxies with different p suffer from the same amount of bias, if all these subsamples are characterized by same m_l and σ_{Mp}. All the plateau data are also selected.

4.1.2 Results.

This method has been applied to B-band TF distances, with line widths corrected for non-circular motions (Bottinelli et al., 1986a) and to H-band TF relation with non-corrected line widths (Bottinelli et al, 1987b). The main results are the following:

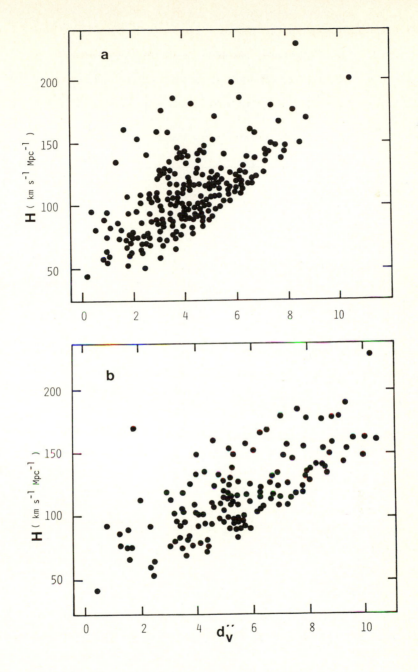

Figure 3: H vs. normalized distance for (a) the B-band sample, cut at the limiting magnitude $B_T^0 = 12$ and (b) the H-band sample, cut at the limting magnitude $H_{-.5} = 10$.

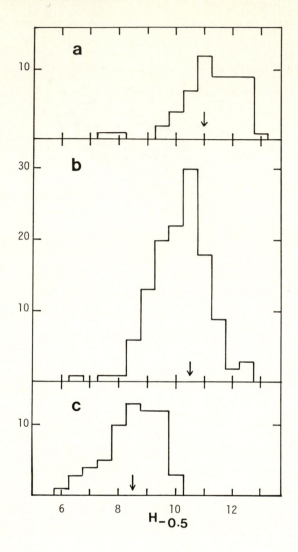

Figure 4: Histograms of $H_{-.5}$ magnitudes in 3 different log V_m ranges (a) $1.9 \leqslant \log V_m < 2.1$ (b) $2.1 \leqslant \log V_m < 2.3$ (c) log $V_m \geqslant 2.3$. The arrows indicate the magnitude up to which the sample can be considered as being complete.

– both samples are affected by a strong bias (Fig 3)

– the plateau data give a mean value of the Hubble constant, in de Vaucouleurs local scale, $H_0 = (72 \pm 3)$ km s^{-1} Mpc^{-1} in both B-band and H-band, which is significantly different from the values of H_0 previously determined from similar samples (de Vaucouleurs et al., 1981, Aaronson and Mould, 1983)

4.1.3. Discussion.

Giraud (1985) claims that he has brought to light a **type effect** in the zero point of the TF relation. When plotting H against kinematic distance, he finds a segregation of the morphological types: at a given distance, H is larger on the mean for late type (Scd, Sd) galaxies than for early type ones (Sab, Sb). In fact, this has nothing to do with a type effect, but is actually expected from the bias, because late type galaxies are, on the mean, less luminous than early type ones and thus expected to suffer from a larger bias. (Bottinelli et al., 1986b). Moreover, this so-called type effect is not conspicuous in H-band data; it should however not be concluded that this has something to do with a better quality of the H-band data. This comes essentially from the constitution of the samples (fig. 4) where the limiting magnitude is larger for low luminosity galaxies (small p). Thus the two different effects, of m_l and p respectively, on the resulting bias compensate each other

Some authors (Tammann, 1986; Giraud, 1986) have considered the bias arising in the whole sample (including all p) without any normalization; it has been seen previously that the bias is much difficult to put in evidence and that the unbiased data are not so easily recognized. Moreover, Giraud has tried to compute the expected bias , he comes to the conclusion that there remains an intrinsinc increase of H with kinematic distance after correcting the bias, which should explain only one third of the effect. In fact this method is a step backward in comparison to the method using the normalized distance and relies on a strong assumption concerning the luminosity function. The separate curves of fig. 2 indicate how galaxies in different $\log V_m$ intervals populate the diagram. In order to calculate the average dependence of the biased H on d one must know how crowded each curve is by galaxies. The expected bias depends on the mean absolute magnitude and σ_M of the global luminosity function of galaxies, which is assumed to be gaussian. When dealing with a subsample of galaxies with same value of $\log V_m$, the mean unbiased absolute magnitude of the sample is known from the TF relation. It is not the case here, where it is determined by Giraud from apparent magnitudes and (**biased !**) distances. Moreover the choice of the limiting magnitude, on which the bias is strongly dependent, has not been discussed, by Giraud. For all these reasons, it is difficult to give any credit to his conclusions.

4.1.4 Conclusions

The main conclusions are thus the following:

(1) the bias is the predominant effect. It is much more important than the compared accuracies of B_T^0 or $H_{-.5}$ magnitudes or the effects of non circular motions.

(2) the bias does not depend only or predominantly on the observed dispersion of the relation but also strongly on the limiting magnitude of the sample. It results that, contrary to a common statement, when comparing different distance criteria, the best one is not necessaryly characterized by the smallest scatter and the bias at a given distance is not necessaryly stronger for a criterion with larger dispersion

(3) Is the determination $H_0 = 72 \pm 3$ in de Vaucouleurs primary calibration only local (because of possible local motions) or global ? There are two ways for answering this question. The first one is to increase the sample, thus m_l and the plateau threshold. The second relies on cluster data.

4.2. Cluster incompleteness bias.

Contrary to a general statement, a bias is also expected within a cluster.

4.2.1 Method.

The bias arising when determining distances from a magnitude limited sample of galaxies within a cluster, using relation (4), has been studied by Teerikorpi (1987). It is illustrated in fig. 5. The bias expected at small p is larger because the luminosity function of galaxies with same value of p is more severely cut. A bias decreasing with increasing p and negligeable at large p (plateau region) is thus expected.

4.2.2. Results.

The bias arising in a sample of 10 clusters with velocities ranging from 4000 to 11 000 km s^{-1} plus Virgo cluster has been studied both in B-band and in H-band (Bottinelli et al., 1987 a, b). A normalized log V_m taking into account the clusters distances and limiting magnitudes has been used in order to put all the cluster data together. The data (fig. 6) shows clearly the trend expected. The plateau data in B-band lead to $H_0 = 73 \pm 4$ in de Vaucouleurs primary calibration, adopting an infall velocity of 220 km s^{-1} for the Local group. This result is in remarkable agreement with the value obtained from the field

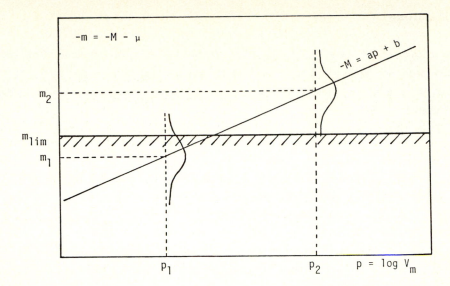

Figure 5: apparent magnitude vs. $p = \log V_m$ diagram in a cluster having a distance modulus m and observed up to a limiting magnitude m_l (hatched line). The straight line stands for the TF relation. It is seen that the luminosity function is more severely cut at p_1 than at p_2 ($p_1 < p_2$).

sample. The limiting magnitude of the $H_{-.5}$ sample is bright, leading to a very small number of plateau data. Using a sample of 19 galaxies including biased data near the threshold and an iterative method for computing the bias, a similar value of H_0 is obtained, in the range 70 – 75.

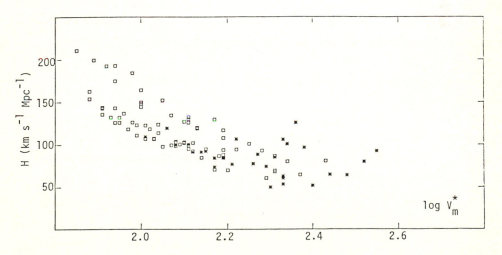

Figure 6: H vs. normalized $\log V_m$ for 10 clusters. The Virgo cluster points (*) have been added, adopting an infall velocity of the Local group equal to 220 km s^{-1}.

5. CONCLUSIONS

(1) The impact of IR data on the primary calibration is quite obvious. The same conclusion does not apply to the TF secondary calibration, where the effect of the Malmquist bias and the cluster population incompleteness bias have been strongly underestimated or even ignored.

What we need before all, are large samples which are complete up to a **large** limiting magnitude. Under this respect the B_T^0 system is presently the best one.

(2) As a general comment, any physical property obtained statistically from a biased sample must be considered with great care. Examples are:

- type effect in TF relation (in fact a differential Malmquist bias)
- particular physical properties of cluster galaxies giving a different slope of the TF relation (the cluster population incompleteness bias predicts a shallower slope)
- all the kinematic studies performed with biased distances (de Vaucouleurs and Peters, 1986; Aaronson et al., 1986).

REFERENCES

Aaronson, M., Bothun, G., Mould, J., Schommer, R.A., Cornell, M.E., 1986 Astrophys. J. **312**, 536

Aaronson, M., Huchra, J., Mould, J., 1979, Astrophys. J. **229**, 1

Aaronson, M., Mould, J., 1983, Astrophys. J. **265**, 17

Bothun, G.D., Mould, J., 1986, Astrophys. J. **313**, 629

Bottinelli, L., Gouguenheim, L., Paturel, G., de Vaucouleurs, G., 1983, Astron. Astrophys. **118**, 4

Bottinelli, L., Gouguenheim, L., Paturel, G., Teerikorpi, P., 1986a, Astron. Astrophys. **156**, 157

Bottinelli, L., Gouguenheim, L., Paturel, G., Teerikorpi, P., 1986b, Astron. Astrophys. **166**, 393

Bottinelli, L. Fouqué, P., Gouguenheim, L., Paturel, G., Teerikorpi, P., 1987a, Astron. Astrophys. **181**, 1

Bottinelli, L., Gouguenheim, L., Teerikorpi, P., 1987b, Astron. Astrophys, submitted

Cook, H.K., Aaronson, M., Illingworth, G., 1986, Astrophys. J. **301**, L45

De Vaucouleurs, G., de Vaucouleurs, A., Corwin, H., 1976, Second Reference Catalog of Bright Galaxies, University of Texas Press (RC2)

De Vaucouleurs, G., Peters, W.L., Bottinelli, L., Gouguenheim, L., Paturel, G., 1981, Astrophys. J. **248**, 408

De Vaucouleurs, G., Peters, W.L., 1986, Astrophys. J. **303**, 19

Freedman, W.L., Grieve, G.R., Madore, B.F., 1985, Astrophys. J. Suppl. Ser. **59**, 311

Giraud, E., 1985, Astron. Astrophys. **153**, 125

Giraud, E., 1986, Astron. Astrophys. **174**, 23

Mc Gonegal, R., Mc Laren, R.A., Mc Alary, C.W., Madore, B.F., 1982, Astrophys. J., **257**, L33

Paturel, G., 1984, Astrophys. J. **282**, 382

Pritchett, C.J., van den Bergh, S., 1987a, Astrophys. J. **316**, 517

Pritchett, C.J., van den Bergh, S., 1987b, Astrophys. J. **318**, 507

Rowan-Robinson M., 1986, The Extragalactic Distance Ladder W.H. Freeman and Co.

Tammann, G. A., 1986, IAU Symp. n° 124, p. 151

Teerikorpi, P., 1984, Astron. Astrophys. **141**, 407

Teerikorpi, P., 1987, Astron. Astrophys. **173**, 39

Tully, R.B., FisherJ.R., 1977, Astron. Astrophys. **54**, 661

Welch, D.L., Mc Alary, C.W., Mc Laren, R.A., Madore, B.F., 1985, in "Cepheids, Theory and Observations", ed. B.F. Madore, Cambrige Univ. press

CONFERENCE SUMMARY

Martin Harwit

Astronomy Department
Cornell University
Ithaca, New York 14853-6801
and
National Air and Space Museum, Washington, D.C. 20560

ABSTRACT

To do justice to so many interesting contributions, both in the form of papers presented as talks and posters represented only by titles in these proceedings, will be difficult. Rather than attempting to list contributions from the individual areas in a representative fashion, I will attempt to see how a few of the striking contributions fit into, or alter, our views on major questions we have been trying to answer during the past few decades – questions dealing with the structure and evolution of the universe, the formation of galaxies and stars, and the origins of the solar system, in short everything from Comets to Cosmology – though I will reverse the order, starting here with cosmological questions and ending up with comets, or rather with zodiacal dust.

1. INTRODUCTION

Joseph Silk and George Efstathiou in their papers pointed to what may be the most fundamentally pressing cosmological problem to which observations almost assuredly will provide answers, if we only work hard enough. It concerns the formation of galaxies, clusters and other large-scale condensations in the universe. Of course, there are many other cosmological problems of equal interest, particularly questions relating to the very earliest epochs. But it is not as clear there how those can ultimately be tackled through observations.

The problem of large-scale condensations in this: The isotropy of the microwave background does not seem to allow significant condensations to have occurred before the era during which radiation decoupled from matter, in a standard expanding cosmological model. At least these isotropy observations don't permit radiation to have significantly coupled to such condensations. Otherwise, the microwave background radiation would show greater spatial fluctuations.

In order to make headway, one then postulates conditions in the early universe, consistent with the uniformity of the background, but able to evolve into the structures seen today. So far such models are not entirely successful. If and when they succeed, they also may turn out not to be unique. But ultimately they may provide us with a hint about the makeup of the early universe, and we might then be able to apply physical arguments to decide whether or how such a cosmological model might make sense. To do that, however, we will need to make further observations with the Cosmic Microwave Background Explorer (COBE) and with a whole series of other future facilities, particularly the Infrared Space Observatory (ISO) and the Space Infrared Telescope Facility (SIRTF) – for, much of the information needed can be sought only through sensitive infrared observations. Such observations include number counts and statistical approaches such as those described at this meeting, by Davis, Houck, Longair, Rowan-Robinson and de Zotti, on behalf of their co-authors.

2. THE BACKGROUND RADIATION

Matsumeto (these proceedings) has noted a background radiation feature at 2μm which he attributes to redshifted Ly-α radiation, possibly from early stages of galaxy formation at $z \approx 20$. He and Paul Richards (also these proceedings) similarly refer to an excess submillimeter flux in bands at 15 and 22 cm^{-1} comprising roughly 10 percent of the total microwave energy density. We take these two background fluxes up in turn. It is interesting to see the extent to which such fluxes put a strain on plausible sources of energy in the universe:

1) The background due to young galaxies was studied by Partridge and Peebles (1967). If an amount of energy $\Delta\epsilon$ is generated at red shift, z, in what today is measured as a unit volume, then today's radiation energy density will be $\Delta\epsilon/(1+z)$ and the integrated flux-crossing unit surface per solid angle will be

$$I = \frac{c}{4\pi} \frac{\Delta\epsilon}{(1+z)}$$

Taking today's baryonic mass density to be ρ_o, a fraction $f = \Delta\epsilon/\rho_o c^2 \eta = 4\pi I(1+z)/\rho c^3 \eta$ of the mass density would have had to be converted into radiation at epoch z, to provide the observed flux, provided the mass was converted into energy with efficiency η. Let ρ_o correspond to $\Omega = 0.1$ where Ω is the baryonic density in units of the closure density $3H^2/8\pi G$, where G is the gravitational constant, and $H = 75$ km/sec Mpc $= 2.5 \times 10^{-18}$ sec^{-1}. Then $\rho_o \sim 10^{-30}$g cm^{-3}. For hydrogen-to-helium conversion $\eta = 7 \times 10^{-3}$. We can then write

$$I = \frac{f\rho_o\eta c^3}{4\pi(1+z)}$$

$$\sim \frac{3\,H^2 f\Omega\eta c^3}{32\pi^2 G(1+z)} \sim \frac{0.5}{(1+z)}\left(\frac{f}{0.05}\right)\left(\frac{\Omega}{0.1}\right)\eta \text{ erg cm}^{-2}\text{sr}^{-1}\text{sec}^{-1}$$

and

$$I \sim \frac{10^{-3}}{(1+z)}\left(\frac{f}{0.05}\right)\left(\frac{\Omega}{0.1}\right)\left(\frac{\eta}{7\times10^{-3}}\right) \text{ erg cm}^{-2}\text{sr}^{-1}\text{sec}^{-1}$$

For $(1+z) \sim 20$ as Matsumato's curves suggests for the 2μm flux which they give as 5×10^{-5}erg cm^{-2}sr^{-1}sec^{-1},

$$I = 5\times10^{-5}\left(\frac{f}{0.05}\right)\left(\frac{\Omega}{0.1}\right)\left(\frac{\eta}{7\times10^{-3}}\right)\left(\frac{20}{1+z}\right) \text{ erg cm}^{-2}\text{sr}^{-1}\text{sec}^{-1}$$

2) If the FIR flux has been redshifted by a factor $(1+z)\sim 10$ from 60 to 600 μm and we have one-tenth the integrated microwave flux in this component, then

$$0.1 \left(\frac{\sigma T^4}{\pi} \right) = 10^{-4} \text{erg}^{-2} \text{ cm}^{-2} \text{sr}^{-1} = 10^{-4} \left(\frac{f}{0.05} \right) \left(\frac{\Omega}{0.1} \right) \left(\frac{\eta}{7 \times 10^{-3}} \right) \left(\frac{10}{1+z} \right)$$

From these two estimates we see that roughly one-fifth the helium believed to have been formed by now, namely about 5% of the hydrogen mass, could have sufficed to produce either flux provided Ω is as high as 0.1. If it is less, f would have to correspondingly rise.

That would be true if the heating had been produced indirectly through inverse Compton scatter as well, as long as the ultimate source of energy was hydrogen-to-helium conversion. Production at a more recent epoch $(1+z) < 10$ would also alleviate the difficulty of accounting for so much energy. Annihilation of matter in the formation of recent black holes could also help, since then η can be of order 10% instead of 7×10^{-3}. The distortion of the submillimeter spectrum, however, cannot have been produced at decoupling since the efficiency of thermalization then was high.

We see then that if either or both of these fluxes are cosmological – a question which, as Stefan Price pointed out in discussion, is not clear, since the near-infrared flux conceivably could still somehow be caused by zodiacal dust scatter or emission – the helium content of the cosmos needs to be re-evaluated. The standard cosmological models suggest that $\sim$24% of the hydrogen was converted into helium during primordial times. That appears reasonably consistent with observations of the atmospheres of early stars, though admittedly such measurements are difficult. If as much as another 10% of the hydrogen had been converted into helium since then, could we have failed to notice it, could it still be bottled up in low-mass stars, or would it have been returned to the interstellar medium by massive evolved stars?

It is too early to tell, but we do know that stars have been shining for a long time; and that energy must now be contributing to a background flux at some wavelengths, perhaps precisely where Matsumoto and Richards were telling us.

3. EXTREMELY LUMINOUS FAR-INFRARED SOURCES (ELFS)

A number of traits seem common to extragalactic sources whose 60μm luminosities exceed $10^{11} L_\odot$.

a) They predominantly appear to be colliding or at least peculiar looking sources.

b) Their optical luminosities often are orders of magnitude below the FIR luminosities.

c) Their radio fluxes are proportional to the FIR fluxes and the proportionality ratio is the same as for galaxies of far lower luminosities and comparable to that of HII complexes.

d) Their molecular hydrogen contents are high $\sim 10^{10}$ $M_\odot$, both as judged from CO, 2.6 mm data and from dust emission at millimeter wavelengths (Young et al., 1986; Krügel et al., 1987).

e) The infrared emission, at least at 10 and 25μm where it can be spatially better resolved through ground-based observations (cf Becklin's contributions in these proceedings) tends to come from the center of the sources. That appears to be true of the CO emission and hence the H_2 concentration as well. The regions in question are perhaps ≤ 1 kpc in diameter, though reliable data exits for only a few of the brighter sources (Becklin and Wynn-Williams, 1987; Scoville et al., 1986).

f) The luminosity function of these sources is known and drops steeply from 10^{11} toward 10^{13} $L_\odot$.

Other frequent, but less consistent traits, have been summarized by Harwit *et al.* (1987).

Four different models have been proposed to account for the ELFS:

1) Those who emphasize the remarkable consistency of trait (c) argue that one is dealing with star formation on a very massive scale; and these views are so widely accepted that the ELFS are often indiscriminately called starburst galaxies.

2) Those who note the extreme compactness of the emitting region and its central location have argued that an active nuclear source must play a role. Trait (b) might be consistent with that as well.

3) Finally, a view Harwit *et al.* (1987) have argued emphasizes traits (a), (b), (d) and (e) and insist that collisions and high gas content must be given primacy, because ELFS without collisions and the presence of gas do not appear to exist.

4) Finally Burbidge (these proceedings) has argued for a model composed of a highly dense aggregate of η Carinae-like stars.

Quantitatively these views lead to the following concerns.

1) Star formation in colliding galaxies is possible, and perhaps is an efficient converter of mass into energy. The sources can persist, and in fact must persist for many millions of years; but the luminosity function of stars must then be quite different from that in well-studied galaxies. Since L $\sim 10^{11}$ to 10^{12} $L_\odot$ and M $\sim$ $M_{gas} \sim 10^{10}$ $M_\odot$, the mass-to-luminosity ratio typically needs to be ~ 100, which is very difficult to explain, though Beichman (this meeting) suggests, L/M $\sim 10 \rightarrow 25$ is characteristic of the most massive molecular clouds. In addition, the stars are sufficiently long-lived and because of their required mass and luminosity would produce sufficiently strong stellar winds, that one would expect any enshrouding dust to be blown away before the stars ceased to shine. Where then is the missing optical luminosity (trait (b)) one would expect? Moreover, with as much gas as is seen in these sources now, even more gas must originally have been present from which the stars formed. That seems far-fetched.

2) The argument for a central active source is plausible, but if traits (a) and (d) are to be incorporated, the active nucleus, presumably a black hole, needs to be fed by in-falling gas.

Two points need to be made about this model. First, the luminosity is strongly dominated by the mass of the black hole, since the gaseous disk at the center of the approaching galaxy probably has quite normal density. The luminosity function for ELFS, which is very steep, should then reflect the mass distribution of the black holes in normal galaxies – a distribution currently not known. Second, this model also fails to explain the high correlation between luminosity and gas content, since primarily gas density rather than mass is crucial.

3) The collisional model attributes the observed luminosity to the dissipation of kinetic energy of colliding gas masses. Harwit *et al.* (1987) have shown that the peak-observed luminosities can be generated that way, and Harwit and Fuller (1987) have shown that the luminosity function can be surprisingly well filled with a model of two colliding disks. This, perhaps, is the model's greatest strength, as well as the circumstance that any collision of two gas-rich galaxies requires the dissipation of kinetic energy as a minimal requirement. One might argue, therefore, that the

dissipative model should at the very least be a feature of models (1) and (2) as well as being possible in isolation.

Two further points can be made. First, Model 3 is less efficient in producing energy than Model 2 since the entire gaseous disk kinetic energy must be dissipated to produce the highest observed luminosities, while only a fraction of the gas in Model 2 needs to fall into the active center. Second, it is possible that only one galaxy needs to be gas-rich in Models 1 and 3 – as well as in Model 2. In Models 1 and 3, gas can collide with itself in the accretion wake of a nucleus and rather high densities and energy dissipation could arise there without the requirement of indigenous gas in both galaxies.

4) Model 4 requires further investigation. Too little is known of the circumstances that lead to the formation of objects like η Carinae.

4. RADIO AND INFRARED EMISSION FROM INTERSTELLAR DUST

Broadbent, Haslam and Osborne told us about the remarkable similarity of maps obtained at 11 cm radio and 60 μm infrared wavelengths. So strong is that similarity that deviations from it in the form of highly intense radio spots apparently can be reliably identified with supernova remnants. Recognition that the infrared and radio fluxes generally are proportional to each other in Galactic sources dates back to papers of the early 1970s, as exemplified by the work of Harper and Low (1971). More recently such authors as Boulanger, Perault and Puget have done a great deal of work along these lines, but a clear-cut rationale for the relationship, which appears to hold for a wide range of luminosities in extragalactic sources as well, is still wanting.

In this connection it may be worth reconsidering an old idea proposed by Hoyle and Wick-ramasinghe which runs something like this. In equilibrium:

$$mr^2\omega^2 \approx kT$$

from which we obtain

$$\nu = \frac{\omega}{2\pi} = \frac{1}{2\pi}\left(\frac{kT}{mr^2}\right)^{1/2} \approx 2.6x10^9 \left(\frac{10^{-21}g}{m}\right)^{1/2} \left(\frac{10\text{Å}}{r}\right) \left(\frac{T}{20°K}\right)^{1/2} \text{Hz}$$

corresponding to a 11 cm wavelength. Here ω is the radial frequency and ν the rotation frequency, while m is the grain's mass and k the Boltzmann constant. If the grain has a dipole moment d = ηer, where $\eta \leq 1$, r is the radius and e is the electron charge, then the grain radiates a power

$$I(11cm) \approx \frac{2\,|\,\ddot{d}\,|^2}{3c^3} \approx \frac{2}{3}\eta^2 e^2 r^2 \omega^4 c^{-3} \approx 4x10^{-24}\eta^2 \text{erg sec}^{-1}$$

At this meeting, Puget told us to expect $2.4x10^{-20}$ erg sec^{-1} to be emitted in the infrared, per carbon atom, or for a polyaromatic hydrocarbon molecule with mass 10^{-21} g, an emission of roughly 10^{-18} erg sec^{-1}. Hence, the expected ratio of infrared–to–radio emission for such grains might be

$$I(60\mu m)/I(11cm) \approx 2.5x10^5 \eta^{-2}$$

where, as already noted $\eta \leq 1$. The actual ratio from Broadbent, Haslam and Osborne's data is

$$[\nu F(\nu)]_{60\mu m} / [\nu F(\nu)]_{11cm} \approx 6x10^5$$

which is rather close to what the model of small grians with reasonable dipole moments predicts. Such dipole moments are quite likely to exist because the attachment of a single metal atom or halogen atom to a hydrocarbon molecule is quite likely to produce significant displacement of charges and result in a dipole moment of the required order.

The reports, at this meeting by Deul that the emission ratio at $60\mu m/100\mu m$ is often enhanced in regions where 21-cm hydrogen line widths are broad, suggests that grains in those regions are heated, at least in part, by the gas rather than by starlight. If grain temperature is the only criterion determining the ratio of infrared-to-radio emission, constancy for that ratio for all regions of one and the same color temperature would be a sign in favor of this emission model, but the ratio of infrared-to-radio emission could vary somewhat as the color temperature changed. The radio emission then would rise roughly as T^2. In all this, however, spectral data will also be required since some of the far-infrared emission may well be contributed by [OI] emission at $63\mu m$, [OIII] emission at 52 and $88\mu m$ and so forth (Harwit, Houck and Stacey, 1986).

5. PERSONAL RECOLLECTIONS

This meeting and its particular emphasis on certain topics has brought back a number of memories dating back more than 20 years; and it may be of some interest to cite them just to see how far we have come. The first day of our meeting dealt largely with the zodiacal dust cloud and how to model it: In particular, IRAS had mapped the zodiacal emission in a relatively narrow range of elongation angles, in order to see how the computed emission could be subtracted from the observeed emission, if one wished to recover information on the diffuse Galactic and extragalactic flux.

Here, it was a pleasure to see that the mid-infrared data obtained by Soifer, Houck and Harwit in 1971 agree so well with IRAS and other modern data – as summarized, for example, by Salama *et al.* (1987). Those observations were obtained with the first successfully flown, rocket-borne, liquid-helium-cooled telescope – the culmination of about seven years of rather difficult work in which a number of us had invented and successively built (*cf* Harwit, Houck and Fuhrmann, 1969), a series of gradually improving designs, the forerunners of the cooled telescopes we now take for granted, IRAS, ISO, SIRTF, and so on. With the same rocket telescope, Judy Pipher, also at Cornell, observed the diffuse Galactic emission at one elongation from the Galactic center; and that value (Pipher, 1973) also is consistent with modern data. I mention this only because both these observations were largely ignored at the time, mainly, I believe, because there were no other experiments which could independently verify them.

Similarly ignored, though for somewhat different reasons perhaps, was a paper prepared for the Liège symposium of 1963 and published in its proceedings (Harwit, 1964). There I had tried to provide a first model for the emission to be observed from the zodiacal dust cloud. And even though I took too low a value for the emissivity of grains – a failing that first became apparent through the Soifer *et al.* 1987 rocket observations which showed the grains to be very dark – I was able to show that the zodiacal emission would make any other, fainter, diffuse glow rather difficult to untangle, and would perhaps be an ultimate limitation on efforts of that kind. That prediction seems to be borne out now at this meeting; but at the time, 24 years ago, nobody really cared:

Since far-infrared astronomical observations didn't even exist, why should one worry about what would ultimately limit them? It didn't make too much sense to worry about such matters.

Perhaps the lesson to be learned from all this – if there is any lesson there at all – is that scientific work needs to be not only right, if it is to be appreciated. It also needs to be done at a time when an appreciative set of colleagues can make use of it.

ACKNOWLEDGEMENTS

It is a pleasure to acknowledge partial support from this meeting's sponsors to defray expenses to attend the meeting. My research has been supported by NASA/JPL Contract 957701 and NASA grant NAGW-761.

References

Becklin, E. E. and Wynn-Williams, G. C., "Star Formation in Galaxies," ed. C. Persson, U.S. Government Printing Office, Washington, D.C., 1987.

Harper, D. A. and Low, F. J., *Astrophys. J.*, **165**, L9 (1971).

Harwit, M., Congress + Coll. de l'Université de Liège, **26**, 506 (1964).

Harwit, M., Houck, J. R. and Stacey, G. J., *Nature*, **319**, 646 (1986).

Harwit, M. *et al.*, *Astrophys. J.* **315**, 28 (1987).

Harwit, M. and Fuller, C. (1987), submitted to *Astrophys. J.*

Harwit, M., Houck, J. R., and Fuhrmann, K., *Applied Optics*, **8**, 473 (1969).

Krügel, E., Chini, R., Kreysa, E. and Sherwood, W., "Mm-Observations of Markarian Galaxies," to be published, *A. A.* (1987).

Partridge, R. B. and Peebles, P. J. E. (1967) *Ap. J.*, **148**, 377.

Pipher, J., I.A.U. Symposium #52 (1973).

Salama, A., *et al.*, *Astrophys. J.*, **92**, 467 (1987).

Scoville, N. J. *et al.*, *Astrophys. J.*, **311**, L47 (1986).

Soifer, B. T., Houck, J. R. and Harwit, M., *Astrophys. J.*, **168**, L73 (1971).

Wickramasinghe, N. C. and Hoyle, F., *Nature*, **227**, 473 (1970).

Young, J. *et al.*, (1986), *Astrophys. J.*, **311**, L17.

POSTER PAPERS

D.K. Aitken, C.H. Smith, and S. James. Arp 220 and NGC 6240 : active nuclei with a starburst component.

L.G. Balazs and M. Kun. Sequential star formation in Cep OB2.

P.Belfort, R.Mochkovitch, and M.Dennefeld. IRAS galaxies and starburst events.

F.Berrilli, C.Ceccarelli, D.Lorenzetti, P.Saraceno, and L.Spinoglio. IRAS observations of HH exciting sources.

C.Birkett, A.Fitzsimmons, and I.P.Williams. Near nucleus dust studies in Comet Halley.

M.F.Bode, E.R.Seaquist, D.Frail, J.A.Roberts, D.C.B.Whittet, A.E.Evans, and J.S.Albinson. Extended far-infrared emission around the old nova GK Persei.

L.Bottinelli. Neutral and molecular gas in IRAS galaxies.

L.Bottinelli, M.Dennefeld, L.Gougenheim, A.M.Le Squeren, J.M.Martin, and G.Paturel. HI, OH–18 cm and continuum study of IRAS luminous galaxies.

K.Brink. Far-infrared radiation from optically bright galaxies.

D.P.Carico, D.B. Sanders, B.T. Soifer, J.H.Elias, K.Mathews, and G. Neugebauer. The IRAS Bright Galaxy Sample III : 1–10μm observations and co–added IRAS data for galaxies with $L_{IR} \geq 10^{11} L_\odot$.

F.Casoli, Ch.Dupraz, and M.Gerin. Dynamically induced star bursts in interacting galaxies.

T.Chester. The search for brown dwarfs in the IRAS data base.

J.T.Clarke. Near-infrared instruments for the Hubble Space Telescope.

R.G.Clowes et al. A north-south comparison of extragalactic IRAS point sources.

M.J.Coe, L.Bassani, N.Mandolesi, L.Spinoglio, and B.Partridge. 1652+395 – a previously unidentified faint galaxy ?

R.J.Cohen, E.E.Baart, and J.J.Joncas. OH masers associated with IRAS far–infrared sources.

P.Cox and A.Leene. Observational constraints on the carriers of the ultraviolet extinction bump as derived from IRAS data.

J.K.Davies and B.Stewart. Minor Planet 3200 Phethon; no evidence of cometary characteristics.

F.X.Desert, D.Bazell, and F.Boulanger. The nearby molecular clouds : a complete survey.

S.A.Eales, C.G.Wynn-Williams, and W.D.Duncan. A search for cold dust in IRAS galaxies.

S.Eales, D.Depoy, and K.Arnaud. X-ray observations of Mkn 231 and Arp 220, the relationship between X-ray and far-infrared emission, and the contribution of spiral galaxies to the X-ray background.

R.Edelson and M.Malkan. Infrared emission from active galaxies.

R.Edelson and M.Malkan. Far-infrared variability in galactic nuclei.

E.F.Erickson. SOFIA - Stratospheric Observatory for Infrared Astronomy.

R.Fong and L.R.Jones. IRAS cirrus and HI at the SGP.

K.I.Fricke, J.Hellwig, and W.Kollatschny. OPtical spectra and FIR fluxes of barred spirals.

A.Fitt, P.Alexander, and M.J.Cox. The radio–infrared correlation : interpretation in terms of a two temperature model.

T.N. Gautier III and F.Boulanger. A study of small–scale structures in infrared cirrus.

D.Gezari and L.Blitz. Comparison of 1.0 mm continuum, CO molecular line, and IRAS data on the NGC 6334 complex.

R.Giovanelli and G.Helou. The luminosity function of faint IRAS galaxies.

S.K.Gosh, K.V.K.Iyengar, T.N.Rengarajan, S.N.Tandon, R.P.Verma, and R.R.Daniel. Observations of southern HII regions in 120 – 300μm band.

D.T.Gregorich, J.R.Gunn, T.Herter, J.R.Houck, G.Neugebauer, and B.T.Soifer. A deep far-infrared survey.

L.Haikala, R.Laureys, and J.T.Armstrong. Extended bipolar molecular outflow in L1228.

D.Halls. Second Generation ST instrument ”HIMS”.

M.G.Hauser, J.C.Mather, C.L.Bennet, and G.F.Smoot. The Cosmic Microwave Background Explorer mission.

J.Harnett, V.Klein, and E.Wunderlich. Radio continuum and far–infrared emission from the spiral galaxy NGC 6946

C.Heiles, W.Reach, and B.-C.Koo. Grain sizes and interstellar shocks.

G.Joncas and C.Kompe. IRAS data analysis of the HII region Sharpless 142.

M.Kessler. The Infrared Space Observatory (ISO).

V.Klein, J.I.Harnett, and E.Wunderlich. Radio continuum and far–infrared emission from the edge on galaxy NGC 4361

S.G.Kleinmann, E.E.Becklin, S.A.Eales, K.D.Kuntz, C.G.Wynn-Williams, W.C.Keel, and D.Hamilton. The most luminous IRAS galaxy.

T.B.H.Kuiper, J.W.Fowler, W.Rice, and J.B.Wheelock. Molecular clouds in the southern galaxy.

J.Loveday. Angular cross-correlation of IRAS and optical galaxy catalogues.

A.Lawrence, M.Rowan-Robinson, W.Saunders, G.Efstathiou, N.Kaiser, R.Ellis, and C.Frenk. The 60μm luminosity function for IRAS galaxies.

K.J.Leech, A.Lawrence, M.Rowan-Robinson, and M.V.Penston. The nature of high IR luminosity galaxies from spectroscopic observations.

A.Leger and L.d'Hendecourt. The physics of IR emission by PAH molecules.

S.K.Legget, R.G.Clowes, M.Kalafi, H.T.McGillivray, P.J.Puxley, A.Savage, and R.D.Wolstencroft. An infrared–optical study of IRAS point sources in the Virgo region.

T.Liljestrom, K.Mattila, and P.Friberg. Interaction between IRAS 04325–1419 and surrounding CO gas in the nearby high latitude cloud L1642.

R.J.Laureijs, G.Chelewicki, P.R.Wesselius, and F.O.Clarke. Evidence for a multimodal distribution of dust grains in IRAS observations of diffuse interstellar clouds.

S.R.Marley, P.Chaloupka, and P.L.Marsden. Mapping of the North American Nebula using IRAS observations.

A.P.Marston and R.J.Dickens. Far–infrared observations of Cen A.

E.J.A.Meurs and R.T.Harmon. An IRAS search for galaxies near the galactic plane.

C.Moss and M.Whittle. Star formation in cluster spirals.

M.de Muizon, L.B.Hendecourt, and T.B.Geballe. Near–infrared spectroscopic observations of Galactic IRAS sources and their implications for the PAH hypothesis.

M.de Muizon, R.Papoular, and B.Pegourie. Statistical analysis of the IRAS spectra.

C.D.Murray. Distribution of meteor orbits : evidence of resonant structure.

G.Needham, A.Lawrence, and M.Rowan-Robinson. The clustering of IRAS galaxies.

Nguyen–Q–Rieu, N.Epchtein, Truong–Bach, and M.Cohen. New CO and HCN sources associated with IRAS carbon stars.

H.L.Nordh, S.G.Olofsson, and P.Modigh. A large scale infrared study of the W3—W5 regions.

B.Payne and M.Coe. The infrared energy distribution of Galactic binary X-ray sources.

A.Prestwich. Near–infrared spectroscopy of Seyfert galaxies.

P.J.Puxley, C.M.Mountain, T.G.Hawarden, and S.K.Legget. The radio continuum emission from an IRAS selected sample of spiral galaxies.

P.Puxley, R.F.Hawarden, C.M.Mountain, and S.K.Legget. Near–infrared spectroscopy of enhanced star formation in barred spiral galaxies.

P.J.Richards and L.T.Little. The Galactic distribution of IRAS sources associated with the early stages of stellar evolution.

W.L.Rice, C.J.Persson, G.Neugebauer, B.T.Soifer, and E.L.Kopan. IRAS catalogue of large optical galaxies.

M.Rowan-Robinson. A physical model for infrared emission from zodiacal dust.

A.Savage, H.T.McGillivray, R.G.Clowes, S.K.Legget, and R.D.Wolstencroft. The IRAS identifications in the Southe Galactic Pole – a catalogue covering 2200 square degrees.

S.Sembay, C.Hanson, M.J.Coe, and R.Clement. A mid–to–far–infrared variability study of a sample of active galactic nuclei.

C.J.Skinner and B.Whitmore. IRAS and mass loss from M supergiants.

C.J.Skinner and B.Whitmore. Infrared excess and chromospheres in M supergiants.

C.J.Skinner and B.Whitmore. IRAS and a new way of calculating cool stellar mass loss.

L.Staveley–Smith, R.J.Cohen, J.M.Chapman, L.Pointer, and S.W.Unger. A systematic search for OH megamasers.

M.Srinivasan, S.R.Pottasch, K.C.Sahu, P.R.Wesselius, and J.N.Desai. Evidence for star formation in cometary globule 22.

M.A.Strauss. The peculiar velocity flow field in space as predicted from the distribution of IRAS galaxies.

M.Tapia, M.Roth, P.Persi, M.Ferrari–Toniolo, and L.Spinoglio. Near infrared observations of IRAS sources coincident with HII regions.

R.I.Thompson. A near infrared instrument for the Hubble Space Telescope.

A.C.Van den Broek. Optical ground–based follow–up studies of an infrared complete sample of extreme IRAS galaxies.

K.Vedi, B.Whitmore, D.Walker, and M.Rowan-Robinson. An all–sky map of the infrared background at $12—100\mu m$ with zodiacal emission subtracted.

M.Werner. SIRTF : The Space Telescope Infrared Facility.

H.J.Walker. Nearby IRAS $100\mu m$ cirrus clouds and their HI counterparts.

H.J.Walker and M.Cohen. An investigation of the IRAS colours of several types of stars – normal and odd.

B.Whitmore and C.J.Skinner. The circumstellar properties of catalogued carbon stars and newly identified carbon stars with the LRS database.

J.G.A.Wouterlot, C.M.Walmsley, and C.Henkel. NH_3 and CO near IRAS sources.

M.G.Wolfire, D.Hollenbach, and A.G.G.M.Tielens. Physical properties of the interstellar medium in nuclei of luminous infrared galaxies.

M.C.H.Wright, J.E.Carlstrom, J.M.Jackson, K.Y.Lo, and P.T.P.Ho. Aperture synthesis observations of starburst galaxies.

F.Werter, L.Magnani, and E.Dwek. Cirrus associated with high latitude molecular clouds.

P.R.Wesselius and R.Assendorp. Star formation in Cha T1.

G.S.Wright, R.D.Joseph, R.Wade, J.R.Graham, I.Gatley, and A.H.Prestwich. Infrared spectroscopy of interacting and spiral galaxies.

C.Xu, G.de Zotti, A.Franceschini, and L.Danese. Optical and far–ir luminosity function of Markarian galaxies.

C.Y.Zhang, R.J.Laureijs, and F.O.Clark. IRAS study of a star forming region : the Serpens molecular cloud.

H.Zinnecker, I.S.McLean, C.Aspin, I.M.Coulson, J.T.Rayner, and M.J.McCaughrean. Near-IR 2D imaging of IRAS dense cores in dark clouds : discovery of a double source in L1495 with IRCAM at UKIRT.

SUBJECT INDEX

AUTHOR CITATION INDEX

OBJECT INDEX

CONFERENCE PARTICIPANTS

SURNAME	INITIALS	INSTITUTE	COUNTRY
A'Hearn	M.	University of Maryland	U.S.A.
Abolins	J.	Rutherford Appleton Laboratory	U.K.
Adams	F.	U.C. Berkeley	U.S.A.
Ade	P.A.R.	Queen Mary College, London	U.K.
Adamson	A.	Lancashire Polytechnic	U.K.
Aiello	A.	University of Florence	ITALY
Aitken	D.	University College of NSW	AUSTRALIA
Alexander	P.	MRAO Cambridge	U.K.
Allen	A.J.	Queen Mary College, London	U.K.
Appleton	P.N.	Lancashire Polytechnic	U.K.
Arnaud	K.	Smithsonian Astrophysical Obs.	U.S.A.
Ashman	K.	Queen Mary College, London	U.K.
Bailey	M.E.	University of Manchester	U.K.
Ball	R.	Caltech	U.S.A.
Baxter	D.A.	University of Cardiff	U.K.
Becklin	E.E.	University of Hawaii	U.S.A.
Beichman	C.	IPAC-Caltech	U.S.A.
Belfort	P.	Institut D'Astrophysique, Paris	FRANCE
Benson	R.	IPAC-Caltech	U.S.A.
Bernard	J.P.	Ecole Normale Superieure, Paris	FRANCE
Berry	D.S.	University of Manchester	U.K.
Bhattacharjee	S.	Queen Mary College, London	U.K.
Birkett	C.M.	Royal Observatory Edinburgh	U.K.
Bode	M.F.	Lancashire Polytechnic	U.K.
Bottinelli	L.	Observatoire de Paris	FRANCE
Boulanger	F.	IPAC-Caltech	U.S.A.
Broadbent	A.	University of Durham	U.K.
Burbidge	G.	U.C. San Diego	U.S.A.
Carico	D.	Caltech	U.S.A.
Carr	B.	Queen Mary College, London	U.K.
Casoli	F.	Lab. de Physique de l'ENS, Paris	FRANCE
Caux	E.	CESR/CNRS Toulouse	FRANCE
Ceccarelli	C.	Frascati	ITALY
Chaloupka	P.	University of Leeds	U.K.
Chester	T.	IPAC-Caltech	U.S.A.
Chlewicki	G.	University of Groningen	NETHERLANDS
Clark	F.	University of Groningen	NETHERLANDS
Clarke	J.T.	NASA Goddard	U.S.A.
Clegg	P.E.	Queen Mary College, London	U.K.
Clement	R.	University of Southampton	U.K.
Clowes	R.G.	Royal Observatory Edinburgh	U.K.
Clube	V.	University of Oxford	U.K.
Coe	M.	University of Southampton	U.K.
Cohen	R.J.	Jodrell Bank	U.K.
Cohen	M.	NASA Ames	U.S.A.
Cong	X.	SISSA Trieste	ITALY

Conrow	T.	IPAC-Caltech	U.S.A.
Costa	V.	University of Granada	SPAIN
Cox	P.	MPI Bonn	WEST GERMANY
Crawford	J.	Queen Mary College, London	U.K.
Cruikshank	D.	University of Hawaii	U.S.A.
d'Hendecourt	L.	GPS Paris	FRANCE
Davidsen	A.F.	Johns Hopkins University	U.S.A.
Davies	S.	University of Southampton	U.K.
Davies	J.	Royal Observatory Edinburgh	U.K.
Davies	M.	U.C. Berkeley	U.S.A.
De Graauw	T.	University of Groningen	NETHERLANDS
de Jong	T.	University of Amsterdam	NETHERLANDS
De Jonge	A.	University of Groningen	NETHERLANDS
de Muizon	M.	Sterrewacht Leiden	NETHERLANDS
De Zotti	G.	University of Padova	ITALY
Dennefield	M.	Institut D'Astrophysique, Paris	FRANCE
Dermott	S.F.	Cornell University	U.SA.
Desert	F.X.	NASA Goddard	U.S.A.
Deul	E.R.	Sterrewacht Leiden	NETHERLANDS
Devereux	N.	University of Hawaii	U.S.a.
Donnison	J.R.	Goldsmith's College, London	U.K.
Dunford	E.	Rutherford Appleton Laboratory	U.K.
Dupraz	C.	Lab. de Physique de l'ENS, Paris	FRANCE
Edelson	R.	Caltech	U.S.A.
Edgington	J.A.	Queen Mary College, London	U.K.
Efstathiou	A.	Queen Mary College, London	U.K.
Efstathiou	G.	IOA Cambridge	U.K.
Elfhag	T.	Stockholm Observatory	SWEDEN
Ellis	G.	Queen Mary College, London	U.K.
Ellis	K.	Queen Mary College, London	U.K.
Elmegreen	B.	Watson Research Center, IBM	U.S.A.
Emerson	J.P.	Queen Mary College, London	U.K.
Encrenaz	T.	Observatoire de Meudon	FRANCE
Epchtein	N.	Observatoire de Meudon	FRANCE
Erickson	E.F.	NASA Ames	U.S.A
Farina-Busto	L.	Queen Mary College, London	U.K.
Fitt	A.	MRAO Cambridge	U.K.
Fitzsimmons	A.	University of Leicester	U.K.
Fong	R.	University of Durham	U.K.
Franceschini	A.	University of Padova	ITALY
Fuller	G.	University of California	U.S.A.
Gautier	T.N.	IPAC-Caltech	U.S.A.
Gerin	M.	Lab. de Physique de l'ENS, Paris	FRANCE
Gezari	D.	NASA Goddard	U.S.A.
Gil	J.	University of Kentucky	U.S.A.
Gillett	F.	Kitt Peak National Observatory	U.S.A.
Giovanelli	R.	Arecibo Observatory	PUERTO RICO
Gondhalekar	P.	Rutherford Appleton Laboratory	U.K.
Gougenheim	L.	Observatoire de Paris	FRANCE
Gregorich	D.	IPAC-Caltech	U.S.A.
Griffin	M.	Queen Mary College, London	U.K.

Habing	H.	Sterrewacht Leiden	NETHERLANDS
Haikala	L.	University of Cologne	WEST GERMANY
Harmon	R.	IOA Cambridge	U.K.
Harnett	J.I.	University of Sydney	AUSTRALIA
Harries	J.E.	Rutherford Appleton Laboratory	U.K.
Harwit	M.	Cornell University	U.S.A.
Hauser	M.	NASA Goddard	U.S.A.
Hawarden	T.G.	Royal Observatory Edinburgh	U.K.
Hilton	J.	Goldsmith's College, London	U.K.
Houck	J.	Cornell University	U.S.A.
Hughes	D.H.	Lancashire Polytechnic	U.K.
Hughes	J.	Queen Mary College, London	U.K.
Iyengar	K.V.K.	Tata Institute, Bombay	INDIA
Jennings	R.E.	University College, London	U.K.
Joncas	G.	University of Laval, Quebec	CANADA
Joseph	R.D.	Imperial College, London	U.K.
Joubert	M.	Marseille	FRANCE
Kailey	W.E.	University of Arizona	U.S.A.
Kaiser	N.	IOA Cambridge	U.K.
Kalmus	P.I.	Queen Mary College, London	U.K.
Kessler	M.F.	E.S.T.E.C.	NETHERLANDS
Kester	D.	University of Groningen	NETHERLANDS
Kollatschny	W.	University of Gottingen	WEST GERMANY
Kopan	G.	IPAC-Caltech	U.S.A.
Kuijken	K.H.	IOA Cambridge	U.K.
Kuiper	T.	Jet Propulsion Laboratory	U.S.A.
Kun	M.	Konkoly Observatory, Budapest	HUNGARY
Kwok	S.	University of Calgary	CANADA
Lahav	O.	IOA Cambridge	U.K.
Lamb	S.	U. Illinois at Urbana-Champaign	U.S.A.
Laureys	R.J.	University of Groningen	NETHERLANDS
Law	S.	Queen Mary College, London	U.K.
Lawrence	A.	Queen Mary College, London	U.K.
Leech	K.	Royal Greenwich Observatory	U.K.
Leger	A.	GPS Paris	FRANCE
Levasseur-Regourd	A.C.	University of Paris	FRANCE
Liljestrom	T.	University of Helsinki	FINLAND
Longair	M.S.	Royal Observatory Edinburgh	U.K.
Lorenzetti	D.	Frascati	ITALY
Loveday	J.	IOA Cambridge	U.K.
MacCallum	M.A.	Queen Mary College, London	U.K.
Malawi	A.	Jodrell Bank	U.K.
Malkan	M.	U.C. Los Angeles	U.S.A.
Marley	S.	University of Leeds	U.K.
Marsden	P.L.	University of Leeds	U.K.
Matilla	K.	University of Helsinki	FINLAND
Matsumoto	T.	Nagoya University	JAPAN
Meurs	E.J.A.	IOA Cambridge	U.K.
Mezger	P.G.	MPI Bonn	WEST GERMANY
Mikulskis	D.	Goldsmith's College, London	U.K.
Moles	M.	University of Granada	SPAIN

Moss	C.	IOA Cambridge	U.K.
Murray	C.D.	Queen Mary College, London	U.K.
Needham	G.	Queen Mary College, London	U.K.
Nordh	L.	Stockholm Observatory	SWEDEN
Norman	C.	STScI Baltimore	U.S.A.
Osborne	J.L.	University of Durham	U.K.
Papaloizou	J.	Queen Mary College, London	U.K.
Papoular	P.	CEN-Saclay	FRANCE
Parker	Q.A.	Royal Observatory Edinburgh	U.K.
Payne	B.	University of Southampton	U.K.
Persson	C.	IPAC-Caltech	U.S.A.
Prusti	T.	University of Helsinki	FINLAND
Puget	J.-L.	Ecole Normale Superieure, Paris	FRANCE
Puxley	P.	Royal Observatory Edinburgh	U.K.
Reach	W.T.	U.C. Berkeley	U.S.A.
Rephaeli	Y.	Tel Aviv University	ISRAEL
Rice	W.	IPAC-Caltech	U.S.A.
Richards	P.L.	U.C. Berkeley	U.S.A.
Richards	P.J.	Rutherford Appleton Laboratory	U.K.
Richardson	K.	Queen Mary College, London	U.K.
Rigault	M.F.	Observatoire de Meudon	FRANCE
Rodriguez-Espinosa	J.M.	ESO Munich	WEST GERMANY
Rowan-Robinson	M.	Queen Mary College, London	UK.
Roxburgh	I.W.	Queen Mary College, London	U.K.
Ryter	C.	CEN Saclay	FRANCE
Salama	A.	MPI Heidelberg	WEST GERMANY
Sanderson	C.	Queen Mary College, London	U.K.
Saunders	W.	Queen Mary College, London	U.K.
Savage	A.	U.K. Schmidt Unit, Siding Springs	AUSTRALIA
Scherrer	R.J.	Center for Astrophysics & Q.M.C.	U.S.A./U.K.
Sellgren	K.	University of Hawaii	U.S.A.
Silk	J.	U.C. Berkeley	U.S.A.
Soifer	T.	Caltech	U.S.A.
Squibb	G.	IPAC-Caltech	U.S.A.
Staveley-Smith	L.	Jodrell Bank	U.K
Stewart	B.	Rutherford Appleton Laboratory	U.K.
Strauss	M.	U.C. Berkeley	U.S.A.
Sykes	M.	University of Arizona	U.S.A.
Tandon	S.N.	Tata Institute, Bombay	INDIA
Tapia	M.	U.N.A.M.	MEXICO
Tavakol	R.K.	Queen Mary College, London	U.K.
Tedesco	E.	Jet Propulsion Laboratory	U.S.A.
Thompson	R.I.	University of Arizona	U.SA.
Truong-Bach		Observatoire de Paris	FRANCE
Tuffs	R.J.	MPI, Heidelberg	WEST GERMANY
Vader	J.P.	Yale University	U.S.A.
Van Den Broek	A.C.	University of Amsterdam	NETHERLANDS
Vedi	K.	Queen Mary College, London	U.K.
Verter	F.	NASA Goddard	U.S.A.
Wainscoat	R.J.	NASA Ames	U.S.A.
Walker	H.	NASA Ames	U.S.A.

Wang	C.	Royal Observatory Edinburgh	U.K.
Ward-Thompson	D.	University of Durham	U.K.
Werner	M.	NASA Ames	U.S.A.
Wheelock	S.	Caltech	U.SA.
White	G.J.	Queen Mary College, London	U.K.
Whitmore	B.	Queen Mary College, London	U.K.
Wilkinson	W.	Queen Mary College, London	U.K.
Williams	I.	Queen Mary College, London	U.K.
Williams	P.G.	Queen Mary College, London	U.K.
Wolfendale	A.W.	University of Durham	U.K.
Wolfire	M.	NASA Ames	U.S.A.
Wolstencroft	R.	Royal Observatory Edinburgh	U.K.
Wouterloot	J.	MPI Bonn	WEST Germany
Wright	G.	UKIRT Hawaii	U.S.A.
Wright	M.C.H.	U.C. Berkeley	U.S.A.
Wunderlich	E.	MPI Bonn	WEST GERMANY
Wynn-Williams	G.	University of Hawaii	U.S.A.
Zhang	C.-Y.	University of Groningen	NETHERLANDS
Zinnecker	H.	Royal Observatory Edinburgh	U.K.

Lecture Notes in Mathematics

Lecture Notes in Physics

Why do things half-way?

ASTRONOMY AND ASTROPHYSICS

A European Journal

Recognized as a "Europhysics Journal" by the European Physical Society

Astronomy and Astrophysics is the most important journal in its field to be published outside North America. Established in 1969, it is the result of the merging of six renowned European journals in astronomy and astrophysics. **Astronomy and Astrophysics** presents papers on all aspects of astronomy and astrophysics – theoretical, observational, and instrumental – regardless of the techniques employed – optical, radio, particles, space vehicles, numerical analysis, etc. Letters to the editor, research notes and occasional review papers are also included.

Astronomy and Astrophysics is divided into thirteen sections:

1. Letters
2. Cosmology
3. Extragalactic astronomy
4. Galactic structure and dynamics
5. Stellar clusters and associations
6. Formation, structure and evolution of stars
7. Stellar atmospheres
8. Diffuse matter in space (including H II regions and planetary nebulae)
9. The Sun
10. The solar system
11. Celestial mechanics and astrometry
12. Physical and chemical processes
13. Instruments, data processing, and computational methods

Astronomy and Astrophysics is edited by an international staff of scientists.

Editors-in-chief: F. Praderie, Meudon, France; M. Grewing, Tübingen, Germany, Federal Republic

Letter-Editor: S. R. Pottasch, Groningen, The Netherlands

Springer-Verlag
Berlin Heidelberg New York
London Paris Tokyo

Springer